Uncertainty in Mechanical Engineering II

Edited by
Peter F. Pelz
Peter Groche

Uncertainty in Mechanical Engineering II

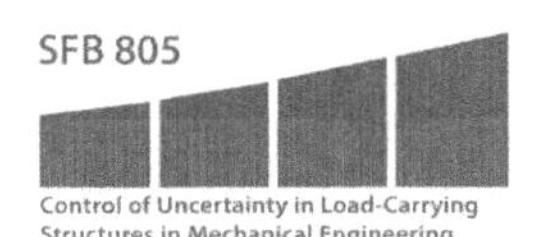

Selected, peer reviewed papers from the
2nd International Conference on
Uncertainty in Mechanical Engineering
(ICUME 2015),
November 19 - 20, 2015, Darmstadt, Germany

Edited by

Peter F. Pelz and Peter Groche

Trans Tech Publications Ltd
Churerstrasse 20
CH-8808 Pfaffikon
Switzerland
http://www.ttp.net

Volume 807 of
Applied Mechanics and Materials
ISSN print 1660-9336
ISSN cd 1660-9336
ISSN web 1662-7482

Full text available online at *http://www.scientific.net*

***Distributed** worldwide by*

Trans Tech Publications Ltd
Churerstrasse 20
CH-8808 Pfaffikon
Switzerland

Fax: +41 (44) 922 10 33
e-mail: sales@ttp.net

and in the Americas by

Trans Tech Publications Inc.
PO Box 699, May Street
Enfield, NH 03748
USA

Phone: +1 (603) 632-7377
Fax: +1 (603) 632-5611
e-mail: sales-usa@ttp.net

Preface

In each phase of the development, production and usage of systems, engineers use *models* for an approximation. Examples for a model would be a rod that is only able to conduct normal forces, an evolution equation for the wear and tear on a machine as well as a usage scenario or a load history. It is generally agreed that a model is the only way to make a system calculable or predictable, even though it is merely a copy of reality and not reality itself. The creation of a model always means to leave out a large portion of reality.

The word a*bstraction* is used to describe this process of consciously ignoring *the part of reality that is irrelevant* to answering the question at hand. However, for cases when the part of reality that is significant to answering the question is ignored in an unconsciously way, the Collaborative Research Centre CRC 805 coined the specific term *unknown uncertainty* or lack of knowledge. Both abstraction and lack of knowledge are always dangerously close to one another. In contrast to empirically based models, abstraction is more absolutely necessary in axiomatically based models. The latter adhere to the law of parsimony (Einstein: "…as simple as possible, but not simpler") and has been a mainstay principle of the natural sciences and engineering since the time of Galileo. The other extreme is the *law against the poverty* or the law against miserliness ("…it is vain to do with fewer what requires more") formulated by Karl Menger. In fact, these two paradigms are not in opposition, but rather complement each other. The simultaneous processing of large amounts of data and nonlinearities which can be modelled adequately by using the law of parsimony represents one research topic. Einstein's and Menger's quotes both reflect the struggle for obtaining valid models on one hand and dealing with lack of knowledge on the other hand.

Lack of knowledge is the most extreme and most significant source of model uncertainty, as its consequences are incalculable and unpredictable. One example is the Challenger accident, when the temperature history was not taken into consideration. Another example is the intercity train accident in Eschede, Germany - here the dynamic material fatigue of elastomers in the railway wheels was unconsciously ignored. An important part of abstraction is to separate the chain of events within a process into individual steps and/or to divide a system into individual components. The abstraction must be carried out at the interfaces. As abstraction and lack of knowledge always go hand in hand, interfaces are a significant source of uncertainty.

Keeping in mind that data are the basis of empirical models, it is assumed below that the creation of status quo models (not future models) has already been accomplished. This implies the existence of either a speculative or a verified and validated functional framework model. In this case, the data are part of the model. We term a model with deterministic parameters and a clear solution *disregarded uncertainty*. *Stochastic uncertainty* takes into account a determinate variability in the form of density functions, in contrast to *unknown uncertainty.*

Uncertainty of a mechanical engineering product as mentioned above, occurs in all phases of the product lifetime – development, production and use. It critically influences process properties and consequently product properties. Incorrect assessment due to uncertainty may have catastrophic consequences in terms of safety, particularly in case of products with load-carrying functions. Thus, it negatively impacts the product's profitability.

In the development process, a finished design is derived from the initial idea. To address uncertainty at this stage, advanced methods of robust design and corresponding mathematical optimisation methods as well as mathematical models for the combination of active and passive load-carrying components within a system network are developed. Furthermore, mathematical algorithms are used to analyse the propagation of uncertainty, and appropriate information models are used to represent and visualise uncertainties. Scaling method-based size range development and dimensional analysis describe the evaluation of uncertainty and modularisation. In addition, new assessment methods and combinations of scaling methods as well as discrete optimization are developed to control uncertainty in the whole system.

In the production process, the physical end product is produced by using raw material. The process chains are optimised with the help of the mathematical methods described above. Metal-forming and metal-cutting methods are rendered in a more flexible way as long as the level of production quality remains consistent. In order to get a complete insight into the individual product lifetime, functional materials for active and sensory components are integrated at an early stage. During the utilization of the load-bearing system, new usage monitoring methods ensure the permanent acquisition of actual loads. At the same time, advanced mechatronic and adaptronic or adaptive technologies stabilise and attenuate the load-carrying structure. Finally, structure or property relationships derived from the utilization process may provide information on the quality and suitability of the product during actual utilization with feedback to the development and production.

From practical experience, causes of uncertainty are usually unclear responsibilities, an increased technical complexity, rising development speed due to increased competition, higher requirements for cross-company quality assurance and an increased cost pressure. In order to address these causes, it is necessary to describe, evaluate and eventually control the uncertainty both in product lifetime and in the whole system by means of appropriate methods. This is also the focus of the research conducted by the Collaborative Research Centre CRC 805, host of the 2nd International Conference on Uncertainty in Mechanical Engineering ICUME 2015 and funded by the Deutsche Forschungsgemeinschaft DFG.

The Organizing Committee of the Second International Conference on Uncertainty in Mechanical Engineering – ICUME is pleased to present several works from an international community and from the CRC 805 giving an academic and industrial perspective to describe, evaluate and control uncertainty in general mechanical engineering and the nine special topics of the mini symposia.

The editors hope to meet the interests of a broad readership with the selection of the following contributions and motivate further investigations.

Peter F. Pelz and Peter Groche

Committees

Local Organizing Committee

E. Abele	Production Engineering and Cutting Machine Tools, Technische Universität Darmstadt, Germany
R. Anderl	Computer Integrated Design, Technische Universität Darmstadt, Germany
R. Bruder	Institute for Ergonomics and Human Factors, Technische Universität Darmstadt, Germany
P. Groche	Production Engineering and Forming Machines, Technische Universität Darmstadt, Germany, *Conference Co-Chair*
H. Kloberdanz	Product Development and Machine Elements pmd, Technische Universität Darmstadt, Germany
M. Kohler	Stochastics, Technische Universität Darmstadt, Germany
U. Lorenz	Department of Technology Management, Universität Siegen, Germany
T. Melz	System Reliability and Machine Acoustics, Technische Universität Darmstadt and Fraunhofer Institute of Structural Durability and System Reliability LBF, Germany
P. Pelz	Chair of Fluid Systems, Technische Universität Darmstadt, Germany, *Conference Chair and Head of SFB 805*
M. Pfetsch	Discrete Optimization, Technische Universität Darmstadt, Germany
R. Platz	Fraunhofer Institute of Structural Durability and System Reliability LBF, Germany
S. Ulbrich	Nonlinear Optimization, Technische Universität Darmstadt, Germany

International Scientific Committee

S. Cogan	Franche-Comté Electronique Mécanique Thermique et Optique - Sciences et Technologies, Université de Franche-Comté, France
S. Donders	Siemens Industry Software NV, Belgium
D. Ewins	Department of Aerospace Engineering, University of Bristol, Great Britain
E. Kostina	Institute for Applied Mathematics, University of Heidelberg, Germany
P. Martins	Instituto de Engenharia Mecanica, Universidade Técnica de Lisboa, Portugal
D. Moens	Production Engineering, Machine Design and Automation Section, Katholieke Universiteit Leuven, Belgium
F. Nobile	Calcul Scientifique et Quantification de l'Incertitude, École polytechnique fédérale de Lausanne, Switzerland
K. Sato	Institute of Design, Illinois Institute of Technology, USA
R. Schultz	Faculty of Mathematics, University of Duisburg-Essen, Germany
K. Schützer	Instituto Educacional Piracicabano, Universidade Metodista de Piracicaba, Brazil
D. Vandepitte	Production Engineering, Machine Design and Automation Section, Katholieke Universiteit Leuven, Belgium

Table of Contents

Chapter 1: Uncertainty in Mechanical Engineering

Chapter 2: Uncertainty of Structural Dynamic Improvements in Light Weight Design

Chapter 3: Modular Design and Scaling for Reduced Uncertainties in the Design Process

Chapter 8: Optimization under Uncertainty

Chapter 9: Binary Decisions under Uncertainty

CHAPTER 1:

Uncertainty in Mechanical Engineering

Applied Mechanics and Materials Vol. 807 (2015) pp 3-12
© (2015) Trans Tech Publications, Switzerland
doi:10.4028/www.scientific.net/AMM.807.3
Submitted: 2015-04-29
Accepted: 2015-08-04

Statistical Consideration of Uncertainties in Bolted Joints of the Drive Train of Sheet-fed Offset Printing Presses

Nicklas Norrick[1,a]

[1]Heidelberger Druckmaschinen AG, Alte Eppelheimer Straße 26, 69115 Heidelberg
[a]nicklas.norrick@heidelberg.com

Keywords: uncertainties, Monte Carlo simulation, bolted joints, sheet-fed, offset, printing press, non-linear system.

Abstract. This paper outlines a design process for the bolted joints of the drive train of sheet-fed offset printing presses incorporating statistical data and methods. Sheet-fed offset printing presses are driven by a continuous geared drive train along the length of the press. The bolted joints of the drive train connecting the gears to the cylinders of the press are subjected to high loads, especially during emergency stops. A nonlinear mechanical model of a printing press implemented in Matlab/Simulink is presented which is used to calculate the occurring loads. Measurements of linear and nonlinear system response are presented to support the quality of the mechanical model. The bolted joints between the main drive train gears and cylinders are designed according to current standards. Statistical information based on experimental data is considered during the application of the standardized method. Using the Monte Carlo technique, a more exact description of the joint's strength is made possible. In this way, the maximum tolerable load for the screw connection is 16% higher than the same result from a standard worst-case calculation.

Introduction

The high print quality of sheet-fed offset printing machines is obtained by the exact transfer of a sheet of paper through the press from one inking unit to the next [1]. The exact handover of the sheets at high speed (up to 18000 sheets/hour, equivalent to 7.6 m/s sheet velocity for the machine type regarded in this paper) is attained by cam-controlled grippers and a continuous geared drive train along the length of the press. The cylinder layout of a modern large format press can be seen in the cutaway view in Fig. 1. The drive train is powered by a single electric motor. In this way, the sheets can be passed through machines over 30 m long from feeder to delivery with a reproducible accuracy of the order of 10 μm.

In this paper we focus on the bolted joints between the main drive train gears and cylinders and show how the statistical consideration of uncertainties enables us to design these joints. In the following section, we will present the general mechanical model used to describe the machine dynamics. Next, we will discuss the maximum load for the bolted connection: the emergency stop. Then we will take a look at the design of the bolted joints and describe the process used to quantify the uncertainties in detail and discuss the results.

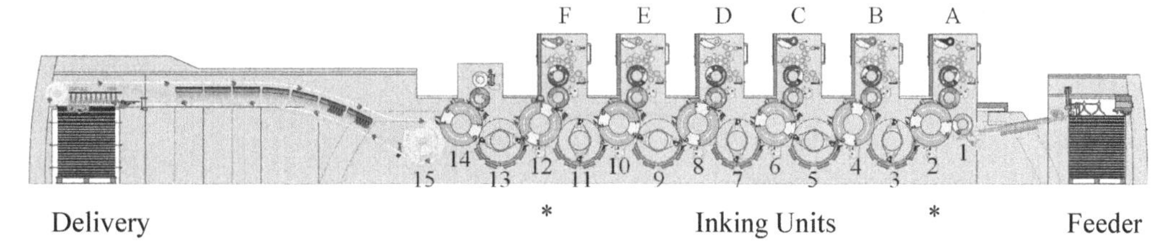

Fig. 1: Cutaway view of a Heidelberg Speedmaster XL162-6+L press with six inking units (marked A to F) and degrees of freedom (marked 1 to 15). Measurement points are marked with a *.

Machine Model

The drive train of the press is modeled using discrete rotational degrees of freedom q_n for every cylinder in the machine that participates in the paper transfer process as shown in Fig. 1. The discrete rotational inertias are connected via nonlinear spring-damper systems which describe the rotational stiffness and damping between each degree of freedom (DoF) including gear tooth clearance (backlash). The nonlinear system equation can be written in its general form

$$M\ddot{q} + B(\Delta q)\,\dot{q} + K(\Delta q)\,q = T\,. \tag{1}$$

The rotary inertia matrix M is constant. Due to backlash, the rotational damping matrix B and the rotational stiffness matrix K are nonlinear and dependent on the relative angles of the cylinders Δq. The excitation vector T is composed of the periodic reaction torques induced by the cam-controlled grippers at each cylinder $T_{cam}(q)$, the machine rotational friction T_{fric} at each degree of freedom as well as the drive torque of the motor T_{mot}, given by

$$T = T_{cam}(q) + T_{fric}(\dot{q}) + T_{mot}(q, \dot{q})\,. \tag{2}$$

The modeling of the complete press including the control loop of the drive motor is done in Matlab/Simulink. The system equations are solved numerically in the time domain for the vector of rotational degrees of freedom $q(t)$. Buck et. al. present this type of nonlinear modeling and diverse measurements for the verification of such models in [2].

For the linearized case when machine vibrations are small and no backlash occurs, the system transfer function between the drive motor and the relative cylinder rotation can be measured. In Fig. 2 we present the amplitude and phase angle of measured transfer functions of seven different presses with the same machine configuration compared with the simulated results. From the coherence function from $n = 7$ samples γ^2 as defined in [3] it can be seen that the system behavior is reproducible from one machine to the next. The natural frequencies and damping of the model fit very well to the measured results.

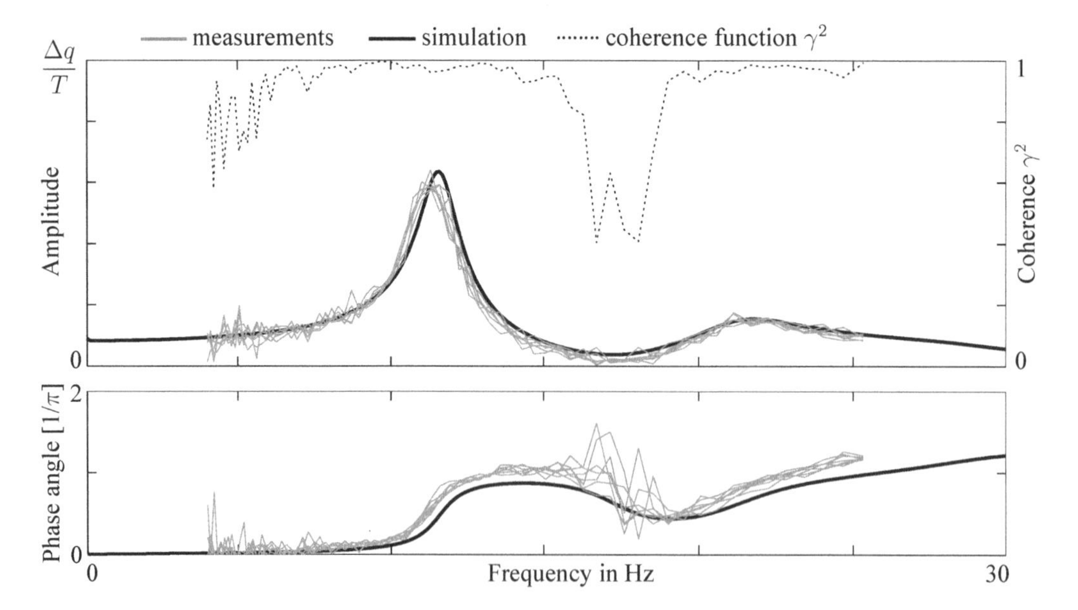

Fig. 2: Comparison of seven measured transfer functions (printing presses with the same configuration) with simulated results (amplitude and phase) and coherence function γ^2.

Emergency Stop

When an emergency stop is triggered, the main drive motor must stop the rotation of the machine within a given time, which is generally dictated by national or international accident prevention regulations [4]. There are several different emergency stop cases to be accounted for, depending on the type of stop: for example the stopping process during a power outage is fundamentally different from the case that the kill switch is pressed. Most importantly, the time sequence of the braking motor torque T_{mot} is fundamentally different. In addition, several other factors are important, such as the operating speed and temperature of the press as well as the machine angle at which the deceleration of the machine begins. There are also uncertainties such as local variations in mass, stiffness and damping parameters as well as gear backlash to be accounted for. It is possible to analyze the effect of stochastic variation of these parameters using Monte Carlo simulations in the manner shown by [5, 6]. Previous studies have shown that these variations only have a small effect on the global machine dynamics, so that in this study these parameters are considered to be fixed.

An active reduction of the mechanical loading of the gear-cylinder connections is technically possible. Nevertheless, for certain emergency stop procedures, e. g. in the case of a power outage, the active control of the motor torque is not possible. Due to safety concerns, we must rely on the passive safety of the bolts.

Fig. 3 shows a comparison of the measured and simulated rotation difference Δq between degrees of freedom 12 and 2 (these are the printing cylinders of the first and last printing unit, thus the whole length of the machine is represented) during emergency stop as well as the effective motor torque T_{mot}. For the sake of clarity in this diagram, the rotation difference is multiplied by the machine radius r to generate displacement values

$$\Delta q\, r = (q_{12} - q_2)\, r \tag{3}$$

on the cylinder circumference. Due to the gear backlash, a substantial overshoot in the displacement values is visible in both measurements 1 and 2 as well as in the simulated values (Fig. 3a). This overshoot correlates with peak acceleration values during emergency stop throughout the press. Because the experimental variation of the input conditions for emergency stops of each machine configuration would take exceedingly long, we resort to simulations. The set of varied input conditions for the emergency stop simulations are listed in Table 1.

Table 1: Input conditions for emergency stop simulations

Condition	Range
Machine angle at the beginning of the stop	$0 < \varphi_0 < 2\pi$
Operating speed $\dot{\varphi}$	equivalent to $3000 < V < 18000$ sheets/hour
Rotational friction (machine warm or cold)	$T_{fric,min} < T_{fric} < T_{fric,max}$
Delay time Δt of the mechanical motor brake	$0 < \Delta t < 200$ ms
Type of emergency stop	time sequence of the braking motor torque T_{mot}

These input ranges result in about 600 to 1000 simulation runs, from which the maximum forces for each gear-cylinder connection are extracted: From one simulation run of an emergency stop we obtain the time series results for the torques between each gear coupling in the machine. From these torques and the gear tooth geometry, we calculate the forces acting on each drive train gear. The vector sum of these forces on the gear in the y-z-plane is then determined. This is the load the bolted joint has to carry through static friction in the surface between the drive gear and the cylinder as can be seen in the sketch in Fig. 4. The force in x-direction is much smaller than the lateral force. For the focus of this study, the dominant load is in the y-z-plane.

Fig. 5 shows the calculated displacement difference $\Delta q\, r$ between degrees of freedom 12 and 2 during emergency stop from four different operation speeds. Relatively small variations of displacement difference across the whole machine result in large variations of lateral force on the joint.

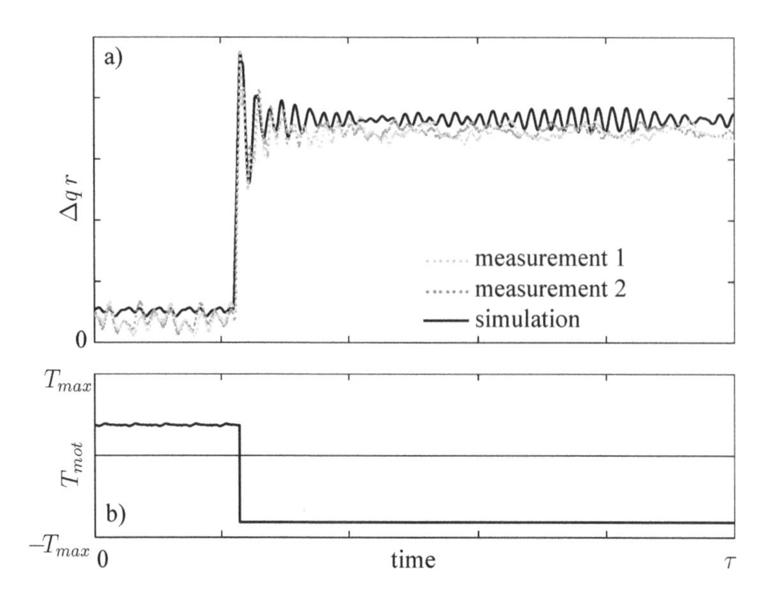

Fig. 3: Comparison of measured and simulated displacement difference $\Delta q\,r$ between degrees of freedom 12 and 2 during emergency stop (a) for the motor torque sequence (b).

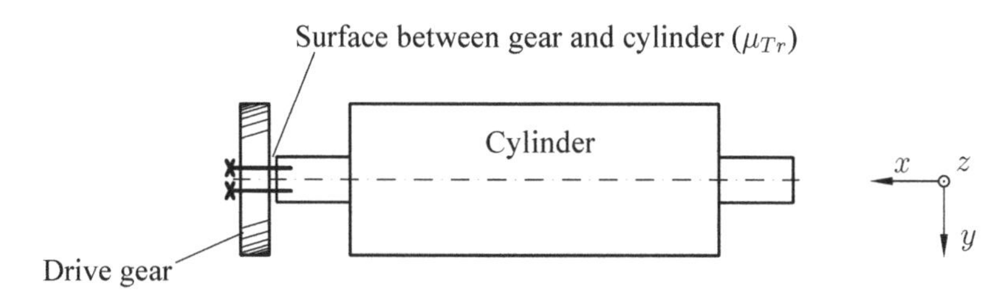

Fig. 4: Sketch of the gear-cylinder connection and coordinate system.

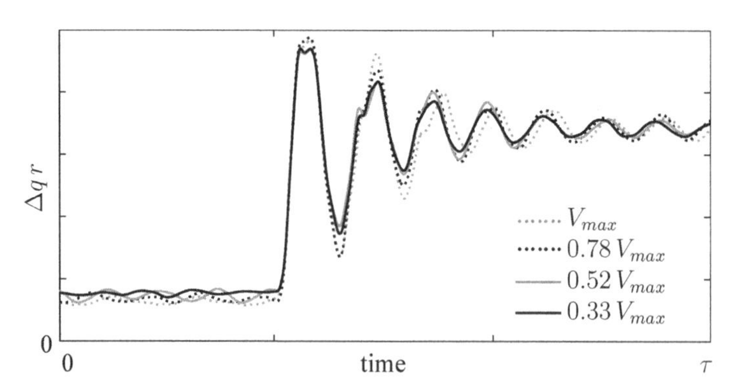

Fig. 5: Comparison of simulated displacement difference $\Delta q\,r$ between degrees of freedom 12 and 2 during emergency stop from four different operation speeds.

Fig. 6 shows the distribution of the maximum lateral force in kN from $N = 864$ simulation runs for a typical machine configuration for three gear-cylinder connections: Front of press (DoF 2), center of press (DoF 8), and end of press (DoF 14). Due to the complexity of the nonlinear system response, it is not feasible to fit the resulting distributions with standard probability distributions. In addition, for some of the input factors it is difficult to make quantifiable predicitions of the actual frequency of occurence. For example, an emergency stop due to power outage may never happen during the life of a certain machine in Central Europe, while in a developing country this type of stop may happen several times a day. For these reasons, a conservative approach is chosen and the absolute maximum forces are the loads F_{max} used in the calculation of the bolted joints according to [7].

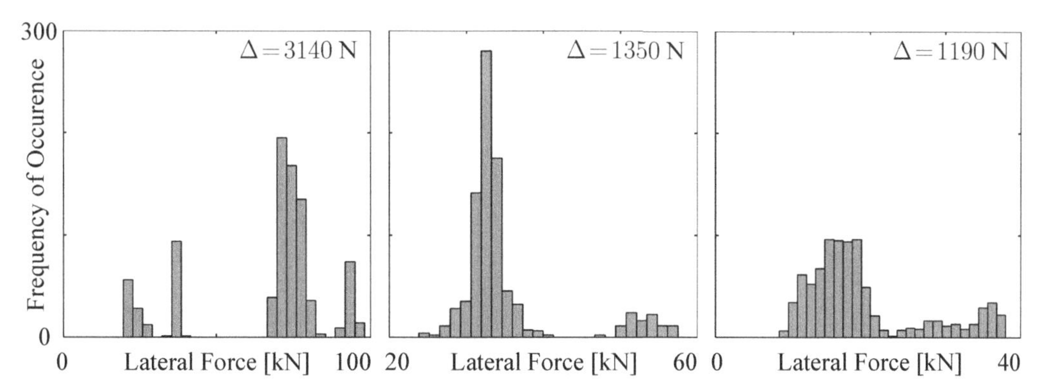

Fig. 6: Histograms of the frequency of occurence of maximum lateral forces for degrees of freedom 2 (left), 8 (center), and 14 (right), calculated from $N = 864$ simulation runs, variation of input conditions according to Table 1, bin-width Δ.

Statistical consideration using the Monte Carlo method

A standard design procedure for a bolted joint calls for the comparison of a maximum operational load F_{max} with a maximum tolerable load F_{tol}, resulting in the factor of safety

$$\text{FoS} = \frac{F_{max}}{F_{tol}} \geq 1\,, \tag{4}$$

which needs to be larger than one for the design to hold and is generally set to a higher value to have an extra margin of safety. If the attained FoS is insufficient, the design of the bolted joint (number of screws, material quality, etc.) must be changed.

The systematic calculation of bolted joints according to VDI 2230 calls for many input factors. Uncertainties in these input factors will result in uncertain output results. Reiff shows a method of calculating fastener torque including statistical tolerancing [8]. The focus is on setting a tightening torque which does not result in loading a bolt beyond its yield strength. The consequence is an 8% increase in average bolt tension with a 32 ppm chance that the bolt will fail.

The statistical approach used in this paper is similar to the method described in [8], but the focus is not on the screw connection but rather on the load carrying capacity of the connection perpendicular to the screw axis (corresponding to the y-z-plane in Fig. 4).

Some of the input parameter uncertainties can be quantified quite exactly, for example the geometrical parameters of the screw connection in question (e.g. size and number of the screws, thread parameters). Other parameters, especially the coefficients of static friction, exhibit large variations from one specimen to the next and also have a profound effect on the calculation results. For this study, we are going to take a closer look at the coefficients of static friction in the screw thread (μ_G), under the screw head (μ_K) as well as the surface between gear and cylinder (μ_{Tr}). In addition, we will discuss the uncertainties in the tightening torque M_A.

For each of the coefficients of static friction we resort to separate sets of measurements, since the conditions differ greatly from one value to the next. For example, the static friction in the screw thread μ_G is dependent on the thread-locking fluid used, while this factor obviously does not affect μ_K and μ_{Tr}. For each specific material combination historical measurements of μ carried out at *Heidelberger Druckmaschinen AG* exist. These measurements exhibit a probability distrubtion which can be fitted with a standard normal distribution. The fitting is done so that the experimental data fit the analytical probability distribution optimally in a least-squares sense. An example histogram of a set of 36 measurements for a specific material combination and the fitted continuous probability density function is shown in Fig. 7. This process results in three normal distributions $\mu_G(\bar{\mu}_G, \sigma_G)$, $\mu_K(\bar{\mu}_K, \sigma_K)$ and $\mu_{Tr}(\bar{\mu}_{Tr}, \sigma_{Tr})$ as shown schematically in Fig. 8.

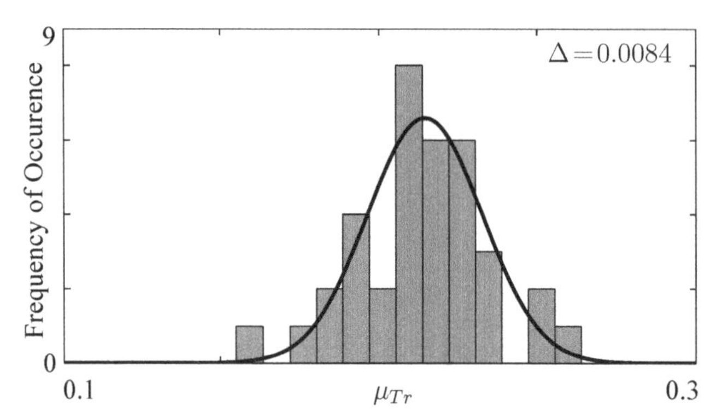

Fig. 7: Histogram of the measured frequency distribution of μ_{Tr} for a specific material combination and fitted normally distributed probability density function, bin-width $\Delta = 0.0084$.

For the tightening torque M_A, we know the nominal value of the torque wrench and assume a conservative tolerance of $\pm 10\,\%$ resulting in a lower and upper boundary M_{min} and M_{max}, respectively. Since we do not have reliable measurements for this value, we assume a uniform distribution of the torque in accordance with a conservative approach. Furthermore, we assume that for a set of s screws belonging to one gear-cylinder connection the torque is the same for all screws. This simulates the practical case when one mechanic using the same torque wrench tightens all screws in succession.

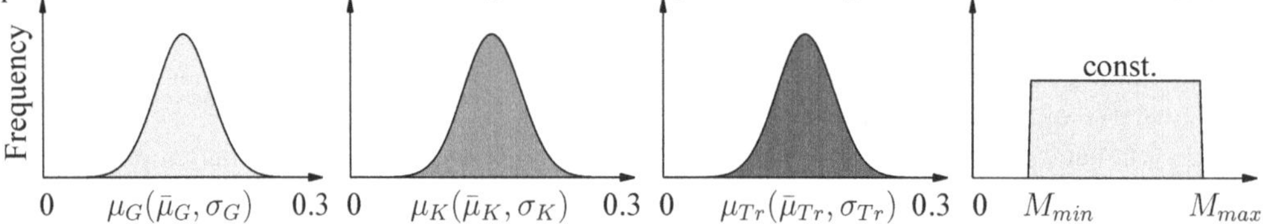

Fig. 8: Schematic diagrams of the chosen density functions for the uncertain parameters μ_{Tr}, μ_G, μ_K and M_A.

For the case that all screws participating in one connection have the same parameters, the maximum tolerable load for the joint in the y-z-plane in simplified form can be given as

$$F_{tol} = s\,\mu_{Tr}\,F_v\,, \tag{5}$$

where F_v is the remaining screw preload after setting. The preload loss due to setting is F_z. The screw preload is calculated as

$$F_v = \frac{M_A}{\frac{D_p}{2}\tan(\beta+\rho) + \mu_K\,\frac{D_m}{2}} - F_z \tag{6}$$

with the abbreviations

$$D_m = \frac{D_k + D_h}{2}, \quad \beta = \frac{P}{D_p \pi}, \quad \text{and} \quad \rho = \arctan\left(\frac{\mu_G}{\cos(\alpha/2)}\right), \tag{7}$$

where α is the thread angle, D_p is the pitch diameter, D_k is the head diameter, D_h is the hole diameter, and P is the thread pitch as labeled in Fig. 9.

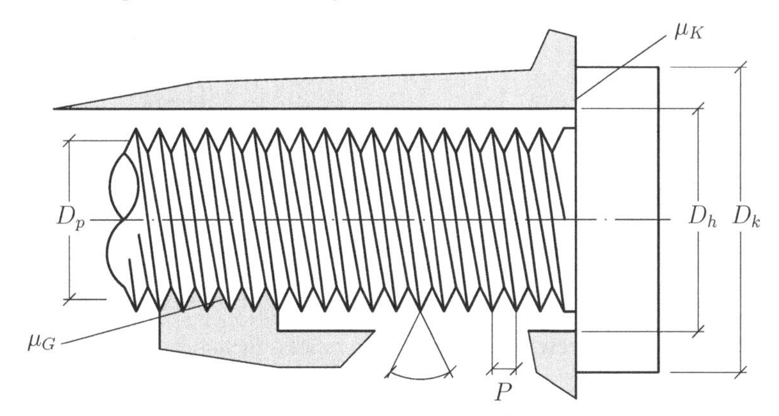

Fig. 9: Detail of the bolt geometry and location of the friction parameters μ_G and μ_K (Note: this sketch does not represent the actual joint design and is not to scale).

In a standard design procedure, we use these parameters and worst-case assumptions for μ_{Tr}, μ_G, μ_K and M_A to calculate the maximum tolerable load F_{tol} (Eq. 5) and compare the load with the maximum design load F_{max} according to Eq. 4.

To incorporate the statistical knowledge we have about μ_{Tr}, μ_G, μ_K and M_A, we use Monte Carlo simulations, calculating a sufficiently large number (N) of individual joints with randomly picked properties, producing a set of N tolerable loads F_{tol}, which form a set of values F_n. Since computing power has become readily available, the Monte Carlo method has developed into a standard tool for the numerical analysis of stochastic uncertainty in computational engineering [9].

The Monte Carlo procedure is implemented in Matlab using the *randn* command to generate the normally distributed pseudorandom numbers according to the Mersenne Twister algorithm [10], enabling the calculation of $N = 10^6$ joints (corresponding to about 10^7 individual screws depending on the cylinder type) within seconds. Fig. 10 shows a flowchart of the process to help clarify the steps involved.

Results

The resulting distribution of the tolerable loads F_n from Monte Carlo simulations and a normal distribution using the mean and standard deviation calculated as

$$\bar{x} = \frac{1}{N}\sum_{n=1}^{N} x_n \quad \text{and} \quad \sigma = \frac{1}{N-1}\sum_{n=1}^{N}(x_n - \bar{x})^2 \tag{8}$$

can be seen in Fig. 11a. The smoothness of the calculated curve is due to the large value of $N = 10^7$ used for the figure. A χ^2-test is used as a standard method for testing the distribution hypothesis [11]. The null hypothesis that the values of F_n are normally distributed holds true if

$$\chi^2 \leq \chi^2_\alpha . \tag{9}$$

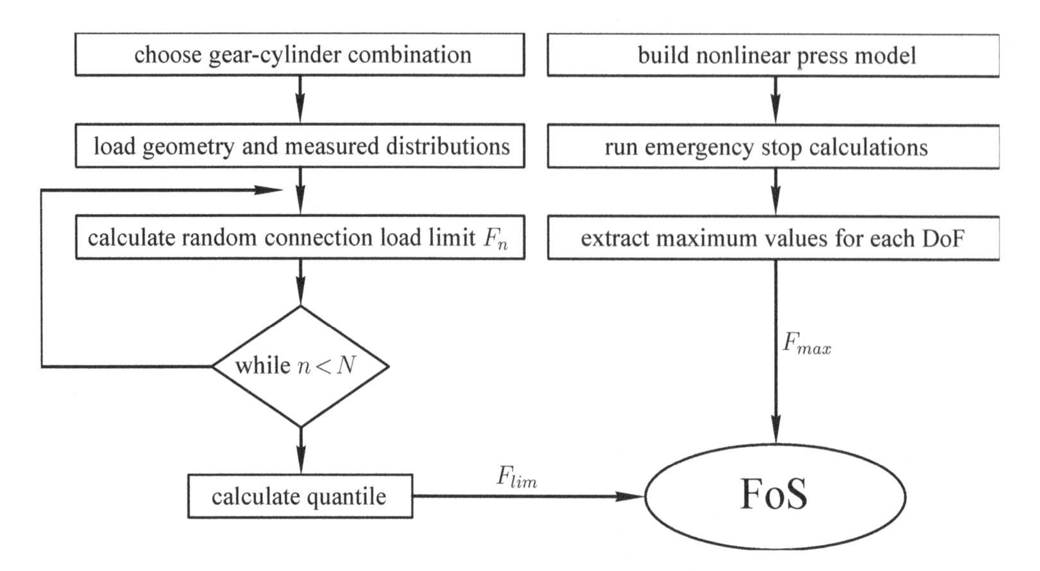

Fig. 10: Flowchart of the screw connection design process including statistical data.

The value of χ^2 can be calculated from the simulated data by sorting the observed values of F_n into K bins. It follows

$$\chi^2 = \sum_{k=1}^{K} \frac{1}{E_k} (O_k - E_k)^2 , \tag{10}$$

where O_k are the observed frequencies in each bin and E_k are the expected theoretical frequencies in each bin from a normal distribution. χ^2_α is the critical value of the χ^2 distribution with $K-3$ degrees of freedom and a level of significance of $\alpha = 5\%$.

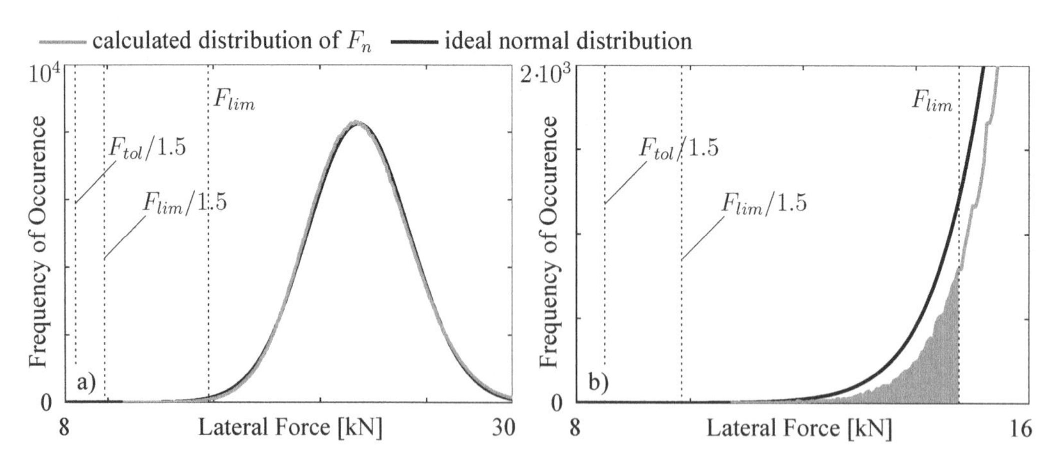

Fig. 11: Resulting calculated frequency distribution of F_n for $N = 10^7$ gear-cylinder connections, ideal normally distributed probability density function, and explicit values for F_{tol} and F_{lim}.

The simulated data yields a value of $\chi^2 > \chi^2_\alpha$, revealing that the resulting distribution is warped and not an ideal normal distribution. It is interesting to note that for smaller values of N of the order of 10^3, the null hypothesis is acceptable, since the amount of data input into the χ^2-test is too small to make the difference noticable.

A connection fails when $F_n < F_{max}$. We set an allowable failure quantile, for example 0.1%. This means that 99.9% of the calculated tolerable loads F_n are higher than the calculated load from the emergency stops F_{max}. This quantile corresponds to an area beneath the distribution curve (marked gray in Fig. 11 on the right). For $N = 10^7$ this corresponds to allowing 10^4 of the calculated connections to have a maximum tolerable load below this value. The dotted line F_{lim} is the extracted load limit from the resulting distribution. Compared to the resulting maximum tolerable load F_{tol} from the worst-case calculation outlined in the previous section the load limit F_{lim} from the Monte Carlo approach is 16% higher. In the figure, the lines $F_{lim}/1.5$ and $F_{tol}/1.5$ mark the resulting load limits when a factor of safety of FoS $= 1.5$ is considered. It can be seen that through the inclusion of the higher FoS, practically no gear-cylinder connections would fail.

Fig. 11b shows a zoomed-in portion of the interesting region of the probability density function, where the residual roughness in the curve due to the chosen value of N is visible. Here the difference between an ideal normal distribution and the calculated distribution, due to the nonlinear terms in the screw calculation, becomes evident. Extracting the 0.1%-quantile from a normal distribution would give an erroneous result for F_{lim}. One option would be to find a different analytical frequency distribution which fits the results better, e.g. Weibull or related distributions. In this paper, we extract the quantile directly from the numerical results. To verify the reliablity of the design process (which is based on pseudorandom numbers) we devise the following convergence test.

To decide which number of calculations N is sufficient for reproducible results, we inspect the standard deviation of our resulting load limit F_{lim} for sets of 20 individual runs for $N = 10^2$ to $N = 10^7$. The result of this convergence test can be seen in Fig. 12. We would like to reliably reproduce the result of F_{lim} within ±10 N, so we choose the point where

$$3\,\sigma(F_{lim}) = 10\,\mathrm{N}\,, \tag{11}$$

in accordance with the three-sigma rule of thumb [12]. At $N = 2 \cdot 10^6$, the standard deviation of the results $\sigma(F_{lim})$ reaches the value $10/3$. Rounding up, $N = 10^7$ is deemed a sufficient number of calculations for this design method and yields an acceptable computing time of 9.3 seconds.

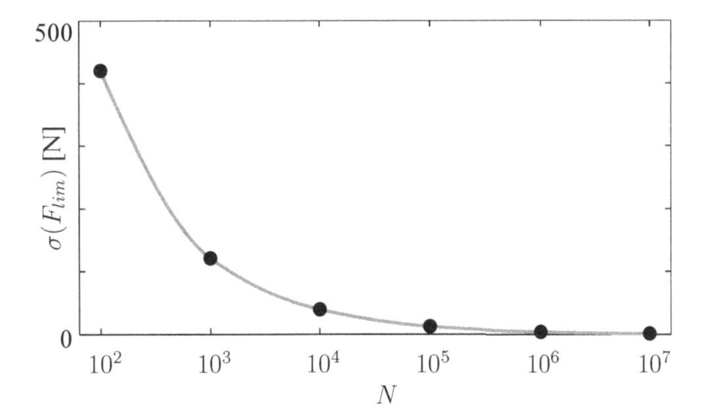

Fig. 12: Standard deviation of resulting load limit $\sigma(F_{lim})$ [N] versus number of calculations N.

Summary

This paper presents a design process for the bolted joints between the gears and cylinders of the main drive train of a printing press. The loading of the connection during emergency stop represents the maximum design load and is calculated from nonlinear simulations of the press dynamics using a detailed mechanical model of the press. The model is implemented in Matlab/Simulink and validated by measurements. By considering statistical information gained from experimental data and using the Monte Carlo method, it is possible to calculate a failure probability for a specific gear-cylinder connection and then set a design criterion. Compared to a standard worst-case calculation, the maximum tolerable load for the example screw connection is 16% higher. This value could be even higher since several conservative assumptions were made in the process, e. g. in the description of the tightening torque M_A. The presented method avoids overengineering and enables a more efficient use of resources. In the future, measurements using sensoric fasteners as presented in [13] could be used to measure the actual preload force F_v, during the manufacturing process or even during operation of the press, resulting in a further reduction of the uncertainty in the design process.

References

[1] H. Kipphan: Handbuch der Printmedien: Technologien und Produktionsverfahren (*Handbook of Print Media: Technologies and Production Methods*), Springer, Berlin Heidelberg, 2000.

[2] B. Buck, E. Knopf, S. Schreiber, and M. Seidler: Nichtlineare Schwingungsphänomene in Bogenoffsetdruckmaschinen (*Nonlinear Vibration Phenomena in Sheet-fed Offset Printing Presses*), VDI Berichte Nr. 1917 (2005) 345–361.

[3] R. Markert: Strukturdynamik (*Structural Dynamics*), Shaker, Aachen, 2013.

[4] BS EN 2210-2 (DIN EN 1010-2) Safety of Machinery - Safety Requirements for the Design and Construction of Printing and Paper Converting Machines - Part 2: Printing and Varnishing Machines Including Pre-Press Machinery, 2011.

[5] R. Platz, S. Ondoua, G. C. Enss, T. Melz: Approach to Evaluate Uncertainty in Passive and Active Vibration Reduction. In: H.S. Atamturktur et al. (eds.), Model Validation and Uncertainty Quantification, Volume 3: Proceedings of the 32nd IMAC, A Conference and Exposition on Structural Dynamics, 2014, Conference Proceedings of the Society for Experimental Mechanics Series, Springer International Publishing (2014) 345–352.

[6] R. Platz, G. C. Enss: Comparison of Uncertainty in Passive and Active Vibration Isolation. In: Proceedings of the 33nd IMAC, A Conference and Exposition on Structural Dynamics, 2015 (awaiting publication).

[7] VDI 2230: Systematic calculation of highly stressed bolted joints, 2014.

[8] J. D. Reiff: A Method for Calculation of Fastener Torque Specifications Which Includes Statistical Tolerancing, Journal of ASTM International, Vol. 2, Issue 3 (2005) 1–12.

[9] T. Müller-Gronbach, E. Novak and K. Ritter: Monte Carlo-Algorithmen (*Monte Carlo Algorithms*), Springer, Berlin Heidelberg, 2012.

[10] M. Matsumoto and T. Nishinura: Mersenne twister: a 623-dimensionally equidistributed uniform pseudo-random number generator, ACM Transactions on Modeling and Computer Simulation, Vol. 8, Issue 1 (1998) 3–30.

[11] I. N. Bronstein, K. A. Semendjajew, H. Mühlig and G. Musiol: Taschenbuch der Mathematik (*Handbook of Mathematics*), Verlag Harri Deutsch, 2000.

[12] J. Hartung: Statistik: Lehr- und Handbuch der angewandten Statistik (*Text- and Handbook of Applied Statistics*), Oldenbourg, Munich Vienna, 1995.

[13] P. Groche and M. Brenneis: Manufacturing and use of novel sensoric fasteners for monitoring forming processes, Measurement, Vol. 53 (2014) 136–144.

Applied Mechanics and Materials Vol. 807 (2015) pp 13-22
Submitted: 2015-04-29
Accepted: 2015-08-04

doi:10.4028/www.scientific.net/AMM.807.13

Localized discrete modelling of contact interfaces to predict the dynamic behaviour of assembled structures under random excitation

Anuj Sharma[1,a*] , Wolfgang Mueller-Hirsch[1,b] , Sven Herold[2,c] and Tobias Melz[2,d]

[1] Divison Automotive Electronics, Department of Design for Reliability and Mechanics (AE/EDE3), Robert Bosch GmbH, Robert Bosch Strasse 2, Schwieberdingen, D 71701, Germany.

[2]Division Smart Structures, Fraunhofer Institute for Structural Durability and System Reliability LBF, Bartningstraße 47, Darmstadt, D64289, Germany.

[a]anuj.sharma2@de.bosch.com, [b]wolfgang.mueller-hirsch@de.bosch.com, [c]sven.herold@lbf.fraunhofer.de, [d]tobias.melz@lbf.fraunhofer.de

Keywords: contact stiffness, contact damping, random vibrations

Abstract. Joints used to fasten different parts are the source of local non-linearity with predominance of contact damping in comparison to inherent material damping. The conventional numerical models can predict the dynamic behaviour to a good accuracy, but their implementation for large systems under real time dynamic excitations like random vibration are encountered with problems of numerical convergence and high computational cost. This paper proposes an approach to model the contact interfaces using discrete elements, with a non-homogeneous definition for the equivalent contact stiffness and damping over the contact interface. The non-homogeneous definition captures the non-linear effects and the local linearisation provides the capability to perform the frequency domain analysis for non-deterministic excitations. The proposed model is validated with experimental results for a test structure excited with random white noise base excitation.

Introduction

Automotive industrial products like Electronic Control Unit (ECU) are mounted on base structures, which are subjected to both deterministic and non-deterministic dynamic excitations from various sources like engine excitation, hydraulic pumps excitation, road profile etc. For an efficient design and operational accuracy, required are numerical models to predict influences of the joints on the dynamical behaviour of the structures. Two stages of modelling is required based on the loadings - the quasi-static loading contact models for assembling the structures and dynamic modelling for the operational conditions.

Experimental investigations for studying the joint dynamical characteristics have been performed by authors like Goodman and Klumpp (1956) [1], Ungar (1973) [2], Gaul et.al [3] and many others. Goodman and Klumpp (1956) concluded that a bolted joint can be designed to yield an optimal pressure with which maximum damping of systems can be achieved. Ungar (1973) performed experiments to define the dissipation in structural members as structural damping, as the dissipation caused due to friction between different planes of the structural material. Gaul et.al studied the isolated joint behaviour, to characterize the local non-linearity induced through jointed interface and influence on the contact dissipation with respect to bolting torque and excitation amplitudes. Authors like Gould and Mikic (1972) [4], Ziada and Abd (1980) [5] studied the pressure distribution over the contact interface due to the bolted joints. The contact pressure is parabolic in nature with circular influence zone of 3.5 times diameter of bolt [4, 5]. Apart from the experimental investigation many analytical and numerical contact models have been developed to capture the observations from experiments - like Hertz (1881), Mindlin (1949), Greenwood and Williamson (1966), Dahl (1968) etc.

Although the conventional contact models predict the dynamic behaviour to good accuracy, their implementation for large complex structures under real time excitations is computationally inefficient. This paper presents a new contact modelling approach to define the equivalent contact stiffness and damping based on the local contact pressure. This leads to a definition of continuous non-homogeneous contribution with local linearisation of contact non-linear forces. The paper describes the important constitutive equations of contact models in the first section followed by description of the proposed new modelling and finally an experimental validation of the proposed model in last section.

Theory: Constitutive equations

The constitutive equations for governing contact laws are described as foundation for the developing new model. The normal and tangential contact laws based on analytical models like Hertz(1881), GREENWOOD AND WILLIAMSON (1966) [6], COULOMB and DAHL [7] are briefly discussed.

Normal contact

The contact between any two bodies is initiated with a normal contact problem, wherein the two bodies experience a normal displacement with the forces perpendicular to their respective surfaces. One of the primitive analytical solutions for the normal contact behaviour was proposed by HERTZ (1881). The Hertzian contact assumes frictionless contact condition, which is justified for normal contact problems having only the normal loading. For a continuous Herztian pressure distribution over the contact interface between two elastic spheres, the normal force F_N is obtained as a function of the normal displacement δ_N [8] and is represented as

$$F_\mathrm{N} = \frac{4}{3} E^* R^{*1/2} \delta_\mathrm{N}^{3/2}, \tag{1}$$

wherein E^* and R^* are the composite elastic modulus and the radius of spheres in contact respectively. These are calculated based on the elastic properties of the two spheres in contact as

$$\frac{1}{E^*} = \frac{1-\nu_1^2}{E_1} + \frac{1-\nu_2^2}{E_2} \quad \text{and} \quad \frac{1}{R^*} = \frac{1}{R_1} + \frac{1}{R_2}. \tag{2}$$

The normal contact stiffness for two elastic spheres in contact is calculated as the derivative of the normal force with respect to the normal relative displacement and is obtained as

$$k_\mathrm{N} = \frac{dF_\mathrm{N}}{d\delta_\mathrm{N}} = 2E^* R^{*1/2} {\delta_\mathrm{N}}^{1/2} = 2E^* r_\mathrm{C}, \tag{3}$$

wherein the contact radius is $r_\mathrm{C} = \sqrt{\delta_\mathrm{N} R^*}$. The stiffness is dependent on the composite elastic constant and the contact radius. The elastic modulus is a constant value but the contact radius varies with loading. The formulations are presented for contact between two elastic spheres, but required are also the formulations for large nominally flat surfaces in contact. All engineering surfaces posses surface roughness, which should be considered while formulating the constitutive equations. Many theoretical and empirical models to accommodate the influence of surface roughness have been studied in detail by authors like ARCHARD [9], GREENWOOD AND WILLIAMSON [6], BOWER AND JOHNSON [10] and many others.

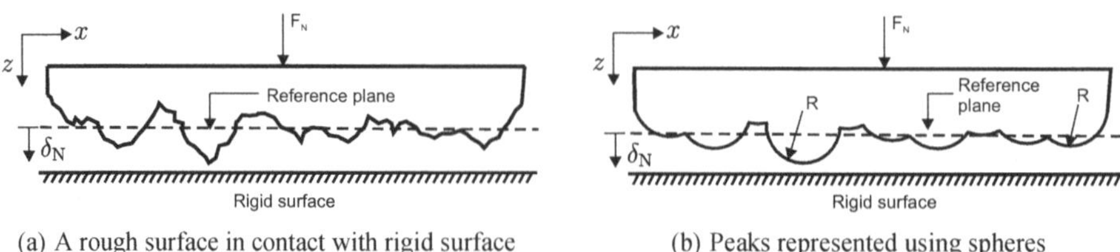

(a) A rough surface in contact with rigid surface (b) Peaks represented using spheres

Fig. 1: Modelling of surface roughness peaks as spherical asperities of same radius based on GREENWOOD AND WILLIAMSON model [6].

GREENWOOD AND WILLIAMSON [6] approximated the surface roughness with a Gaussian normal distribution. An assumption of all asperities possessing a spherical geometry with same radius of curvature is taken, see Fig. 1. The probability of initiation of contact will be significantly at the peaks of the surface roughness [6]. This allows the choice of probability density function to be taken as exponential distribution, where the predominant effects due to the peak-asperities is captured sufficiently. The main advantage of using exponential distribution is the possibility to have analytical closed form solutions of the integral formulation, which is not possible for Gaussian distribution. A modified exponential distribution ϕ_{mexp} [11, 12] to have better correlation to the Gaussian formulation for the predominance at the peaks is represented as

$$\phi_{\mathrm{mexp}}(z) = c \exp\left(-\lambda_c \frac{z}{\sigma_r}\right), \tag{4}$$

wherein c and λ_c represent the dimensionless constant coefficients. Also, σ_r is the standard deviation of the surface roughness. For all practical purposes, the values of $c = 17$ and $\lambda_c = 3$ are feasible [12]. The total normal force is calculated as the summation of local contact areas and is expressed as

$$F_{\mathrm{N}} = \frac{c\pi^{1/2} n E^* R^{1/2} \sigma_r^{3/2}}{\lambda_c^{5/2}} \exp\left(-\lambda_c \frac{\delta_{\mathrm{N}}}{\sigma_r}\right), \tag{5}$$

where n is the number of asperities and R is the radius of the asperity. The normal contact stiffness is calculated as the gradient of the normal contact force with respect to the normal relative displacement.

$$k_{\mathrm{N}} = \frac{dF_{\mathrm{N}}}{d\delta_{\mathrm{N}}} = \frac{\lambda_c}{\sigma_r} F_{\mathrm{N}}. \tag{6}$$

The normal contact stiffness is linearly dependent on the normal force. The basis formulations related to normal force-relative displacement will be used as the foundation in implementation of contact modelling of interfaces in succeeding section.

Tangential contact

Tangential contact problems are governed by the loadings leading to tangential relative displacements between the two bodies in contact. For a large tangential loading, the bonding between the two bodies is broken and leads to slip between the bodies- referred as the sliding contact. COULOMB formulated the equation for the dynamic contacts- defining the tangential force F_{T} in the sliding state as the product of the coefficient of friction μ and normal force F_{N}. For tangential forces F_{T} less than the magnitude of limiting force friction ($F_{\mathrm{C}} = \mu F_{\mathrm{N}}$) value, there will be no sliding but the traction force exists. The state before the sliding occurs is referred to as stick state. The pictorial description of the two states of contact under tangential loading is shown in Fig. 2(a).

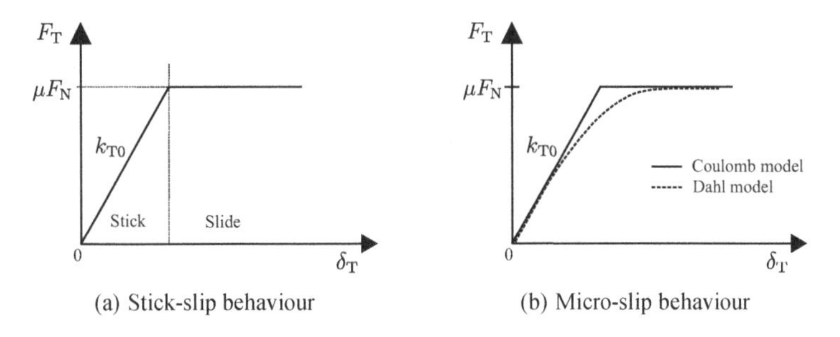

(a) Stick-slip behaviour (b) Micro-slip behaviour

Fig. 2: Stick and sliding region for the tangential loading with transition from stick to slide region at point where the tangential loading exceeds the limit of friction μF_{N}.

The tangential force acting over the contact interface in the stick state under uniform stress distribution is obtained as function of the relative tangential displacement [8]:

$$F_{\mathrm{T}} = 8G^{*}r_{\mathrm{C}}\delta_{\mathrm{T}}, \tag{7}$$

where the composite shear modulus G^{*} is calculated as $\frac{1}{G^{*}} = \frac{2-\nu_1}{G_1} + \frac{2-\nu_2}{G_2}$. From Eq.7, it is clear that stick state of contact has a linear relation between tangential force and relative displacement. The tangential stiffness of the stick state can be calculated by differentiating the tangential force with respect to the tangential relative displacement and is obtained as

$$k_{\mathrm{T0}} = \frac{dF_{\mathrm{T}}}{d\delta_{\mathrm{T}}} = 8G^{*}r_{\mathrm{C}}. \tag{8}$$

Coulomb model describes two states of contact - the stick and sliding states. Before the complete sliding between the bodies occur, there exists a pre-sliding state also refereed as micro-slip state. DAHL [7] proposed a model to include the micro-slip effect to the conventional Coulomb friction model. The tangential contact stiffness for defining all states of stick, micro-slip and sliding is given by

$$\frac{dF_{\mathrm{T}}}{d\delta_{\mathrm{T}}} = k_{\mathrm{T}} = k_{\mathrm{T0}}\left(1 - \mathrm{sgn}\left(\dot{\delta_{\mathrm{T}}}\right)\frac{F_{\mathrm{T}}}{F_{\mathrm{C}}}\right). \tag{9}$$

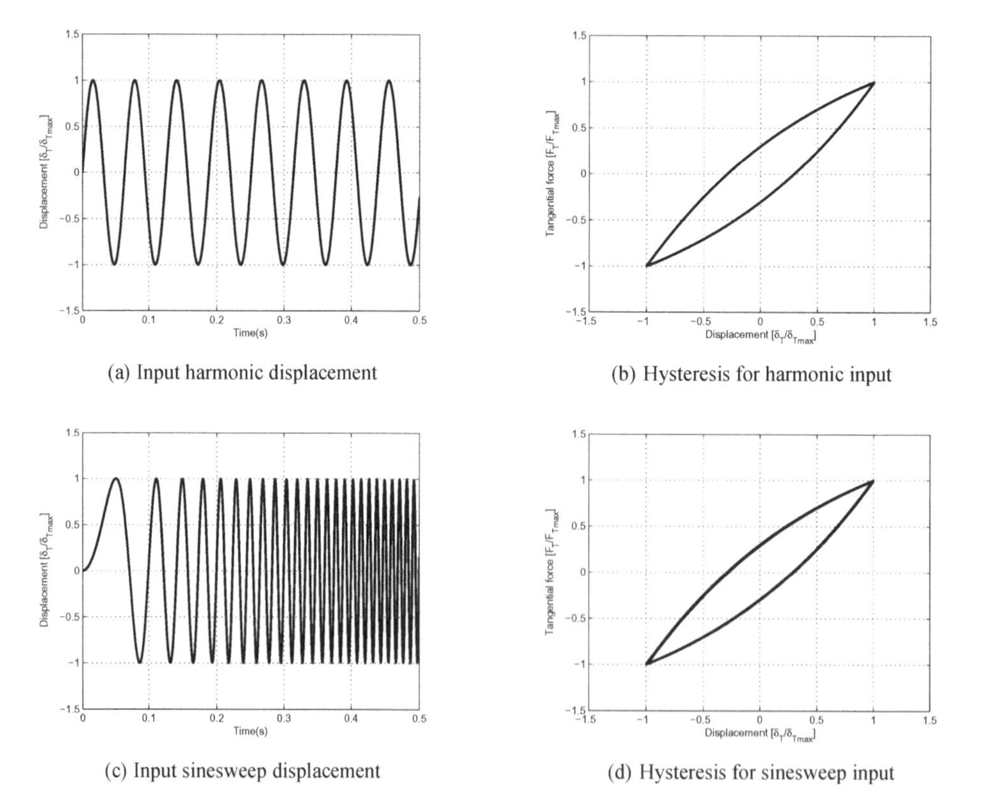

(a) Input harmonic displacement

(b) Hysteresis for harmonic input

(c) Input sinesweep displacement

(d) Hysteresis for sinesweep input

Fig. 3: The generation of hysteresis due to steady state excitation based on the Dahl model for dry frictional contacts.

The contact stiffness in the stick, micro-slip and sliding state has a linear constant value, non-linear decreasing value and zero value respectively, see Fig.2(b). The Dahl formulation shown in Eq.9 leads to generation of the hysteresis between the tangential force and relative displacement. The area under the hysteresis is associated with the frictional loss or dissipation due to the relative displacement between the contact interfaces. Fig. 3 shows the hysteresis generated with steady state excitations inputs. It can be observed that for both harmonic and sine-sweep excitations, there is no influence on the hysteresis. Hence, it can be concluded that the contact dissipation for dry contacts is independent of the frequency of excitation. However, for the random input the generated hysteresis has chaotic behaviour, see Fig.4. This shows that contact formulations for the non-deterministic excitations lead to very complicated behaviour, which cannot be easily used for complex large systems.

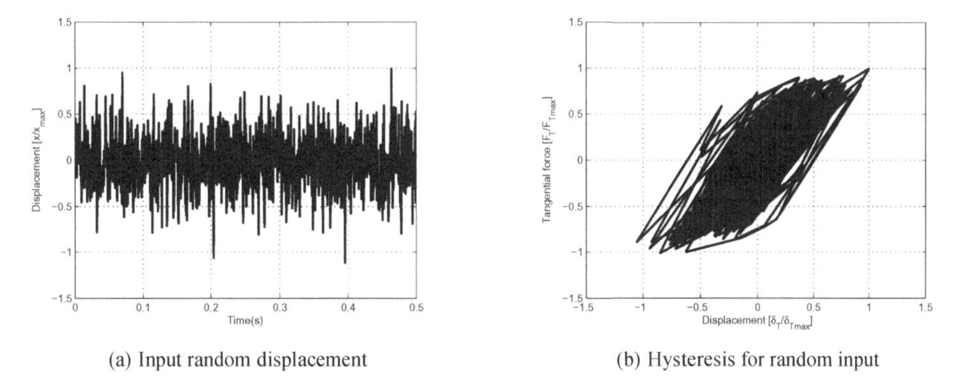

(a) Input random displacement

(b) Hysteresis for random input

Fig. 4: The generation of chaotic hysteresis due to random excitation based on the Dahl model for dry frictional contacts.

Damped Pressure Dependent Joint

As discussed in preceding section, the contact interfaces are disintegrated into normal and tangential contact behaviour. For random excitations, the time domain representation for the contact kinetic leads to chaotic behaviour with high computational cost. Hence, required is an equivalent representation of non-linear contacts in frequency domain to predict the dynamical behaviour of structures under random vibrations. The modelling approach is to define the non-linear forces due to the contact as the summation of the equivalent contact stiffness and contact damping. The equivalent contact stiffness and damping are proposed to be defined based on the contact pressure and hence referred as the Damped Pressure Dependent Joint (D-PDJ) model. The contact interfaces are discretized into point to point coupling, with each contact pair characterized using the D-PDJ model based definitions of the equivalent contact stiffness and damping.

Contact stiffness

The contact stiffness is calculated based on the force-displacement relation. The modified exponential distribution (see Eq. (4)) is used to formulate a generalized relation between the contact pressure and the normal deformation. The resultant exponential pressure-penetration law [13, 14] is presented in below equation as

$$P_{\mathrm{N}} = P_{\mathrm{N0}}\, e^{\lambda\left(\delta_{\mathrm{N}}-\delta_{\mathrm{N0}}\right)}, \qquad \lambda = \frac{1}{\sigma_{\mathrm{r}}}. \tag{10}$$

The constants used in above equation - P_{N0} and δ_{N0} - are the initial contact pressure and displacement when no loading is applied. The important parameter used is λ, which governs the curvature of the exponential curve and is idealized as the inverse of the standard deviation σ_{r} of the surface roughness. Each local nodal point in contact will be governed through this non-linear exponential pressure-penetration for the normal contact description. Due to the unbounded behaviour of the exponential function, very large magnitudes of pressure are required to produce very small deformation after certain saturation deformation limit. This often leads to problems in the numerical convergence. This problem is handled by defining a transition from non-linear to linear behaviour for higher pressure values. The pressure at the transition point is referred to as saturation pressure P_{N1} and the pictorial representation of non-linear and linear regions is depicted in Fig. 5(a). The region before the saturation pressure uses the proposed exponential pressure-penetration formulation and the region after the saturation pressure uses a constant slope with a linear relation between pressure and deformation.

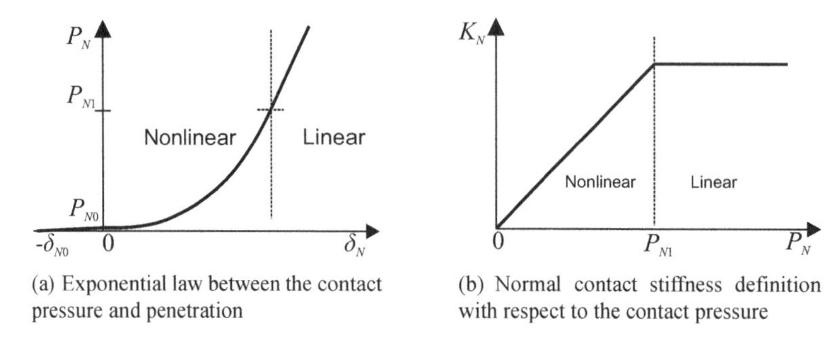

(a) Exponential law between the contact pressure and penetration

(b) Normal contact stiffness definition with respect to the contact pressure

Fig. 5: The governing normal contact law based on exponential relation between pressure-penetration with divisions of the non-linear and linear regions at saturation pressure P_{N1}.

The stiffness density K_{N} is calculated as the derivative of the pressure with respect to penetration. There exists a linear relation between the stiffness density and pressure with slope as curvature parameter λ. However, as the pressure-penetration formulation is modified into regions of non-linear and linear, the stiffness density exhibits two states - linear and constant stiffness, as shown in Fig.5(b).

$$K_{\mathrm{N}} = \frac{dP_{\mathrm{N}}}{d\delta_{\mathrm{N}}} = \begin{cases} \lambda P_{\mathrm{N}} & P_{\mathrm{N}} \leq P_{\mathrm{N1}} \quad \text{Non-linear region.} \\ \\ \lambda P_{\mathrm{N1}} & P_{\mathrm{N}} > P_{\mathrm{N1}} \quad \text{Linear region.} \end{cases} \tag{11}$$

The tangential contact stiffness can be calculated based on the normal contact stiffness density for an assumption of a stick contact status. Dividing Eq.8 by Eq.3, leads to definition of tangential contact stiffness density as

$$K_{\mathrm{T}} = \beta_0 K_{\mathrm{N}}, \tag{12}$$

where the coupling factor $\beta_0 = \dfrac{4G^*}{E^*}$. This coupling factor can be updated for increase in the tangential loading to describe the equivalent tangential contact stiffness for cases of micro-slip or sliding contacts.

Contact damping

As the contact stiffness is defined based on the local contact pressure, similar localized definition of contact damping at each point to point contact is required. A generalized example of bolted joint is used to discuss the local contact dissipation phenomenon at the contact interface. The region near the bolted joint has very high bolting pressure leading to almost no relative motion. This region has very low contact dissipation and is referred as stick region. The region away from the bolted joint has very low or no bolting pressure leading to very large relative motion with opening of contacts. This region too has very low contact dissipation with almost flat-horizontal hysteresis and is referred as slide region. The region between the stick and slide regions is then idealized as the region having optimal pressure leading to the maximum dissipations and is referred as the region of micro-slip [15].

DAMISA et.al (2008) [16] also proved this observation for layered beam with linear pressure variation over the contact interface. Summarizing the behaviour of the contact dissipation over the interface, the contact dissipation increases to maximum value from stick to micro-slip regions and decreases to zero from micro-slip to slide region. This behaviour can be formulated with use of a Rayleigh distribution function having the local contact pressure as the primary variable. Contact Damping definition based on the Rayleigh distribution function [17] satisfies the requirement of a higher damping contribution in the micro-slip region and is given below in equation as

$$d_{\mathrm{loc}}(x,y) = \left(\frac{P_{\mathrm{N}}(x,y)}{P_{\mathrm{m}}}\right) \exp\left(-\frac{1}{2}\left(\frac{P_{\mathrm{N}}(x,y)}{P_{\mathrm{m}}}\right)^2\right) \chi_r, \quad \text{where} \quad P_{\mathrm{m}} = \frac{\mathrm{mean}(P_{\mathrm{N}}(x,y))}{P_{\mathrm{m}}^{\mathrm{loc}}}. \quad (13)$$

Two damping parameters are used in the above expression to define the local contact damping d_{loc}. The first parameter P_{m} is the corresponding pressure at which maximum damping occurs. The parameter P_{m} is normalized to mean contact pressure with use of a locating parameter $P_{\mathrm{m}}^{\mathrm{loc}}$. The second parameter χ_r is the amplification factor to control the limits of damping value and is a non-dimensional quantity. The damping expression in Eq.13 is a dimension less quantity and required to be defined in terms of equivalent damping used in the numerical modelling of dynamic problems.

The local contact damping is defined as an equivalent structural damping loss factor. There are two reasons to justify this statement. Firstly, the loss factor also is dimensionless quantity and can be directly equated as the localized damping value. Secondly, the structural damping is independent of the frequency of excitation which is the same phenomenon observed for dissipation due to dry frictional contacts [2, 18], see Fig.3. Thus, the discrete structural damper elements at the point-to-point coupling can be used with their values calculated based on the formulation of Rayleigh distribution function of Eq.13.

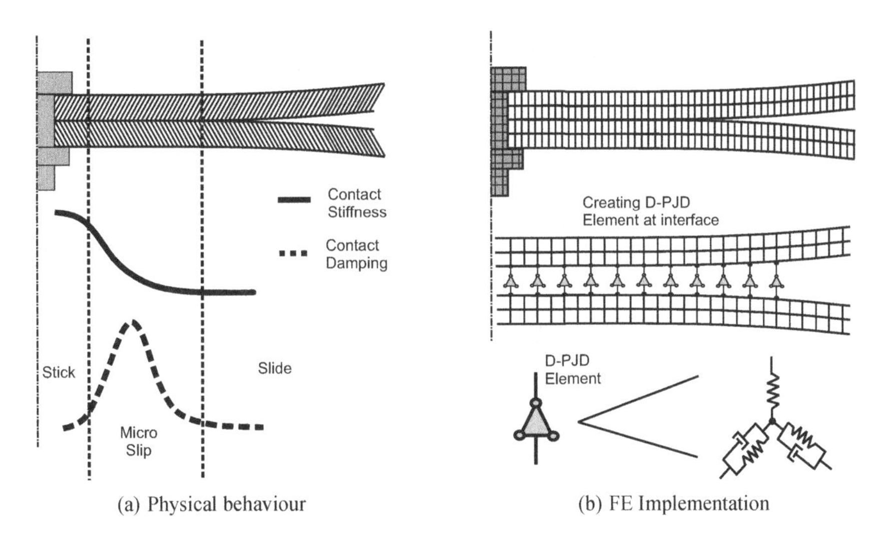

Fig. 6: Division of regions with contact stiffness and damping distribution based on pressure distribution (left). Implementation of D-PJD using finite elements (right).

With localized definition of contact stiffness and contact damping based on the pressure distribution, the D-PDJ model is characterized as a model having local linearised definition with non-homogeneous global description. The local linearisation allows to implement the model in frequency domain and hence the possibility for performing the dynamic simulations for the non-deterministic excitations using the conventional techniques of random Power Spectral Density (PSD) vibrational analysis. The non-homogeneous description of contact stiffness and damping over the contact interface is sufficient to capture the effects of the non-linearity due to the dry frictional contacts. A pictorial representation of D-PDJ model is depicted in Fig.6.

Model validation

For the experiment validation of proposed D-PDJ model, a thin stainless steel plate structure of Airbag ECU is bolted on a large Aluminium base support structure. This test specimen resembles the prototype model of ECU's baseplate mounted on chassis system. The exploded view of the assembly of the plate structure with base structure is shown in Fig.7. Used are M8 size bolts for fastening the plate onto base structure with 8 Nm bolting torque. The contact area is considerable large and hence the dynamic characteristic of the assembled structure gets influenced due to the contact interface. The assembled structure is excited with a PSD value of 1 $(m/s^2)^2/Hz$ in frequency range of 100-2100 Hz at the bottom of the base structure. The material properties of plate structure are optimized to match the modal frequencies from the Experiment Modal Analysis (EMA) of freely hanging plate structure. As the plate structure is obtained from forging process, there exists the discontinuity in mass distribution and local pre-stress effects at the bent regions of the plate structure. The simulation model of plate structure does not include these uncertainties and leads to small errors in estimation of modal frequencies.

Table 1: Modal frequencies comparison between EMA and simulation for the plate structure with Young's modulus $E = 2.17$ GPa, Poisons ratio $\nu = 0.32$ and mass density $\rho = 7850$ kg/m^3.

EMA.	345	1131	1418	1686	1901	2019	2257	2323	2734	2902	3444
Simulation	347	1120	1400	1682	1906	1985	2255	2324	2732	2874	3398
Error (%)	0.6	0.6	-1.2	-0.2	0.3	-1.7	0.1	0.0	-0.1	-1.0	-1.3

A pre-stress modal superposition based - base excitation random PSD analysis is performed in Ansys v15, with modelling of contact interfaces based on the D-PDJ model. The comparison between experiment and D-PDJ simulation are done for measured transmissiblity at the corner point on the plate structure, see Fig.7 for the measurement point.

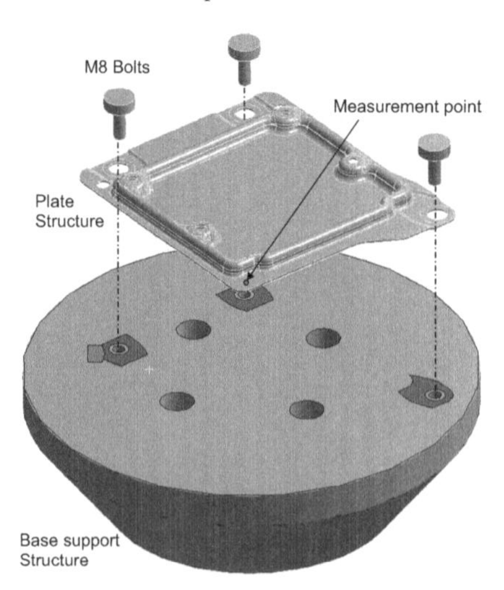

Fig. 7: Exploded view - assembly of plate and base support structure with region of contact interface highlighted. The corner point on plate structure is used for comparing results of experiment and simulation.

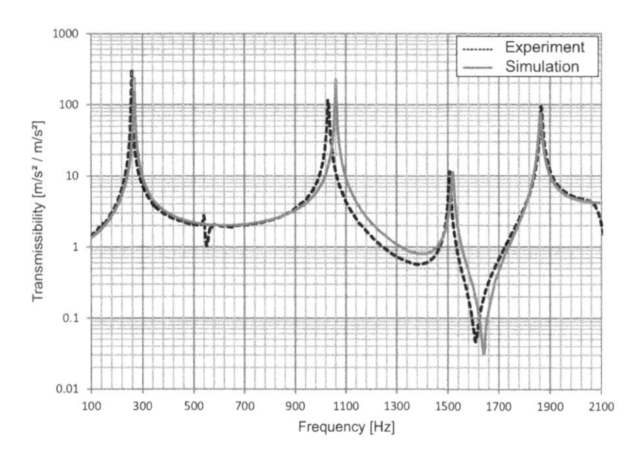

Fig. 8: Comparison of transmissibilty between experiment and D-DPJ simulation for the plate-base support structure assembly in base excitation set-up. The contact parameter used : $\lambda = 4000, \lambda P_{\mathrm{N1}} = 8000, \beta_0 = 1.6, \chi_r = 2, P_{\mathrm{m}}^{\mathrm{loc}} = 10$.

A very good match of transmissibility is achieved between the experiment and D-PDJ simulation for the assembled structure, see Fig.8. The contact parameters used for the D-PDJ simulation are listed in Fig. 8. The contact parameters λ and P_{N1} have influence on the contact stiffness and hence control the resonance frequencies. The coupling constant $\beta_0 = 1.6$ is obtained for the contact between stainless steel and aluminium material, to calculate the equivalent local tangential contact stiffness based on Eq.12. The contact parameter $P_{\mathrm{m}}^{\mathrm{loc}}$ is chosen such that the maximum damping occurs at pressure value equal to 0.1 times the mean contact pressure.

Conclusion

The proposed D-PDJ model characterizes the contact interfaces as assimilation of discrete point to point coupling, where each contact pair describes the contact between the two asperities. Each contact pair is characterised with equivalent contact stiffness and damping based on the local contact pressure. This defines a non-homogeneous contribution over the global contact interface while linearisation at local contact pair. The local linearisation retains the advantages of linear system while the non-homogeneous definition contributes to capturing the non-linear effects from contacts. Implementation of D-PDJ model in frequency domain hence allows to simulate the non-deterministic excitation cases using modal superposition random-PSD vibration analysis. A very good match between the experiment and D-PDJ simulation motivates the use of the model for large complex structures used in industries.

References

[1] L. E. Goodman, J.H. Klumpp, Analysis of slip damping with reference to turbine blade vibration, Journal of Applied Mechanics. 23 (1956) 421.

[2] E. E. Ungar, The Status of Engineering Knowledge Concerning the Damping of Built-Up Structures, Journal of Sound and Vibration. 26(1) (1973) 141-154

[3] L. Gaul, J. Lenz, Nonlinear dynamics of structures assembled by bolted joints, Acta Mechanica. 125 (1997) 169-181.

[4] H. H. Gould, B. B. Mikic, Area of contact and pressure distribution in bolted joints, Transactions of ASME, Journal of Manufacturing Science and Engineering. 94(3) (1972) 864-870.

[5] H. H. Ziada, A. K.Abd, Load pressure distribution and contact areas in bolted joints, Institution of Engineers (India). 61 (1980) 93-100.

[6] J. A. Greenwood, J. B. P. Williamson, Contact of Nominally Flat Surfaces, Proceedings of Royal Society of London. 295 (1966) 300-319.

[7] P. R. Dahl, Solid Friction Damping of Mechanical Vibrations, Journal of Aircraft-American Institute of Aeronautics and Astronautics .14 (1976) 1675-1682.

[8] V. L. Popov, Contact mechanics and Friction, Springer-Verlag Berlin Heidelberg (2010).

[9] J. F. Archard, Elastic deformation and the laws of friction, Proceedings of Royal Society of London. 243 (1957) 190-205.

[10] A. F. Bower, K. L. Johnson, The influence of strain hardening on cumulative plastic deformation in rolling and sliding line contact, Journal of the Mechanics and Physics of Solids .37 (1989) 471-493.

[11] J. A. Greenwood, J. B. P. Williamson, Developments in the Theory of Surface Roughness, In Proceedings of 4th Leeds-Lyon Symposium on Tribology (1977)

[12] A. A. Polycarpou, I. Etsion, Analytical Approximations in Modeling Contacting Rough Surfaces, Journal of Tribology .121 (1999) 471-493.

[13] M. Mayer, Zum Einfluss von Fuegestellen auf das dynamische Verhalten zusammengesetzter Strukturen, PhD thesis, University Stuttgart, Germany (2007).

[14] U. Bittner, Strukturakustische Optimierung von Axialkolbeneinheiten, PhD thesis, Karlsrue Institute of Technology, Germany (2013).

[15] M. Groper, Microslip and macroslip in bolted joints, Experimental Mechanics .25 (1985) 171-174.

[16] O. Damisa, V.O.S. Olunloyo, C.A. Osheku, A.A. Oyediran, Dynamic analysis of slip damping in clamped layered beams with non-uniform pressure distribution at the interface, Journal of Sound and Vibration .309 (2008) 349-374.

[17] A. Sharma, W. Mueller-Hirsch, S. Herold, T. Melz, Non-homogeneous localized Kelvin-Voigt model for estimation of dynamical behaviour of structures with bolted joints, Proc. of 11th World congress on computational mechanics, Barcelona, Spain, (2014).

[18] N. Peyret, J.L. Dion, Energy dissipation by micro-slip in an assembly, analytic and experimental approach, Proceedings of the ASME 2011 International Design Engineering Technical Conferences & Computers and Information in Engineering Conference IDETC/CIE (2011).

Applied Mechanics and Materials Vol. 807 (2015) pp 23-33
© (2015) Trans Tech Publications, Switzerland
doi:10.4028/www.scientific.net/AMM.807.23

Submitted: 2015-04-29
Accepted: 2015-08-04

Variability in composite materials properties

Andy Vanaerschot[1,a] , Stepan Lomov[2,b] , David Moens[1,c]
and Dirk Vandepitte[1,d*]

[1]KU Leuven
Dept. of Mechanical Engineering, Kasteelpark Arenberg 41 - box 2449 — Belgium

[2]KU Leuven
Dept. of Materials Engineering, Kasteelpark Arenberg 44 - box 2450 — Belgium

[a]andy.vanaerschot@kuleuven.be, [b]stepan.lomov@mtm.kuleuven.be, [c]david.moens@kuleuven.be,
[d]dirk.vandepitte@kuleuven.be

Keywords: composite materials, anisotropic properties, orthotropic properties, material identification, variability, correlation length

Abstract. Composite materials are created as a quite complex architecture which includes a fibre reinforcement structure and matrix material. Many material parameters play a role when composite structures are modelled, e.g. in finite element models. In addition to the properties of the raw fibre and matrix materials which are used, also geometrical parameters have a significant effect on structural characteristics. Fibre reinforcement geometry together with material properties of fibre and matrix determine homogenised material properties.

The first part of the paper gives an overview of the most important processes which are used in composites processing industry. The factors which affect variability are also listed, and the effect of variability on material parameters is mentioned as well. The second part of the paper elaborates the identification of geometrical variability of the fibre reinforcement structure which is encountered with one particular type of composite material, namely a twill 2/2 carbon fibre weave with an epoxy matrix.

Introduction

Composite materials are available for several decades already. The many families of composite materials have just one thing in common: they have some kind of fibre reinforcement and some kind of matrix material to hold the fibre reinforcement in place. Apart of that, there are many different variants, which can be identified on three different aspects:

- fibre reinforcement: here again, there are different aspects
  - fibre material is a parameter, with glass and carbon as the most commonly used materials, but natural fibres such as flax and hemp have a very promising future
  - length of fibre: short or long
  - fibre reinforcement architecture: short fibre are usually organised in some random distribution, whereas long or continuous fibres are organised in a well-structured architecture
- materix material: there are two main categories
  - thermoplastic matrix, in which the polymer becomes mouldable above a specific temperature and solidifies again after cooling
  - thermoset matrix, in which a chemical reaction takes place
- processing technology: a wide range of technologies are available, with different characteristics

The three categories which are listed above are not independent. A particular processing technology is appropriate for one or several specific types of fibre reinforcement and for particular types of matrix materials. The particular choice of a fibre type and reinforcement architecture, matrix and processing technology depends on the application, with again a multitude of parameters which play a role: geometry of the component, specifications on structural performance, weight of the component, processing time, size of series, environmental requirements, recyclability, ...Obviously, cost of manufacturing and exploitation of the component is a very important factor.

For most modern products and applications, numerical analysis is done in the design phase to verify the performance of the product. In this paper, the particular case of structural characteristics is taken as a reference. Structural analysis is often done using finite element analysis. In order to be able to conduct a reliable design analysis, representative models are required.

The representation of a composite material requires many parameters, as listed above. Each of these parameters exhibits uncertainty and variability. This paper uses these terms as they are defined by Oberkampf et al. [1]. As many design parameters play a role, likewise many sources of non-determinism have to be taken into account, which implies that many model parameters have to be represented by a relevant non-deterministic model, either in a probabilistic format through a probability density function or in a non-probabilistic format through an interval number or a fuzzy number [2].

Probabilistic methods are used to describe scatter in properties. Probability distribution functions (pdf) can be established for all uncertain parameters, taking into account the correlation between different parameters. The result of the analysis can be interpreted in a statistical sense, and the probability of every output quantity depends on the input probabilities and their correlations. It is important that all these inputs must be validated in order for the result to allow for a statistical interpretation. Unfortunately this fact is often neglected by many scientists, and assumptions are made on the input pdfs [2]. The definition of an input pdf is then subjective, and so is the result. Freudenthal [3] states that "ignorance of the cause of variation does not make such variation random.". The availability of objective and validated data is thus required.

An additional factor is the distribution of a parameter within the structural component. Both structural parameters, such as plate thickness, and material parameters, such as Young's moduli, may not be constant within the entire structure. Physical reality usually implies some relation between a parameter value at a specific position inside the component and another position. In a context of probabilistic analysis, the concept of random fields is well suited to express spatial variability in a component [4]. In a context of non-probabilistic analysis, the concept of interval fields which is proposed by Verhaeghe et al. [5] offers an attractive perspective.

This paper starts with an overview of different processing technologies, all of which are well known. The focus of the discussion is on the sources of uncertainty and variability which are relevant for that technology.

The second part of the paper focusses on the quantification of variability in composite material stiffness properties. The methodology is applied on a typical woven textile composite.

Effects of uncertainty in composites production processes

The processing technique which is used in the manufacturing phase inevitably introduces uncertainty and variability in the mechanical characteristics of the composite product.

Composites technologies This section starts with an overview of the most important processes for composites manufacturing and processing. They are usually a combination of a fibre reinforcement architecture, a thermoplastic or thermoset matrix and a processing technology. Each of these processes has specific quality characteristics, which are discussed below. The term quality refers to the degree by which the product matches the pre-defined specifications. The present paper focusses on geometrical characteristics, and quality expresses how well product geometry matches the ideal geometry, especially when the process is executed repeatedly.

- low or high pressure injection moulding techniques, such as resin transfer moulding, vacuum infusion, RTM-light, injection moulding, reaction injection moulding all inject resin in a liquid state into a closed mould in which the fibre reinforcement is already present. Depending on the precise definition of the process, the fibre volume fraction of the composite may be variable.

  The precise position and shape of the reinforcement is not well known because the resin which flows into the mould may shift or deform it, which implies that the exact orientation of the reinforcement is not guaranteed. An important aspect of this process is the development of the flow front, for which simulations are done to verify the quality of the process.

- high pressure compression moulding techniques, such as SMC (sheet moulding compound), BMC (bulk moulding compound) typically have a low fibre volume fraction. This is manufactured by dispersing long strands of chopped glass fibres or carbon fibres on a bath of resin (commonly polyester resin, vinylester resin or epoxy resin) before compressing them.

  The degree of variability is quite high, because the fibres are randomly oriented and the fibre volume fraction is variable within the volume of the component.

- material configuration techniques such as hand lamination, spray-up, prepreg autoclave curing, automated tape placement, filament winding, …all prepare the fibre reinforcement in a mould or on a mandrel.

  In manual lamination, the operator positions the reinforcement in the mould. The common type of reinforcement with hand lamination is a weave, with a structure which should be quite regular. However, the example in the next section of this paper demonstrates that individual tow paths do exhibit some irregularity, even within one weave. The manual procedure of positioning the layers in the mould introduces additional uncertainty, as the accuracy is limited. Also the thickness of the structure is not precisely known, especially when layers overlap each other.

  In automated tape laying, the same operation is done by a tape laying machine. The tape laying machine is either a gantry type robot or an articulated robot. The fibres are usually arranged in a uni-directional orientation, as they come from the tape which is unrolled from the tape layer. The quality of automated processes is obviously much better controlled than with manual processes. Adjacent tapes are well positioned next to each other, and the automated process guarantees that there are no overlaps. The orientation of the tape is also quite accurate, as it coincides with the moving tape layer head.

  In the filament winding process, each long fibre is placed individually on a rotating mandrel in a continuous process. Position control is thus quite accurate and also the orientation of the fibre is close to the nominal design. Thickness may exhibit some variation because the contact pressure between the fibre and the mandrel is not well controlled. And the amount of resin which is applied on the mould cannot be precisely controlled either.

- continuous production, with technologies such as pultrusion and continuous lamination: fibre reinforcement is applied along the longitudinal direction of the component.

  The fibre is mainly aligned with the production direction of the material, but the precise orientation is not exactly known.

- sandwich construction consists of minimum three layers, with two thin face sheets and a thick and lightweight core in between. The three layers are usually glued to each other. The face sheets have either isotropic in-plane properties (e.g. aluminum foils) or orthotropic properties (e.g. a glass or carbon fibre weave). The core may consist of punctual connections (pole fibres), line connections (corrugations), a full two-dimensional connection (e.g. foam), or a structured line polygonal connection (e.g. a honeycomb core).

Each of the three constituents has its own uncertain characteristics: geometrical arrangement, precise thickness and material properties. On top of that, the glue layers add a substantial degree of uncertainty too, as it may be difficult to precisely control the amount of glue which is applied at each position. Especially with thin foil honeycomb cores, the area of the wetted surfaces between the cell wall or the face sheet and the adhesive in relation to the volume of adhesive is uncertain, affecting the strength of the bond.

A particular aspect of sandwich materials is the multitude of failure modes which may occur (core shear, face sheet buckling, dimpling, ...). It is unconservative to assume that all constituents remain intact, but it is difficult to identify the position when imperfections occur and to estimate the level of degradation which they bring about.

This overview of diverse processing techniques shows that most processes do not have the capability to guarantee the precise position and orientation of the fibre reinforcement. For some processes, the fibre volume fraction is not constant within the volume of the component either.

Effect of parameter scatter on material stiffness properties The designer has many degrees of freedom, including the selection of raw materials for both the matrix and the fibre reinforcement, the architecture of the fibre reinforcement, the fibre volume fraction, the number of layers and the orientation of layers. For the analyst, this large set of design degrees of freedom translates into a wide range of model parameters, and inevitably also a wide range of uncertain or imprecise material data.

Most composite materials with long fibre architectures exhibit orthotropic behaviour, expressed by Eq.1:

$$\epsilon = \begin{Bmatrix} \epsilon_{11} \\ \epsilon_{22} \\ \epsilon_{33} \\ \epsilon_{12} \\ \epsilon_{23} \\ \epsilon_{31} \end{Bmatrix} = \begin{bmatrix} \frac{1}{E_{11}} & -\frac{\nu_{21}}{E_{22}} & -\frac{\nu_{31}}{E_{33}} & 0 & 0 & 0 \\ -\frac{\nu_{12}}{E_{11}} & \frac{1}{E_{22}} & -\frac{\nu_{32}}{E_{33}} & 0 & 0 & 0 \\ -\frac{\nu_{13}}{E_{11}} & -\frac{\nu_{23}}{E_{22}} & \frac{1}{E_{33}} & 0 & 0 & 0 \\ 0 & 0 & 0 & \frac{1}{2G_{12}} & 0 & 0 \\ 0 & 0 & 0 & 0 & \frac{1}{2G_{23}} & 0 \\ 0 & 0 & 0 & 0 & 0 & \frac{1}{2G_{31}} \end{bmatrix} \begin{Bmatrix} \sigma_{11} \\ \sigma_{22} \\ \sigma_{33} \\ \sigma_{12} \\ \sigma_{23} \\ \sigma_{31} \end{Bmatrix} = \mathbf{D}\sigma. \tag{1}$$

With ϵ representing strain and σ stress, E_{11}, E_{22} and E_{33} Young's moduli along the direction 1 (aligned with the warp fibre), 2 (aligned with the weft fibre) and 3 (through the thickness). G_{12}, G_{23} and G_{31} represent the shear moduli in the planes 12, 23 and 31. When the load is applied along orientations x and y which include an angle $\theta \neq 0$ with the orientations 1 and 2, the compliance matrix **D** in the constitutive relation Eq.1 changes into the matrix **D′** as expressed in Eq.2 [6]:

$$\mathbf{D}' = \mathbf{T}\mathbf{D}\mathbf{T}^{-1} \quad \text{with} \quad \mathbf{T} = \begin{bmatrix} \cos^2\theta & \sin^2\theta & 0 & -\sin 2\theta & 0 & 0 \\ \sin^2\theta & \cos^2\theta & 0 & \sin 2\theta & 0 & 0 \\ 0 & 0 & 1 & 0 & 0 & 0 \\ -0.5\sin 2\theta & 0.5\sin 2\theta & 0 & \cos 2\theta & 0 & 0 \\ 0 & 0 & 0 & 0 & \cos\theta & \sin\theta \\ 0 & 0 & 0 & 0 & -\sin\theta & \cos\theta \end{bmatrix}. \tag{2}$$

This relation is used to express the variation of material stiffness constants for a change of orientation of the load. In the application to uncertainty, the misalignment of the fibre orientation with respect to the orientation of loading may be accidental, but the effect is significant, as illustrated by figure 1, which is taken from [7] and [8]. The left hand side of the figure shows the variation of the elastic orthotropic stiffness constants for different orientations of a uniaxially reinforced glass fibre composite lamina with respect to the applied uniaxial tensile load. The graph shows a significant decrease of stiffness with increasing misalignment of the fibre. The right hand side of the figure is valid for a cross-ply

(0°-90°) carbon-epoxy system. The graph shows the variation of the Young's modulus for different alignments of the fibre orientations with respect to the loading direction. The graphs show that the equivalent material stiffness depends strongly on the fibre placement. An imprecise placement of the fibre inevitably leads to a change of stiffness with respect to the nominal values. The left hand side of the graph also shows that the orthotropic elastic constants are inter-related.

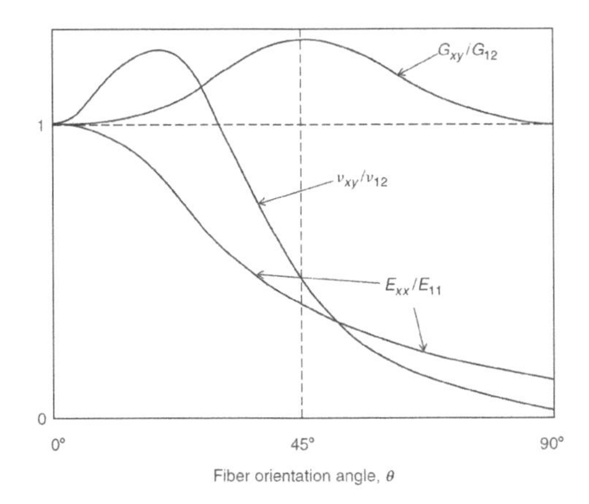

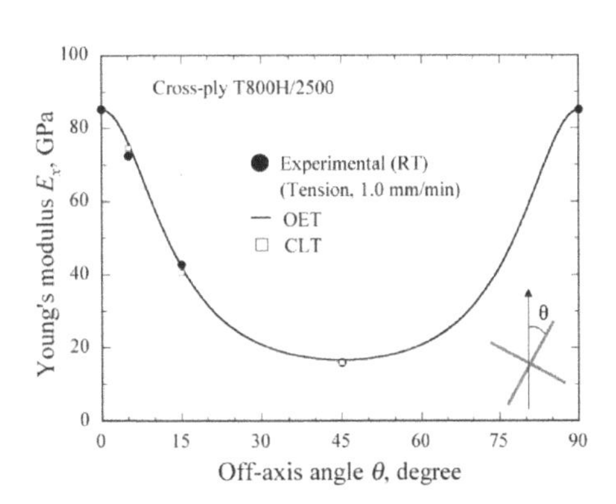

Fig. 1: Dependency of in-plane material parameters on the orientation θ of the major fibre axis to the loading direction; left: variation of the elastic constants of a continuous E-glass fibre lamina [7]; right: variation of the tensile Young's modulus for a cross-ply carbon-epoxy composite [8]

Another geometrical parameter that determines the homogenised stiffness characteristics of a textile composite material is the so-called crimp factor. It is a measure of the waviness of the yarn through the thickness of the panel. A general tendency is that the equivalent modulus of a textile composite increases with decreasing crimp.

Realistic data on material uncertainty and variability

Powerful mathematical tools are available to represent and propagate the effect of uncertainty and variability on finite element models of structural systems. However, there is a severe lack on practical data on real material systems and on concepts to generate reliable and useful data in a format that can be fed into the powerful numerical models that are available. Authors usually content themselves with making assumptions, e.g. on the covariance function of the Young's modulus and the correlation length [2]. They have done sensitivity analyses and they prove that the correlation length has a very large effect on the final result, yet validated data are not available. Charmpis and Schuëller [9] propose two approaches to achieve significant advances in realistic material modelling

1. establishing experimental data on the spatially correlated random fluctuations of uncertain material properties
2. deriving probabilistic information for macroscopic properties from the lower scale mechanical characteristics of materials

Although the need for experimental data was identified already in the 80ies, it appears that no evident step in this direction has been made until a few years ago. The second approach is essentially a multi-scale approach.

Extensive research efforts are currently ongoing to develop a multi-scale modelling procedure at successive scales. Depending on the type of material, the micro-scale describes properties with a reference length in the order of $10^{-6} - 10^{-4}$m, the meso-scale describes properties with a reference length in the order of $10^{-4} - 10^{-2}$m, and entire component structural behaviour is described on the macro-scale, with reference lengths in the order of $10^{-2} - 10^{0}$m and above. The step from a lower level to a higher level is made using homogenisation procedures, that assign overall properties at a higher scale based on lower scale data. So far, these models are mainly deterministic. When these models will be well established, they present an excellent opportunity to introduce variability at the appropriate level, and to predict the propagation of their effect to a higher level, and ultimately to the entire component. Experiments will however always be required to validate these models.

Multi-scale models also have the advantage that spatial variation of homogenised properties can be described based on lower scale characteristics. This presents opportunities for realistic quantification of random fields, for which experimental data are currently missing.

Quantification of variability on composite geometry in carbon fibre weaves

The objective of this section is to develop a consistent modelling strategy which can be used to generate virtual samples of a composite reinforcement architecture. The statistics of the set of samples which are generated should match the statistics of a typical real composite panel.

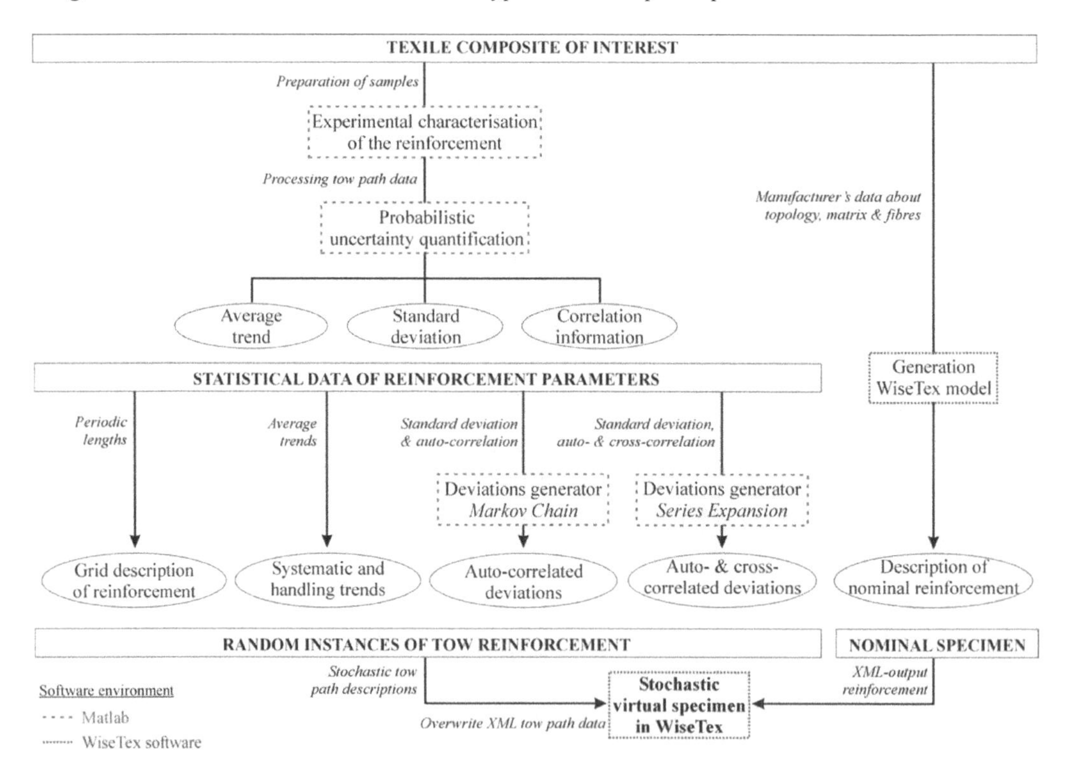

Fig. 2: The general methodology for modelling data uncertainty in composite reinforcement geometry consists of three steps: 1 (top)- acquisition of geometry data; 2 (middle)- processing of statistical data and correlation within one tow and between adjacent tows; 3 (bottom)- generation of virtual tow realisations which have the same statistics as the experimental data set

Figure 2 summarises the entire procedure.

Modelling strategy Realistic woven specimens are acquired that are replicas of experimental samples. Randomness is introduced in the numerical models at the meso- and macro-level; scatter in the matrix and fibre properties are not considered. Variability of each tow path is defined for the centroid co-ordinates (x, y, z), tow aspect ratio AR, tow area A and orientation θ in cross-section which fully describe a woven reinforcement. Figure 2 presents an overview of the multi-scale framework, where three main steps can be distinguished to obtain such random representations:

1. The first step is the collection of experimental data and the subsequent statistical analysis, according to the procedure proposed by Bale et al. [10]. Two length scales are considered, one on the level of the representative volume element using μCT and the other on the scale of multiple unit cells. The former set gives precise positional and cross-sectional data on the meso-level, identifying full details of 3D geometry. The latter data set is required to describe long-scale deviations from the nominal reinforcement pattern.

   (a) Characterisation of the short-range scatter (meso-scale) with samples close to the unit cell size.

   (b) Characterisation of the long-range variation (macro-scale) with samples spanning several unit cells.

   (c) Statistical analysis of the tow path parameters in terms of average trends, standard deviation and correlation lengths. For each parameter, the average trend $\langle \epsilon_i^{j,t,p} \rangle$ and the local deviation $\delta\epsilon_i^{j,t,p}$ are separated:

   $$\epsilon_i^{j,t,p} = \langle \epsilon_i^{j,t,p} \rangle + \delta\epsilon_i^{j,t,p}. \tag{3}$$

   where i refers to the location, j to the tow number, t to the tow genus (warp or weft) and p to the ply or sample.

   Vanaerschot et al. [11] give full details on the analysis procedure and statistical data processing.

2. The second step is the stochastic multi-scale modelling of the reinforcement architecture that is experimentally identified in step 1.

   (a) Definition of systematic and handling trends from the experimental data.

   (b) Generation of zero-mean deviations correlated along the tow path using the Monte Carlo Markov chain method.

   (c) Generation of zero-mean deviations correlated along and between neighbouring tow paths using the cross-correlated series expansion technique.

   For the generation of geometry data, two options are available. When cross-correlation between adjacent tows are not taken into account, the procedure by Blacklock et al. [12] is used. The representation is done using a Markov process. Each tow parameter is generated in an independent way, and for the calibration step, standard deviation and the nearest neighbour correlation information is used. When cross-correlations are taken into account, a cross-correlated series expansion is used, based on Karhunen-Loève decomposition, as proposed by Vorechovský et al. [13, 14]. Vanaerschot et al. [15] give full details on the generation of an approximated random field.

3. Construction of virtual specimens in the WiseTex software

   (a) Simulation of the nominal model with matrix and fibre properties from the manufacturer.

   (b) Redefinition of the reinforcement information with the produced tow paths.

   (c) Recalculations of the path orientation vectors and length, in addition to the updated general unit cell properties.

The WiseTex [17] software suite is developed at the Materials Engineering Department of KU Leuven. It is a pre-processor for the generation of virtual 3D models of a wide range of textile composite architectures. The pre-processor prepares for the generation of finite element models for the mechanical analysis of composite structures.

This step of the analysis generates virtual specimens which have the same statistical characteristics as the samples on which the experimental data acquisition was done. For each virtual sample, a WiseTex model is generated, with the original nominal tow path being overwritten by updated paths. Discrete random tow path realisations are interpolated and accompanied with information on the orientation vectors of each tow cross-section, the path length for each segment between two discrete locations and with new, off-nominal unit cell dimensions and properties.

Vanaerschot et al. [16] present details of the procedure for generating virtual samples.

The entire methodology allows for the generation of a set of statistically relevant samples. In a subsequent analysis step, these models are used for the investigation of the homogenised properties of a composite panel.

When other topologies than woven are considered, additional parameters should be quantified to allow a full description, e.g. the braid angle for braids and the distortion of the z-yarn in case of non-crimp fabrics.

Application to a carbon fibre twill weave

This section presents an application to the methodology which is described above. For a first validation experiment, a composite architecture is selected which is expected to exhibit only a low to moderate degree of geometrical variability.

Sample data The entire procedure is applied on a polymer textile composite with a twill 2/2 woven topology (Hexcel®G0986 Injectex), with 6K carbon fibre AS4C tows. The matrix is an expoxy resin Epikote®828LVEL with Dytek®DCH-99 hardener. The samples are prepared in a Resin Transfer Moulding (RTM) process and the unit cell size is 11.4mm×11.4mm. Some of the samples on which data have been acquired have 7 layers, others have only a single layer.

Geometry data acquisition Two sets of measurements have been conducted on two different types of samples of the composite. For the characterisation of short-range scatter μCT scans have been taken on samples with dimensions close to the unit cell size. These samples have 7 layers. Olave et al. [18] discuss the details of this measurement campaign. For the characterisation of longe-range scatter optical scans have been taken on samples with dimensions equal to several unit cell size. These samples have only one layer.

Statistical processing of tow data Statistical analysis is conducted on the date of each individual tow. The average trend in the figure exhibits a wavy pattern, marking the crossing of one warp tow over or under a weft tow.

Modelling of tow data In the modelling step, models are generated which extend over 10 unit cells, with the length of one unit cell being represented by 32 equi-distant points. Short-range periodic trends and long-range handling trends are combined. Figure 3 shows the average trends for the warp and weft genus, as they are measured in the experimental μCT analysis, for each of the five tow attributes. It was found that simulated trends correspond to experimental trends when cross-correlation is taken into account. Correlation is verified comparing a set of generated tow paths to a set of measured paths. For the out-of-plane and for the in-plane transverse centroid co-ordinates visual comparison is easy. Another element of comparison is a histogramme. Both methods of comparison show a very good match between simulations and experiments.

Generation of virtual models In the final step, 3D models are generated in the WiseTex preprocessor. 3D models allow for straightforward visual comparison, and models are found to actually exhibit the appropriate degree of spatial scatter.

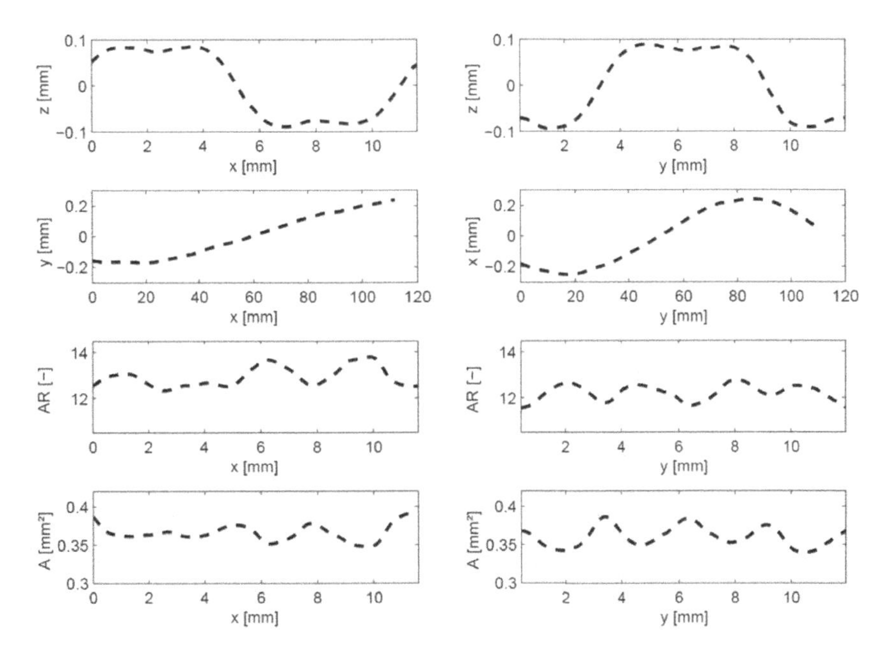

Fig. 3: Trends on each of the tow attributes as they result from the μCT analysis: the left hand graphs apply for the warp genus, and the right hand column applies to the weft genus; the top line graphs show the out-of-plane position of the tow centroid for one unit cell; the second line graphs represent the lateral in-plane deviation of the tow centroid for 10 unit cells; the third and fourth line graphs show the aspect ratio and the cross-section, respectively, each for one unit cell

Conclusions

This research is an effort to characterise uncertainty and variability in composite materials. The methodology is demonstrated for a carbon-epoxy 2/2 twill woven composite. The comparison between experimental and simulated deviation trends is found to be good. Randomly generated virtual specimens possess the target statistical information.

With the multitude of composite fibre architectures and processing methods, a lot of research work is still required to validate the proposed procedure also for other composite materials. A flat 2/2 twill woven composite has in theory a highly regular reinforcement pattern, but a curved geometry or braided or knitted architecture add up substantially to the complexity of the process and probably also to the degree of variability in the product. Another aspect is the size of the product, with parts in the order of several metres having much more variability than the unit cell size and the small components which have been investigated in this work.

Acknowledgement

The authors gratefully acknowledge the funding agencies: FWO-Vlaanderen and IWT-Vlaanderen.

References

[1] W. Oberkampf, S. DeLand, B. Rutherford, K. Diegert, K. Alvin, K., A new methodology for the estimation of total uncertainty in computational simulation, Proceedings of the 40th AIAA/ASME/ASCE/AHS/ASC Conference, AIAA-99-1612, 3061–3083 (1999)

[2] D. Vandepitte, D. Moens, Quantification of Uncertain and Variable Model Parameters in Non-Deterministic Analysis, IUTAM 2009, IUTAM Symposium on the Vibration Analysis of Structures with Uncertainties, 27, 15-28 (2011)

[3] A. Freudenthal, Fatigue sensitivity and reliability of mechanical systems, especially aircraft structures, WADD Technical Report 61-53 (1961)

[4] E. Vanmarcke, Random fields: analysis & synthesis, MIT Press, Cambridge, MA, London (1983)

[5] W. Verhaeghe, W. Desmet, D. Vandepitte, D. Moens, Interval fields to represent uncertainty on the output side of a static FE analysis, Computer methods in Applied Mechanics and Engineering, 260, 50–62 (2013)

[6] C.T. Sun, Mechanics of aircraft structures, John Wiley & Sons, Inc. (2006)

[7] P.K. Mallick, Fiber-reinforced composites: Materials, Manufacturing and Design, 3rd edition, CRC Press - Taylor & Francis Group (2007)

[8] M. Kawai, N. Honda, Off-axis fatigue behavior of a carbon/epoxy cross-ply laminate and predictions considering inelasticity and in situ strength of embedded plies, International Journal of Fatigue, 30 (10-11), 1743–1755 (2008)

[9] D.C. Charmpis, G.I. Schuëller, M.F. Pellissetti, The need for linking micromechanics of materials with stochastic finite elements: a challenge for materials science, Computational Materials Science, 41, 1, 27–37 (2007)

[10] H. Bale, M. Blacklock, M.R. Begley, D.B. Marshall, B.N. Cox, R.O. Ritchie, Characterizing three-dimensional textile ceramic composites using synchrotron x-ray micro-computed-tomography, Journal of the American Ceramic Society, 95(1), 392–402 (2012)

[11] A. Vanaerschot, B. Cox, S.V. Lomov, D. Vandepitte, Stochastic multi-scale modelling of textile composites based on internal geometry variability, Computers & Structures, 122, 55–64 (2013)

[12] M. Blacklock, H. Bale, M. Begley, B. Cox, Generating virtual textile composite specimens using statistical data from micro-computed tomography: 1D tow representations for the Binary Model, Journal of the Mechanics and Physics of Solids, 60(3), 451–470 (2012)

[13] M. Vorechovský, Simulation of simply cross correlated random fields by series expansion methods, Structural Safety, 30(4), 337–363 (2008)

[14] M. Vorechovský, D. Novák, Correlation control in small-sample Monte Carlo type simulations 1: A simulated annealing approach, Probabilistic Engineering Mechanics, 24(3), 452–462 (2009)

[15] A. Vanaerschot, B. Cox, S.V. Lomov, D. Vandepitte, Stochastic framework for quantifying the geometrical variability of laminated textile composites using micro-computed tomography, Composites A, 44, 122–131 (2013)

[16] A. Vanaerschot, B. Cox, S.V. Lomov, D. Vandepitte, Simulation of the cross-correlated positions of in-plane tow centroids in textile composites based on experimental data, Composite Structures, 116, 75–83 (2014)

[17] http://www.mtm.kuleuven.be/Onderzoek/Composites/software/wisetex (2015)

[18] M. Olave, A. Vanaerschot, S.V. Lomov, D. Vandepitte, Internal geometry variability of two woven composites and related variability of the stiffness, Polymer Composites, 33 (8), 1335–1350 (2012)

Applied Mechanics and Materials Vol. 807 (2015) pp 34-44
© (2015) Trans Tech Publications, Switzerland
doi:10.4028/www.scientific.net/AMM.807.34

Submitted: 2015-04-30
Accepted: 2015-08-04

Fluid-Structure Interaction Simulation of an Aortic Phantom with uncertain Young's Modulus using the Polynomial Chaos Expansion

Jonas Kratzke[1,a*], Michael Schick[2,b] and Vincent Heuveline[1,2,c]

[1]Engineering Mathematics and Computing Lab (EMCL), IWR, Heidelberg University, Speyerer Strasse 6, 69115 Heidelberg, Germany

[2]RG: Data Mining and Uncertainty Quantification, Heidelberg Institute for Theoretical Studies, Schloss-Wolfsbrunnenweg 35, 69118 Heidelberg, Germany

[a]jonas.kratzke@iwr.uni-heidelberg.de, [b]michael.schick@h-its.org, [c]vincent.heuveline@uni-heidelberg.de

* corresponding author

Keywords: Uncertainty Quantification, Polynomial Chaos, Aortic Phantom, Fluid-Structure Interaction, Computational Fluid Dynamics

Abstract. To add reliability to numerical simulations, Uncertainty Quantification is considered to be a crucial tool for clinical decision making. This holds especially for risk assessment of cardiovascular surgery, for which threshold parameters computed by numerical simulations are currently being discussed. A corresponding biomechanical model includes blood flow, soft tissue deformation, as well as fluid-structure coupling. Thereby, structural material parameters have a strong impact on the dynamic behavior. In practice, however, particularly the value of the Young's modulus is rarely known in a precise way, and therefore, it reflects a natural level of uncertainty. In this work, we introduce a stochastic model for representing variations in the Young's modulus and quantify its effect on the wall sheer stress and von Mises stress by means of the polynomial chaos method. We demonstrate the use of uncertainty quantification in this context and provide numerical results based on an aortic phantom benchmark model.

Introduction

Highest reliability is a requirement for diagnostic techniques in the field of surgical risk assessment. In its technological design, the diagnostic uncertainty is to be minimized, or, at least, to be quantified for conscientious clinical decision making. In preparation for cardiovascular surgery, individual risk assessment for patients suffering from cardiac conditions can be based on diagnostic parameters defined in international guidelines [1]. In the case of aneurysmatic pathologies, these parameters should i. a. predict the risk of the vessel wall to rupture. However, literature shows an ongoing discussion on a proper choice of reliable indicators [2, 3].

A recent review [4] on methods of finite element analysis pointed out that simulations of aortic biomechanis have the potential to support the individual assessment of risk factors. In particular, occurring low or high wall shear stress, computed by computational fluid dynamics, can indicate conditions of increased risk. As described in [4], several studies showed that, statistically, either remarkable high or low wall shear stress was found in the area of aortic wall ruptures. Wall shear stress influences endothelial cell remodeling by elongation and realignment. Further studies [5] propose to take the stress distribution within the vessel wall under consideration, involving the vessel wall's elasticity and, with that, fluid-structure interaction simulations. Especially in the region of the aortic root and bow, the vessel wall displaces to a significant extent during each cardiac cycle, such that computational fluid dynamics studies with the assumption of a rigid vessel wall may be inadequate. The simulation of blood flow and aortic wall displacement allows to compute the von Mises stress distribution in the aortic wall as a stress measure especially suited for failure analysis. It could serve as an comparative value and alternative risk parameter.

Calibration of cardiovascular numerical simulations is typically based on methods of medical imaging and diagnosis. A number of model parameters can be obtained directly from the respective measurement and, thus, exhibit a relatively small uncertainty with respect to the individual patient. Other parameters can only be determined using highly invasive measurement methods or are even inaccessible. For these parameters, general estimations have to be taken into account, which bear high uncertainty with respect to individual patient data [6]. In this work, we focus on the calibration of aortic fluid-structure simulations. This requires precise knowledge about elastic vessel wall properties, which, in practice, is rarely available and therefore subject to high uncertainty.

We exemplarily consider a model of a prototypical aortic phantom. This silicon phantom has been utilized to evaluate state-of-the-art 4D phase contrast magnetic resonance imaging, as well as to compare fluid flow simulation results to magnetic resonance-measured velocity fields in a previous work [7]. Phase contrast magnetic resonance imaging enables the measurement of a fully spatially resolved and time dependent velocity field of aortic blood flow over one cardiac cycle. The outcome of the previous work suggests to also take the observable displacement and elasticity of the silicon vessel wall into account and hence consider fluid-structure interaction. Our main focus in the present work addresses the general uncertainty with respect to the elastic vessel wall properties and aims to quantify the uncertainty of computed stress distributions. We model the uncertainty of the elastic material in assuming a basic linear constitutive law and describing the Young's modulus by a uniformly distributed random variable. We rely on the (generalized) polynomial chaos expansion [8, 9] in order to model the dependence of quantities of interest with respect to the uncertain Young's modulus. The coefficients of the polynomial chaos expansion are computed numerically by a finite element method. Thereby, we employ a non-intrusive approach based numerical quadrature for discretization of the stochastic space, which allows the re-use of existing software originally developed for deterministic fluid-structure interaction problems. In contrast to classical Monte Carlo methods, this approach exhibits faster convergence and, therefore, only few realizations are required to achieve a high accuracy in approximating the polynomial chaos coefficients. We demonstrate how the uncertainty propagates through the fluid-structure model and evaluate the probability distributions of the wall sheer stress and von Mises stress. In addition, we validate our numerical computations by means of experimental data.

The paper is structured in the following way: In section *Mathematical modeling* we describe the mathematical modeling of the deterministic physics and proceed with an approach of representing uncertainty in section *Uncertainty representation.* In section *Numerical results* a comprehensive discussion on the numerical results is given, followed by a conclusion of this work in section *Conclusions.*

Mathematical modeling

Motivated by a prototypic aortic phantom [7], shown in fig. 1a, we model its essential vessel segment (part 3 in fig. 1a) as a three-dimensional tube with an elastic wall (fig. 2). At a pulsating flow profile over time a blood like fluid is driven through the vessel segment leading to an alternating expansion and contraction of the silicone tube.

In large blood vessels such as the aorta, blood flow can be approximatively modeled as a incompressible Newtonian fluid [10]. In addition, 4D magnetic resonance imaging of the considered phantom showed laminar flow behavior in part 3 of fig. 1a. The corresponding Navier-Stokes equations are stated in equations (1), momentum conservation, and (2), continuity equation. They relate the velocity vector field $v = (v_x, v_y, v_z)^T$ and the pressure field p of fluid flow incorporating the constant parameters density ρ and viscosity ν in the fluid domain Ω_{fluid}:

$$\partial_t v + (v \cdot \nabla)v - \nu \Delta v + \frac{1}{\rho}\nabla p = 0, \qquad \text{in } \Omega_{\text{fluid}}, \tag{1}$$

$$\nabla \cdot v = 0, \qquad \text{in } \Omega_{\text{fluid}}, \tag{2}$$

for all $t \in [0, T]$, for a period time T.

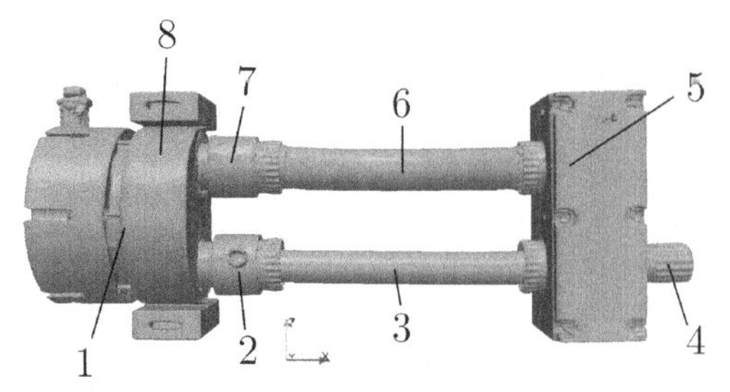

(a) A pulsating flow is induced by an elastic membrane in part (1). Directed by artificial valves (2 and 7), the internal fluid is first driven through the aortic vessel segment (3). The latter is of elastic silicon imitating the elasticity of an aortic vessel wall. Emulating the resistance effect of the capillary system, the flow resistance at the transition from (3) into the fluid reservoir (5) can be adjusted (4). The second tubular segment (6) closes the circuit.

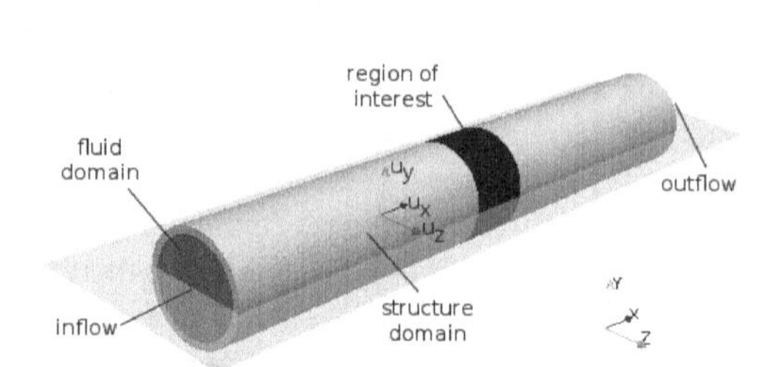

(b) Three-dimensional computational model of the essential vessel segmet (3) shown in fig. 1a. The fluid flow domain is surrounded by the silicone structure domain, which is subject to displacement u. The considered quantities of interest are computed as average values over the region of interest. More details are given in fig. 2 with respect to the indicated cut plane.

Fig. 1: The considered prototypic aortic phantom from [7] forms a closed fluid circuit reflecting basic parts of the human blood circuit.

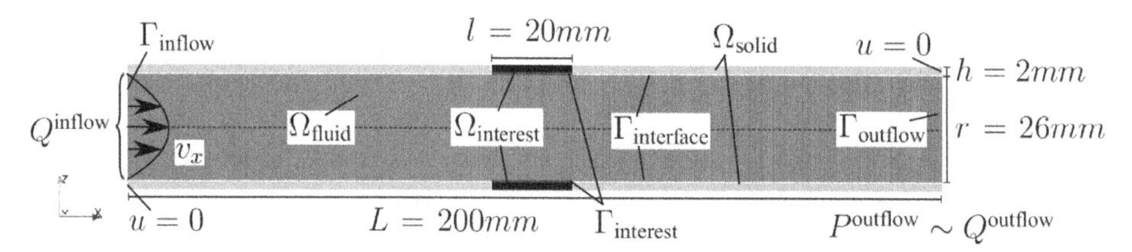

Fig. 2: Cross section as indicated in fig. 1b of the computational domain. The denotations of computational domains and interfaces as well as geometrical dimensions are illustrated. Additionally, in- and outflow boundary conditions for the fluid flow and fixing boundary conditions for the structure displacement are indicated.

The pulsatile inflow rate $Q^{\text{inflow}} = \int_{\Gamma_{\text{inflow}}} v ds$ (fig. 3) is obtained by phase-contrast magnetic resonance measurement data from the prototypical aortic phantom and interpolated by Fourier-interpolation to obtain a smooth periodic inflow rate profile in time. In the computational model, the measured inflow rate is realized with a Poisseuille Dirichlet boundary condition on the inflow boundary Γ_{inflow} (cf. fig. 2).

To emulate the resistance at the outflow of the model we use a linear absorbing outflow boundary condition proposed in [11]. Additionally to the structural material parameters Poisson ratio γ and the Young's modulus Y, it takes the outflow cut-plane area A and the wall thickness h into account. The model imposes a pressure boundary condition at the outflow by means of a linear relation of outflow rate Q^{outflow} and pressure average over the outflow boundary P^{outflow}:

$$P^{\text{outflow}} = \frac{\sqrt{\rho\beta}}{\sqrt{2}A^{5/4}} Q^{\text{outflow}}, \qquad \beta = \frac{\sqrt{\pi} h Y}{1-\gamma^2}. \tag{3}$$

During each cardiac cycle, the silicone wall of the considered vessel segment shows expanding and contracting displacement, which we describe by a displacement vector field $u = (u_x, u_y, u_z)^T$ (cf. fig. 2). Derived deformation measures are given by the deformation gradient tensor $F = \nabla u + I$ and the symmetric, rotationally invariant Green-Lagrange strain tensor $E = \frac{1}{2}(F^T F - I)$. The structure

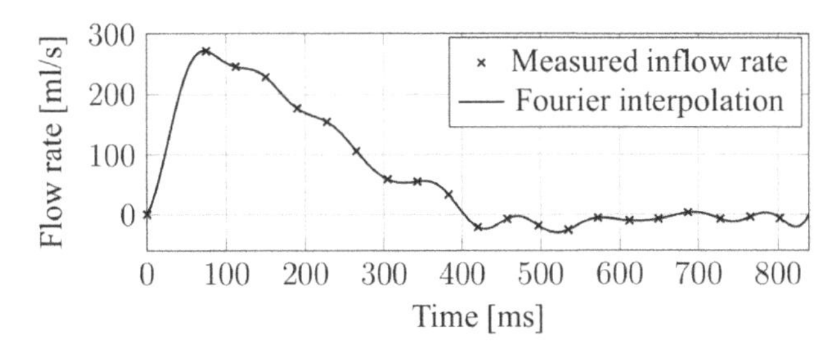

Fig. 3: The inflow rate, measured by 4D magnetic resonance imaging is trigonometrical interpolated to obtain a smooth periodic function defining the computational inflow boundary condition.

of human aortic wall tissue exhibits complex inhomogeneous, nonlinear and anisotropic behavior which is infeasible to assess in living tissue of a patient [12]. To reflect the uncertainty in the general material behavior, we use the Saint-Venant Kirchhoff constitutive law as a basic structural model along with high uncertainty in the stiffness of the model. In addition being also suitable for silicone material, the Saint-Venant Kirchhoff model constitutes a linear relation between the Green-Lagrange strain tensor E and the second Piola-Kirchhoff structural stress tensor $\Sigma = \lambda_1 \mathrm{tr}(E)I + 2\lambda_2 E$. The elastic material parameters Poisson ratio ξ and the Young's modulus Y specify the Lamé constants $\lambda_1 = Y\gamma(1+\gamma)^{-1}(1-2\gamma)^{-1}$, $\lambda_2 = 0.5Y(1+\gamma)^{-1}$ [10]. The resulting elasticity equation of momentum conservation is given for all $t \in [0, T]$ by

$$\partial_t^2 u - \frac{1}{\rho}\nabla \cdot (F\Sigma) = 0, \qquad \text{in } \Omega_{\text{solid}}. \tag{4}$$

According to the configuration of the silicone phantom (fig. 1a), the inlet and the outlet of the vessel segment are fixed by Dirichlet zero boundary conditions for the displacement field.

At the interface $\Gamma_{\text{interface}} = \Omega_{\text{fluid}} \cap \Omega_{\text{solid}}$ between the fluid flow domain and structure volume, physiological coupling conditions are given by the equality of fluid and structure velocity (5) as well as the equality of fluid and structure boundary forces (6) [10]. Hereby, n_{fluid} and n_{solid} denote the outer unit normal vector of the respective domain (cf. fig. 2).

$$v = \partial_t u, \qquad \text{on } \Gamma_{\text{interface}}, \tag{5}$$

$$(\rho\nu(\nabla v + \nabla v^T) - pI) \cdot n_{\text{fluid}} + F\Sigma \cdot n_{\text{solid}} = 0, \qquad \text{on } \Gamma_{\text{interface}}. \tag{6}$$

As described in the *introduction*, stress measures on the aortic vessel wall can be taken into account that may indicate high risk of adverse aortic events. We compute the stress measures averaged over a region of interest Ω_{interest}, which we locate in the middle of the vessel segment. It has a length of $l = 20mm$ and stands representatively for an area where rupture risk conditions are to be investigated. Firstly, we focus on the von Mises stress as a stress measure especially suited for failure analysis

$$\zeta = \frac{1}{|\Omega_{\text{interest}}|}\int_{\Omega_{\text{interest}}} \sqrt{\sigma_{xx}^2 + \sigma_{yy}^2 + \sigma_{zz}^2 - \sigma_{xx}\sigma_{yy} - \sigma_{xx}\sigma_{zz} - \sigma_{yy}\sigma_{zz} + 3(\sigma_{xy}^2 + \sigma_{xz}^2 + \sigma_{yz}^2)}dx, \tag{7}$$

where $\sigma = \det(F)^{-1}F\Sigma F^T$ denotes the Cauchy stress tensor. Secondly, we consider the magnitude of the wall shear stress, which is exerted by the fluid flow field on the fluid-structure interface with normal vector n_{fluid}. It has an influence on endothelial cell remodeling via elongation and realignment [4].

$$\tau = \frac{1}{|\Gamma_{\text{interest}}|} \int_{\Gamma_{\text{interest}}} |\rho\nu(\nabla v + \nabla v^T) \cdot n_{\text{fluid}}| ds. \tag{8}$$

Uncertainty representation

Uncertainty model for Young's modulus. We model the scalar Young's modulus Y as a single random variable with uniform probability distribution. Specifically, we define

$$Y := Y(y) := Y_0 + yY_1, \quad y \sim U(-1,1), \tag{9}$$

where y is a uniformly distributed random variable in the interval $(-1,1)$. Consequently, $Y_0 > 0$ denotes the mean of Y and $Y_1 > 0$ a variation factor. Their specific values will be defined in Section *Numerical results*. Such a probability model is usually referred to as "ignorance model", where we do not assume any preferred values on the probability of Y.

Polynomial Chaos expansion. Polynomial Chaos expansions [8, 9] use a truncated series of orthogonal polynomials in random variables to model the dependence of quantities of interest on the input parameters, in our case the Young's modulus. We illustrate the approach exemplified on the velocity component v. Assuming a finite variance of v, we can express its dependence on y via:

$$v = v(y) = \sum_{i=0}^{\infty} v_i L_i(y), \tag{10}$$

where L_i denote Legendre polynomials in y with polynomial degree $i \in \mathbb{N}$. This approach is commonly referred to as generalized polynomial chaos expansion [9], since the classical formulation employs Gaussian distributed random variables with Hermite polynomials. Legendre polynomials satisfy the orthogonality condition

$$\langle L_i, L_j \rangle := \frac{1}{2} \int_{-1}^{1} L_i(y) L_j(y)\, dy = 0, \quad i \neq j, \tag{11}$$

where the factor $1/2$ corresponds to the probability density function of y. For numerical computation the series in (10) is truncated by prescribing a maximum polynomial degree $P \in \mathbb{N}$, such that

$$v(y) \approx \sum_{i=0}^{P} v_i L_i(y). \tag{12}$$

The "modes" or coefficients v_i need to be computed numerically, which we explain in the following section. Note that the polynomial chaos expansion is employed in a similar way for all other quantities of interest. We assume the same truncation order P for every expansion.

Computation of the polynomial chaos expansion. Many methods exist in order to compute the coefficients in the polynomial chaos expansion. They can be categorized as "intrusive" and "non-intrusive", differing in the fact the first requires the modification of existing software code for the solution of deterministic systems, while the latter only requires to evaluate solutions on prescribed or randomly given parameter values. This makes "non-intrusive" methods very attractive for extending existing software, and it comes with the benefit of independent sampling or independent evaluation points allowing efficient parallel computations. The most popular approaches are Monte-Carlo methods, which, however, typically require a large number of parameter realizations to obtain a good accuracy. In contrast, sparse-grid collocation methods [13] are based on deterministic quadrature rules, which have prescribed parameter evaluation points and show fast convergence for low to moderate sized stochastic problems. Since in our case, we only consider one random variable y, the collocation approach simplifies to a one-dimensional quadrature rule.

We again illustrate the procedure by focusing on the velocity $v = v(y)$. Due to the orthogonality of the polynomial chaos expansion, the corresponding modes can be computed by

$$v_k = \frac{1}{2}\frac{1}{\langle L_k, L_k\rangle}\int_{-1}^{1} v(y)L_k(y)\,dy, \tag{13}$$

for $k = 0, \dots, P$. Using a numerical quadrature rule, the integral in (13) can be approximated by

$$v_k \approx \frac{1}{2}\frac{1}{\langle L_k, L_k\rangle}\sum_{i=1}^{N} w_i v(y_i)L_k(y_i), \tag{14}$$

where $w_i \in \mathbb{R}$ denote the weights and $y_i \in [-1, 1]$ denote the evaluation points of the employed numerical quadrature rule. Therefore, only $v(y_i)$ needs to be computed numerically by solution of deterministic systems associated to y_i in order to compute the polynomial chaos expansions. Since we model v as a polynomial in y, a Gauss-Legendre Quadrature rule would be very efficient in terms of accuracy. However, it is not nested, i.e., increasing N for higher degrees P requires to recompute everything. In contrast, the Clenshaw-Curtis rule requires more quadrature points than a Gauss-Legendre rule, but exhibits a nested structure.

Based on the polynomial chaos expansion, stochastic moments, e.g., mean and standard deviation, can be computed in a straight-forward way due to the orthogonality of the chaos polynomials:

$$\mathbb{E}(v) = v_0, \quad \sigma(v) = \left(\sum_{i=1}^{P} v_i^2\langle L_i, L_i\rangle\right)^{1/2}. \tag{15}$$

The major benefit, however, lies in the fact that a continuous representation of the uncertainty dependency has been obtained. Therefore, the polynomial chaos expansion can be used for example for evaluating the probability density function by means of histograms to get a full characterization of the propagated uncertainty.

Numerical results

Table 1 Summarizes the parameter values employed for our benchmark problem.

Parameter	Symbol	Value
Density (fluid and structure)	ρ	1×10^3 kg m^{-3}
Viscosity	ν	4×10^{-6} m^{2}s^{-1}
Reynolds number	Re	3500
Young's modulus	$[Y_0,\ Y_1]$	$[500,\ 300]$ kPa
Poisson ratio	ξ	0.45
Period time	T	840 ms

Table 1: List of parameters the numerical simulation is calibrated with. The blood flow parameters, namely density and viscosity, are chosen from [10]. Values for the structural parameters, given by Poisson ratio and uncertain Young's modulus, are motivated by [14].

The fluid-structure interaction-problem is solved with a monolithic arbitrary Lagrangian Eularian approach with a fixed reference domain [10]. To stabilize the slightly convection dominated flow problem, we use a simplified residual-based streamline diffusion scheme [15]. Discretization of the model equations is done by the finite element method with a stable combination of quadratic Lagrange elements for displacement and velocity and linear elements for the pressure. Time-stepping is realized with the second-order accurate Crank-Nicolson scheme. We employ a 8th order polynomial chaos expansion for the quantities of interest. For an exact evaluation of all stochastic integrals, this results in 17 collocation points of the Clenshaw-Curtis quadrature rule.

In each cardiac period, the fluid propagates in a pulsatile way through the tube-shaped model. Coming along with the pulsatile inflow rate (fig. 3), a pressure wave traverses the vessel, which induces consecutive displacement of the vessel wall.

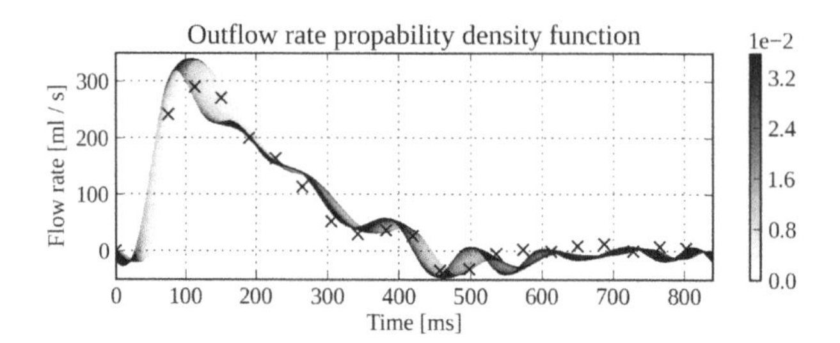

Fig. 4: The resulting probability density function of the outflow rate is shown. The crosses denote outflow measurements of 4D magnetic resonance imaging for comparison.

For the purpose of validation, one can compare the experimental outflow rate data with the corresponding computational results (fig. 4). The computed outflow rate is in relatively well agreement with experimental data, which for the greater part is found within the range of the outflow probability density function. Small deviations can occur due to discretization errors of the deterministic model.

Fig. 5 and fig. 6 show different modes of the polynomial chaos expansion for the Von Mises stress (7) and the wall shear stress (8). A decay in the mode's amplitude can be observed, which reflects the convergence property of polynomial chaos. The amplitudes quickly decrease three orders of magnitude from zeroth mode to higher mode indices (not depicted). In total, only 17 deterministic sample simulations were required to compute the polynomial chaos modes. Thereby, a single run consists of about 1 million spatial degrees of freedom per time step and takes about 1 day of computing time on a 32 core machine. This would render a standard Monte Carlo approach impractical, due to its slow convergence rate $\mathcal{O}(1/\sqrt{N})$ with N denoting the number of samples.

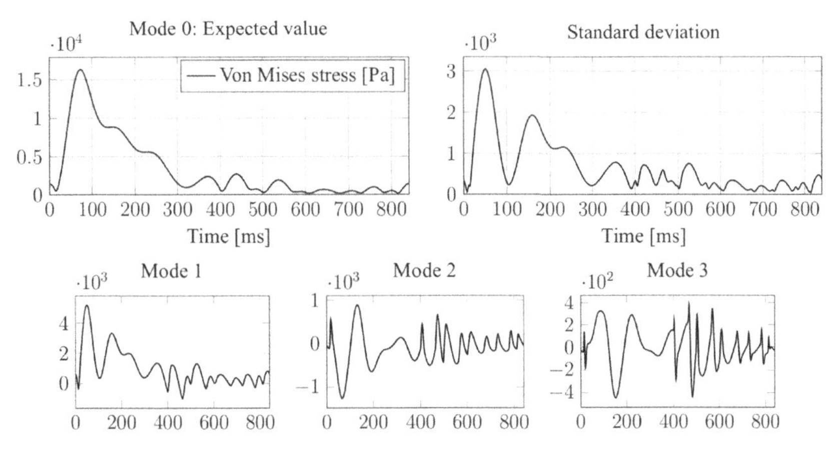

Fig. 5: Quantification of uncertainty of the average von Mises stress (7) in the region of interest (cf. fig. 1). The expected value is highest at the peak of inflow rate (cf. fig. 3) and drops to a small value present in the second half of a period. Standard deviation has its maximum shortly before and shows a second peak during decrease of inflow. The amplitudes of higher modes decrease monotonously.

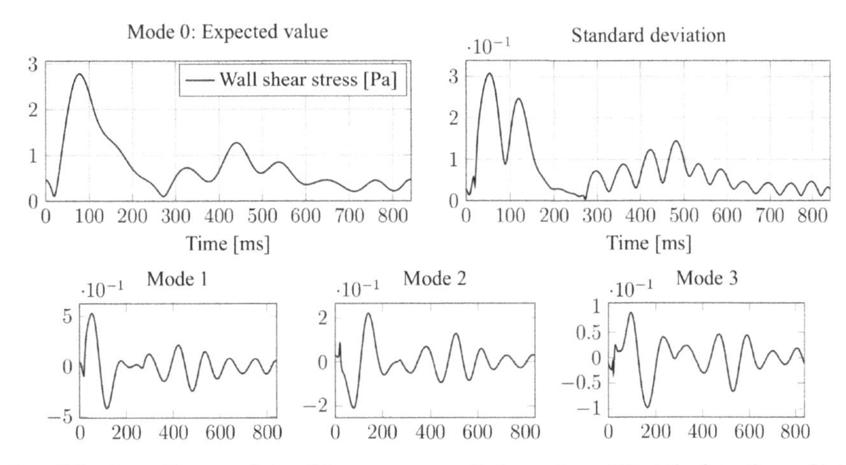

Fig. 6: Quantification of uncertainty of the average wall shear stress (8) at the interface of interest (cf. fig. 1). The expected value is highest at inflow peak (cf. fig. 3). Standard deviation shows two peaks shortly before and after maximal inflow. The amplitudes of higher modes decrease monotonously.

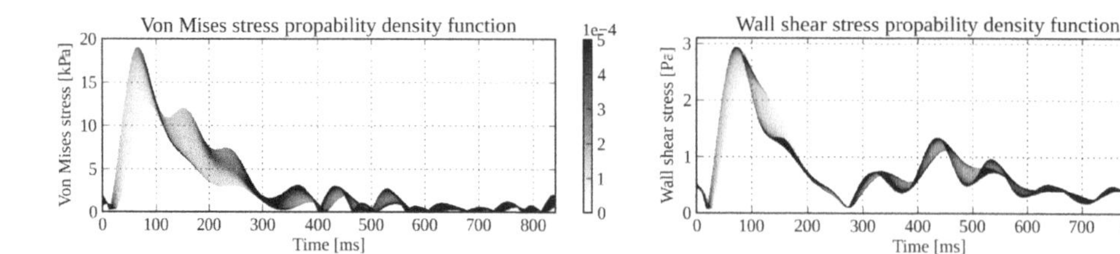

Fig. 7: Illustration of the probability density functions of the average von Mises stress (7) and wall shear stress (8). Both quantities of interest are relatively broadly distributed at highest inflow (cf. fig. 3) and show a tighter distribution in the second half of a period.

Besides the computation of expected value and standard deviation of the quantities of interest (fig. 5 and 6), polynomial chaos enables evaluating the probability density function shown in fig. 7. For both the von Mises stress and the wall shear stress, the expected value of the average over the region of interest has its maximum at highest inflow and flattens to small oscillations subsequently.

The respective standard deviation plots show relatively increased uncertainty at peaks of the respective expected value and oscillations along with oscillations in the expected value. In accordance with the respective standard deviation, the probability density is relatively broadly distributed at times of high standard deviation. The wall shear stress shows less uncertainty at its expected peak than the von Mises stress. This can be due to the direct dependence of the von Mises stress on the Young's modulus. However, the wall shear stress in the first instance depends on the velocity field and just has an indirect connection to the Young's modulus via the fluid-structure interface coupling (5, 6).

Conclusions

In this work, we considered the computation of risk parameters for aortic pre-surgery diagnosis by fluid-structure interaction simulation by the example of a prototypic aortic phantom. We proposed an approach to represent the uncertainty of the hardly assessable stiffness of the aortic wall using polynomial chaos.

Both, the evaluated von Mises stress and wall shear stress show maximal uncertainty at their highest expected value which occurs at maximal inflow rate. However, the wall shear stress appears to be a more reliable parameter than the von Mises stress under uncertain Young's modulus. If the von Mises stress is taken into diagnostic consideration, the results suggest to integrate the dependency on uncertain material properties in the numerical simulation.

One deterministic simulation run of the considered fluid-structure interaction problem showed to be computationally too expensive for standard Monte Carlo methods. However, polynomial chaos in this work approved to give accurate results with few collocation points. With a theoretical convergence rate of $\mathcal{O}(1/\sqrt{N})$ standard Monte Carlo is expected to require orders of more sample realizations than the here used polynomial chaos-based approach with satisfactory convergence using 17 collocation points.

Further developing the proposed approach to the quantification of uncertainty in an aortic fluid-structure simulation, future work may include the reduction of parameter uncertainty based on experimentally measured flow field data we already used in this work for validation. Another aspect can be seen in the application to patient-specific data with a heterogeneous uncertainty distribution of material parameters over a vessel wall area.

Acknowledgement

The authors wish to thank the Transregional Collaborative Research Centre (TCRC) 125 'Cognition-Guided Surgery' of the German Research Foundation (DFG) for providing the setup of the phantom. The authors gratefully acknowledge the computing time granted by the John von Neumann Institute for Computing (NIC) and provided on the supercomputer JUROPA at Jülich Supercomputing Centre (JSC).

References

[1] Hiratzka, L. F., et al., 2010 ACCF/AHA/AATS/ACR/ASA/SCA/SCAI/SIR/STS/SVM Guidelines for the Diagnosis and Management of Patients With Thoracic Aortic Disease: Executive Summary. Journal of the American College of Cardiology 55(14) (2010) 1509–1544.

[2] Pape, L. A., et al., Aortic diameter 5.5 cm is not a good predictor of type a aortic dissection observations from the international registry of acute aortic dissection (irad). Circulation 116(10) (2007) 1120–1127.

[3] Cozijnsen, L., Braam, R. L., Waalewijn, R. A., Schepens, M. A., Loeys, B. L., van Oosterhout, M. F., Barge-Schaapveld, D. Q., and Mulder, B. J., What is new in dilatation of the ascending aorta? review of current literature and practical advice for the cardiologist. Circulation 123(8) (2011) 924-928.

[4] Chung, B. and Cebral, J., CFD for evaluation and treatment planning of aneurysms: Review of proposed clinical uses and their challenges. Annals of Biomedical Engineering 43(1) (2015) 122–138.

[5] Valencia, A., Burdiles, P., Ignat, M., Mura, J., Bravo, E., Rivera, R., and Sordo, J. Fluid structural analysis of human cerebral aneurysm using their own wall mechanical properties. Computational and mathematical methods in medicine, (2013).

[6] Damughatla, Anirudh R., et al. Quantification of aortic stiffness using MR Elastography and its comparison to MRI-based pulse wave velocity. Journal of Magnetic Resonance Imaging 41(1) (2015) 44-51.

[7] Kratzke, J., Schoch, N., Weis, C., Müller-Eschner, M., Speidel, S., Farag, M., Beller, C., Heuveline, V. Enhancing 4D PC-MRI in an aortic phantom considering numerical simulations SPIE: Physics of medical imaging, 9412-47 (2015).

[8] Wiener, N., The homogeneous chaos. American Journal of Mathematics, 60(4) (1938) 897-936.

[9] Xiu, D. and Karniadakis, G. E., The Wiener-Askey polynomial chaos for stochastic differential equations. SIAM J. Sci. Comput. 24(2) (2002) 619-644.

[10] Formaggia, L., Quarteroni, A. M., and Veneziani, A., Cardiovascular mathematics. Milan, Springer, 2009.

[11] Janela, J., Moura, A., and Sequeira, A., Absorbing boundary conditions for a 3d non-newtonian fluidstructure interaction model for blood flow in arteries. International Journal of Engineering Science 48(11) (2010) 1332 - 1349.

[12] Fung, Y. C., Biomechanics: Mechanical Properties of Living Tissues, Second Edition, Springer, New York, NY, 1993.

[13] Nobile, F., Tempone, R. and Webster, C.G., A Sparse Grid Stochastic Collocation Method for Partial Differential Equations with Random Input Data. SIAM J. Numer. Anal., 46(5) (2008) 2309-2345.

[14] Nichols, W., O'Rourke, M., and Vlachopoulos, C. (Eds.). McDonald's blood flow in arteries: theoretical, experimental and clinical principles. CRC Press, 2011.

[15] Brooks, A. N. and Hughes, T. J., Streamline upwind/petrov-galerkin formulations for convection dominated flows with particular emphasis on the incompressible navier-stokes equations. Computer methods in applied mechanics and engineering 32(1) (1982) 199-259.

Applied Mechanics and Materials Vol. 807 (2015) pp 45-54
© (2015) Trans Tech Publications, Switzerland
doi:10.4028/www.scientific.net/AMM.807.45

Submitted: 2015-04-30
Accepted: 2015-08-04

Investigation of uncertainty sources of piezoresistive silicon based stress sensor

Palczynska Alicja[1,a], Schindler-Saefkow Florian[2,b], Gromala Przemyslaw[1,c], Kerstin Kreyßig[2,d], Sven Rzepka[2,e] Dirk Mayer[3,f] and Melz Tobias[3,g]

[1]Robert Bosch GmbH, Reliability Modeling and System Optimization (AE/EDT3), Reutlingen, 72703, Germany

[2]Fraunhofer ENAS, Micro Materials Center MMC, Chemnitz, 09126, Germany

[3]Fraunhofer-Institut für Betriebsfestigkeit und Systemzuverlässigkeit LBF, Darmstadt, 64289, Germany

[a]Alicja.Palczynska@de.bosch.com, [b]Florian.Schindler-Saefkow@enas.fraunhofer.de, [c]PrzemyslawJakub.Gromala@de.bosch.com, [d]Kerstin.Kreyssig@enas.fraunhofer.de, [e]Sven.Rzepka@enas.fraunhofer.de, [f]Dirk.Mayer@lbf.fraunhofer.de, [g]Tobias.Melz@lbf.fraunhofer.de

Keywords: stress sensor, measurement uncertainties, sensitivity analysis

Abstract. The aim of this paper is to get insight into measurement uncertainties for thermomechanical measurements performed using a piezoresistive silicon-based stress sensor in a standard microelectronic package. All used sensors have the same construction, were produced in the same technological processes at the same time, yet the measurement results show significant distribution. The possible causes for this phenomenon are discussed in this paper. Additionally, Finite Element Method (FEM) model is created and validated, what enables a study of sensitive parameters influencing the measurement uncertainties.

Introduction

A piezoresistive silicon based stress sensor has already proven itself as a powerful tool to monitor the stresses inside electronic packages during various production processes e.g. transfer molding [1], post mold cure [2,3] and underfill process [4,5]. It can also be used to monitor thermomechanical loads during thermal cycling. Recently the potential to use it in Prognostics and Health Monitoring (PHM) has been investigated [6,7]. However, after performing a series of measurements with these sensors, a spread in the results was noticed. In order to make conclusions about the state of health based on the measurements conducted with this chip, it is mandatory to know the sources of this spread. Thus, the need to investigate the measurement uncertainties of this chip arose. To realize this goal, statistical analysis of the measurements was conducted to quantitatively evaluate the variance of measurements and possible causes were found. Moreover, a FEM model was created to enable an analysis of influence of specified parameters on the behavior of the package.

Experimental setup

The sensor used in this study is a piezoresistive stress sensor, called IForce [8]. It has the advantage over other types of sensors (such as strain gages) that similarly built components are used in real products. In consequence, the obtained information can be interpreted as the stress state in actual devices. Its specific construction enables the stress measurements with high spatial resolution. In this device the sensing elements are created by the channels of MOSFET transistors in a current mirror circuit as shown in Fig. 1. The current mirror is very sensitive to differences in parameters of the transistors constituting it. The channels of MOSFETs are oriented in such a way that the change in stress is changing their resistivity. Both of these properties are used to measure the stresses with very high sensitivity [9]. By measuring the currents flowing through both branches of the current mirror, one can calculate an in-plane shear stress, σ_{xy}, and the difference in in-plane normal stress components, σ_{xx}-σ_{yy}, from the following relationships [9]:

$$\sigma_{xy} = \frac{1}{\pi_{11}^{(n)} - \pi_{12}^{(n)}} \frac{I_{OUT} - I_{IN}}{I_{OUT} + I_{IN}} \quad , \tag{1}$$

$$\sigma_{xx} - \sigma_{yy} = \frac{1}{\pi_{44}^{(p)}} \frac{I_{OUT} - I_{IN}}{I_{OUT} + I_{IN}} \quad . \tag{2}$$

Where:
π_{11}, π_{12}, π_{44} - piezoresistive coefficients of silicon,
I_{IN}, I_{OUT} – currents measured respectively at the input and at the output of current mirror.

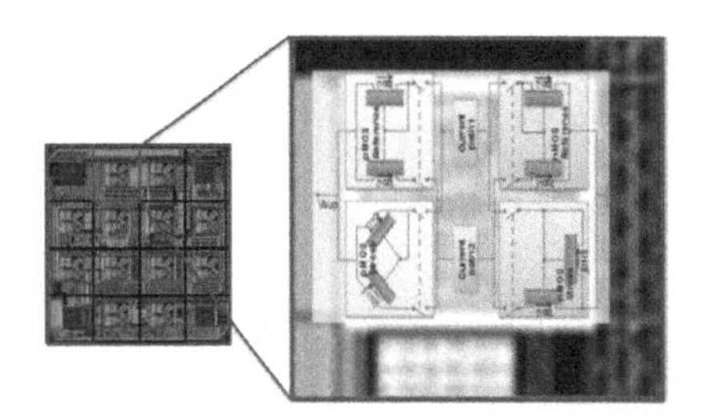

Fig. 1 Construction of IForce stress sensor [9]

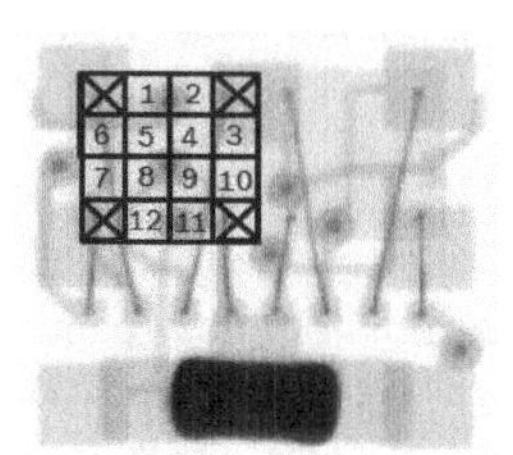

Fig. 2 X-Ray image of applied sensor with 12 sensing cells marked

Each sensor consists of a matrix of sensing cells. The sensor with 12 sensing cells is used in the test, being placed in a 4 × 4 array. Four cells in the corners are inactive as marked in Fig 2 that presents a layout of the sensor. The sensor is packaged in a standard microelectronic LGA package as presented in Figure 3. The silicon die is attached to a PCB using a die attach adhesive. Electrical connections are formed by wire bonds. Additionally, a dummy ceramic component soldered on the PCB. The whole construction is overmolded using commercially available epoxy molding compound. The final dimension of the package is 3 mm × 3 mm × 1 mm.

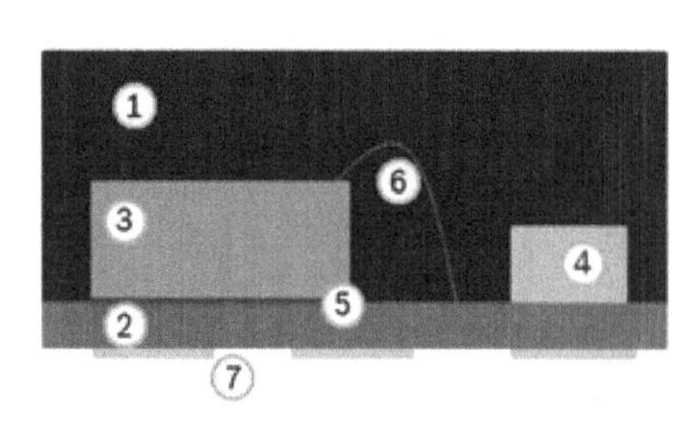

Fig. 3 Construction of LGA Package. 1 – mold, 2 – PCB, 3 – stress sensor, 4 – ceramic, 5 – die attach, 6 – wire bond, 7 – soldering pads

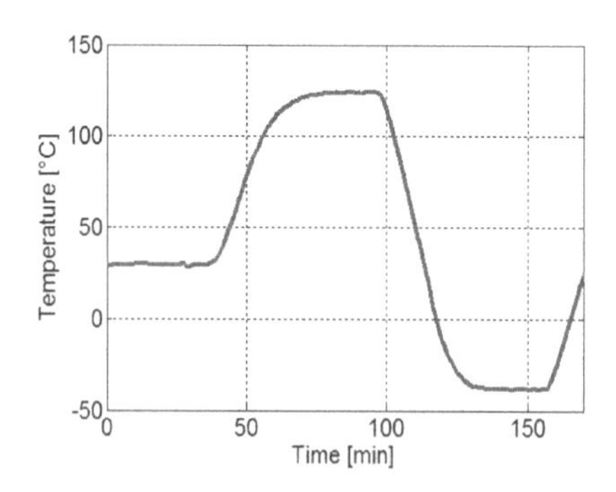

Fig. 4 Passive temperature profile used in the investigation

In this investigation 6 samples were measured during passive thermal cycling. The temperature profile is shown in Fig. 4. Temperature during cycle ranges from -40°C to 125°C with a dwell of

20 minutes at maximal and minimal point. Since all the samples were measured simultaneously in a climate chamber, the errors related to the temperature differences can be omitted assuming uniformity of the temperature inside the chamber. According to the specification, the temperature fluctuation inside a chamber does not exceed ±0.5 K.

Statistical analysis of data

The analysis is conducted using standard measurement uncertainty evaluation tools, it means average, standard deviation and relative standard deviation. In this paper only type A uncertainty will be considered, focusing on the differences between samples. It is reasonable, knowing that the measurement equipment is very accurate and calibrated before every measurement. Furthermore, the uncertainties related to the temperature drift of piezoresistive coefficients and calibration of the sensor were already assessed by Jaeger et al. [10] and are not in scope of this work. The sensor itself is very accurate. In this paper it is investigated, if the discrepancies between different samples can be caused by different stress situation introduced by production parameters distribution.

There are two domains in which the results can be evaluated. First, the behavior of measured values in time with changing temperature can be investigated. Additionally, at the given temperature, the spatial distribution of uncertainties can be considered. It is reasonable to consider the data from every cell separately, not to loose information about spatial distribution of uncertainty. It should be also mentioned, that, to be able to compare the results with simulation, we consider the evolution of the stresses during thermal cycle, not absolute values of stress.

The average shear stress and normal in-plane stress difference in all cells are presented in Fig. 5a) and b). It can be observed that the difference in normal in-plane stress has much bigger amplitude than shear stress, what influences the magnitudes of standard deviation. To make it easier to understand what causes the difference in respective cells, the average spatial distribution of stress is evaluated at temperatures 125°C and -40°C for Panel 1 as are shown in Fig. 6 and Fig 7. The measured difference in normal in-plane stress is smallest at the left edge of the chip and the magnitude of change is the smallest. Shear stress is very small in the most of the area of the chip, only on the borders its value rises. These observations are useful in order to further analyze of the results.

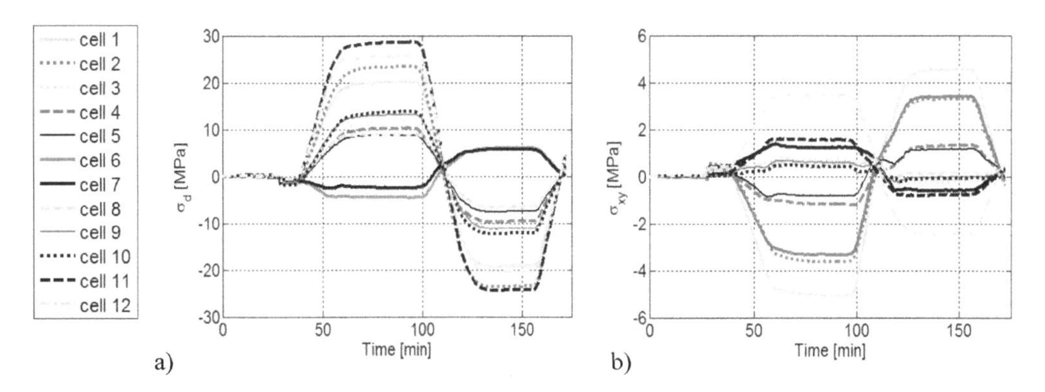

Fig. 5 The average a) normal in-plane stress difference b) shear stress of the Panel 1 population

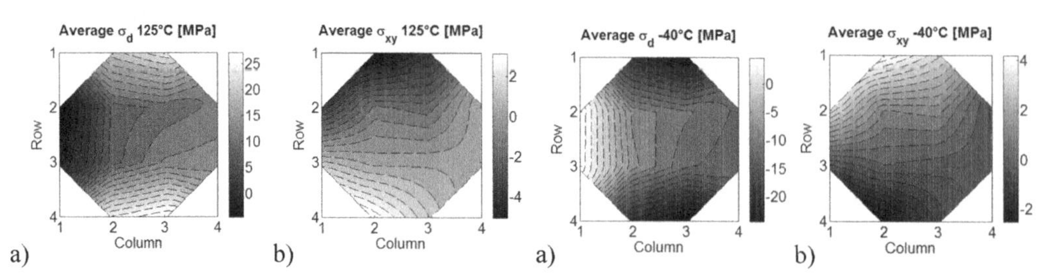

Fig. 6 Average stresses measured in Panel 1 at 125°C a) difference in normal in-plane stresses b) shear stress

Fig. 7 Average stresses measured in Panel 1 at -40°C a) difference in normal in-plane stresses b) shear stress

Uncertainties in time

To evaluate the variation of the data the standard deviation, as well as relative standard deviation is calculated. The standard deviation of both σ_D and σ_{XY} feature time dependency on time as shown in Fig. 8. It seems to change proportionally to the average - it's the smallest at the points where the average is close to zero. The average values of stresses cross zero at some points. Therefore, it is numerically impossible to calculate relative standard deviations in the ranges close to it. Thus the relative standard deviations are evaluated only at -40°C and 125°C and are presented in next section.

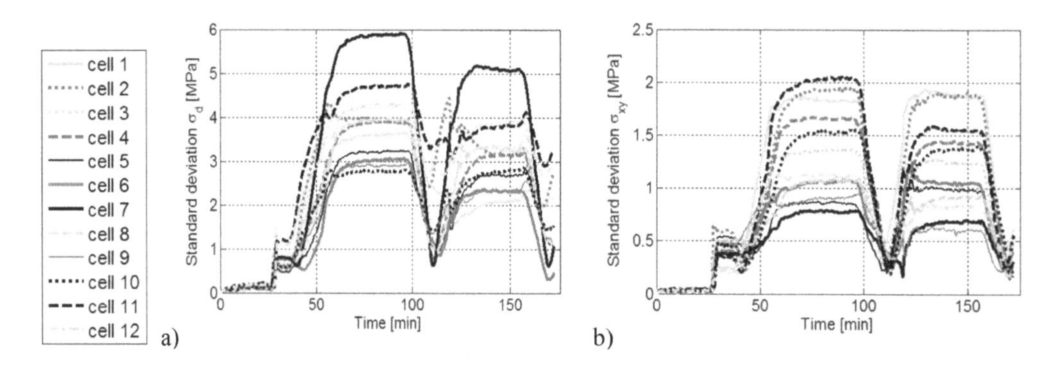

Fig. 8 Standard deviation in time a) normal in-plane stress difference b) shear stress

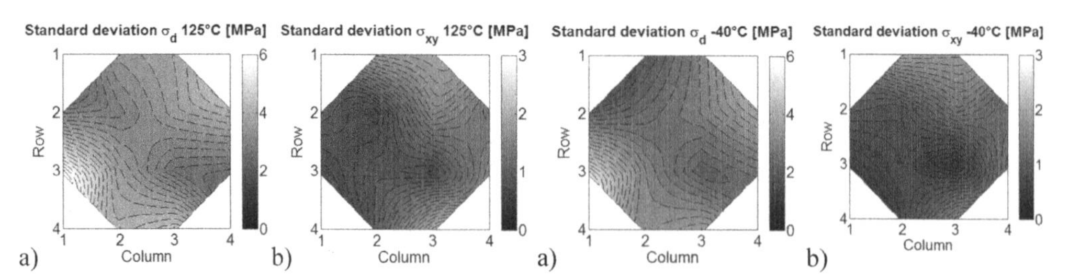

Fig. 9 Standard deviation of stresses measured in Panel 1 at 125°C a) difference in normal in-plane stresses b) shear stress

Fig.10 Standard deviation of stresses measured in Panel 1 at -40°C a) difference in normal in-plane stresses b) shear stress

Uncertainties over the chip

The spatial distribution of standard deviation at -40°C and 125°C are presented in Fig. 9 and Fig.10. The highest uncertainties for difference in normal in-plane stresses are observed in cell 7 in the corner of the chip and for shear stress in cells 1 and 2 in upper part of the chip. It is confirmed for difference in normal in-plane stress by the relative standard deviation presented in Fig. 11 and 12. Comparing the relative uncertainties between normal in-plane stresses and shear stresses, it is noticeable that for shear stresses the values are much higher. This is understandable since the average values of shear stress are significantly lower. The regions in Fig. 11 and 12 where the relative standard deviations are very high are the regions with very small values of stresses measured. To confirm these observations the relative and absolute standard deviations are plotted against average values of stresses as shown in Fig. 12 and Fig. 13. These plots endorse the hypothesis that the big values of relative standard deviations are caused by dividing by small averages. At certain level of measured stress the relative standard deviation stabilize at about 15% and 35% for difference in normal in-plane stresses and shear stresses, respectively. There is no clear trend visible in absolute standard deviation behavior; it reaches 6 MPa for σ_D and 2 MPa for σ_{XY}.

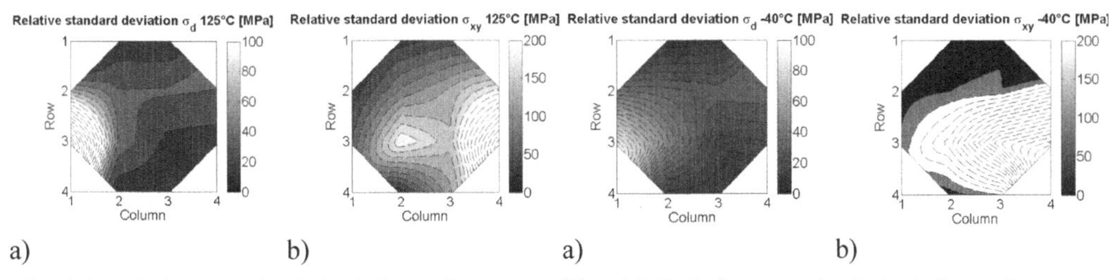

a) b)

Fig. 11 Relative standard deviation of stresses measured in Panel 1 at 125°C a) difference in normal in-plane stresses b) shear stress

a) b)

Fig. 12 Relative standard deviation of stresses measured in Panel 1 at -40°C a) difference in normal in-plane stresses b) shear stress

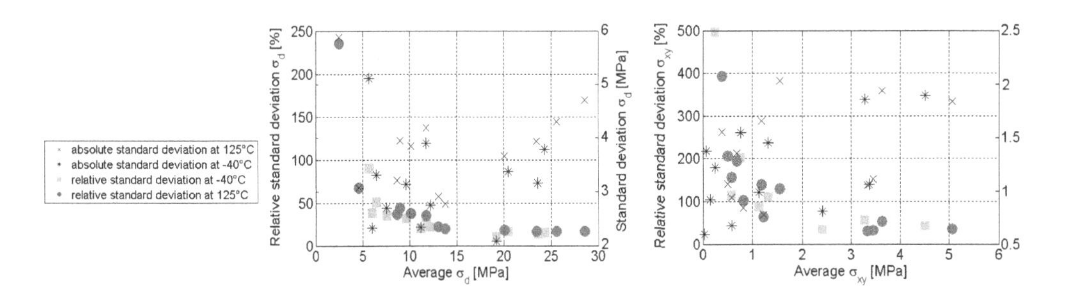

Fig. 12 Relative and absolute standard deviations versus average value of stresses for difference in normal in-plane stresses

Fig. 13 Relative and absolute standard deviations versus average value of stresses for shear stresses

Looking at the evaluated measurement uncertainties, it is obvious that the variations between the samples are substantial. In the following section the possible causes for these differences will be investigated.

Variation sources

The stress state on a chip can be influenced by many parameters related to electronic packaging processes. Possible reasons for differences between the stress levels in investigated sensors are:

Die attach thickness - during the production process the adhesive is deposited on the PCB, then the silicon die is placed on top of it and the curing of glue is performed. The thickness of die attach given in the specification is 10 μm, but it may vary depending on the position in the batch etc. This has influence on the stresses measured in the chip.

Silicon die thickness - in order to encapsulate the sensor in LGA package, the silicon die is thinned to 590 μm. This process could have been imprecise, leading to differences in silicon chip thickness.

Mold coverage – the process of transfer molding is controlled to obtain 200 μm mold coverage. The variances in these values can also have a substantial influence on the stress on the chip.

Position of the chip – looking at the X-Ray images of various sensors, the slight differences in placement of the die can be noticed. For example, in one chip the cell is placed over a soldering pad and due to small shift, in the other chip it is not, it can potentially create the variances in the stress distribution. It is a possible reason for the highest difference in normal in-plane stresses standard deviation in cell 7.

Material parameters – the production processes as well as different material lots can cause the differences in the material properties between the samples, caused for example by inhomogeneous temperature during transfer molding, curing etc.

All the above parameters influence the stress state in a package. The further investigation should be conducted to analyze which of them has the major influence on the stresses and if it is possible to explain the observed variations in measured values taking them into account. This study can be conducted with help of FEA modeling. In the next section the created FEA model will be introduced.

FEA modeling

The numerical model was created using ANSYS software. All the materials were characterized and their thermomechanical properties are summarized in Table 1. The model was verified using moiré interferometry measurements [10]. The discretized model of silicon chip was created in a way that enables extraction of the results from individual cells as obtained from measurements as shown in Fig. 15.

Table 1 Material parameters

Material	Young's modulus [GPa]	CTE [ppm/K]
Silicon	168.9	2.8
Copper	125	17
Solder	50	20
Al	64	25.3
Ceramic	300	8.0

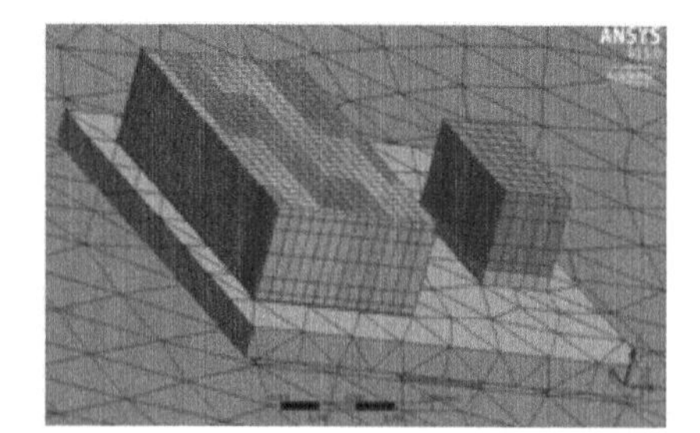

Fig. 15 The details of the stress sensor model

The comparison between results from simulation and measurements is presented in Fig. 16 and Fig. 17. The results obtained from simulation agree with the measurement taking into account measurement uncertainties. Thus, the simulation can be used to perform a sensitivity study for the sources of uncertainties.

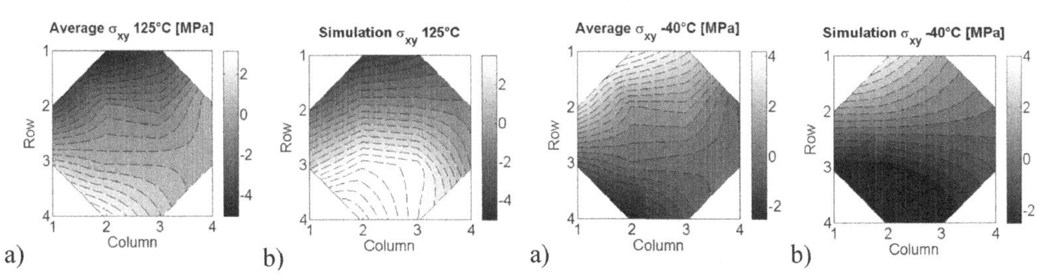

Fig. 16 Comparison between simulation and measurement for shear stress at 125 °C a) measurement b) simulation

Fig. 17 Comparison between simulation and measurement for shear stress at -40 °C a) measurement b) simulation

Sensitivity study

To perform an investigation of the influence of variation of different packaging parameters on the measurement results, two potential sources of uncertainties were chosen. First of all, the thickness of the silicon die was adjusted. The default value of die thickness is 590 μm, but due to manufacturing process it can vary. The assumed variation was ±10 μm, thus simulations with silicon die thickness of 580 μm and 600 μm were performed. It is worth to notice, that the mold coverage was changed accordingly, to maintain a constant total package thickness. The other adjusted parameter was the position of the chip inside the package in x and y direction. The variation of this parameter was noticed looking at X-Ray images of large sample of sensors. The assumed variation of position of the chip was ±150 μm in both x and y directions. The simulations were performed separately for each of these cases. In total 7 simulations were executed as summarized in Table 2.

Table 2 Summary of performed simulations

Case	Silicon thickness [μm]	Relative position of the chip in x-direction [μm]	Position of the chip in y-direction [μm]
Reference	590	0	0
Max. silicon thickness	600	0	0
Min. silicon thickness	580	0	0
Max. shift in x direction	590	+150	0
Min. shift in x direction	590	-150	0
Max. shift in y direction	590	0	+150
Min. shift in y direction	590	0	-150

Silicon die thickness

The results of simulations with varied thickness of silicon die are presented in Fig. 18. The values of change introduced by adjustment of the chosen parameter are plotted in function of the cell number to enable easy evaluation of results. In case of shear stress the introduced change is in range -2 MPa to +1 MPa and there's no clear pattern of distribution of introduced stress. The values in all cells are uniform. In case of difference in normal in-plane stresses, the observed variations are in range of ±5MPa in almost whole chip. Only in cells number 6 and 7, which are on the edge of the chip, the introduced changes are significantly bigger and are reaching 20 MPa. Similar behavior for normal in-plane stress difference was observed in the measurements – the biggest spread of values was observed in cell 7.

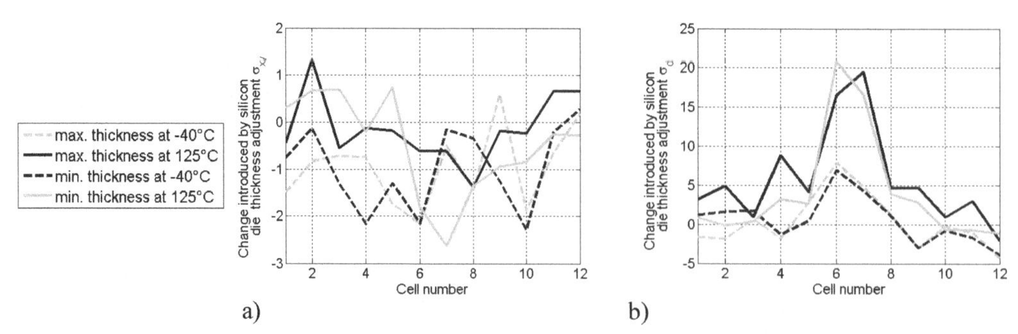

Fig. 18 The differences in the values of stress introduced by adjustment of the silicon die thickness a) shear stress b) difference in normal in-plane stresses

Position of the chip in x and y direction

The results of simulations with varied position of silicon die in x and y direction are presented in Fig. 19 and Fig. 20. For both x and y direction the changes in shear stresses are in range of ±2 MPa. Interesting observation is that this value is biggest in cells 1 and 2, what stays in accordance with the measurement results. It reaches there from -4 MPa to 5 MPa. In case of x-position shift, there is an apparent rise of the shear stress change also in cell 6, what is not observed in measurements. This needs to be further investigated. In the results of simulation of normal in-plane stress difference, the same behavior as in case of adjustment of thickness of silicon die is observed. Mainly, the biggest stress change is introduced in cells 6 and 7. In all other cells the stress variation reaches ±5 MPa for x position change and from -10 MPa to +5 MPa for y position change.

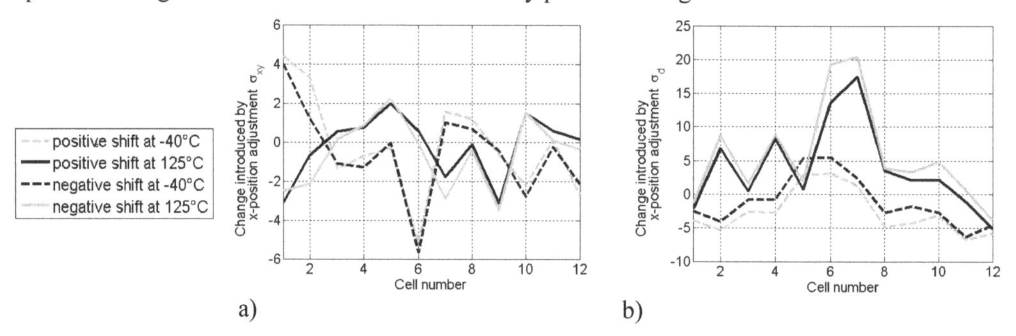

Fig. 19 The differences in the values of stress introduced by adjustment of the position of silicon die in x-direction a) shear stress b) difference in normal in-plane stresses

For checked parameters, the most sensitive cells to the changes in geometry are cell 6 and 7 in case of normal in-plane stress difference and cell 1 and 2 for shear stress. This is consistent with observations from measurements. All these cells are lying on the borders of the chip, what can have a link to observed changes. The influence of investigated parameters changed in a chosen range on the stress state is similar – the observed amplitudes of changes are comparable. All three checked parameters are introducing changes independently from each other. However, to calculate the joint uncertainty introduced by them it would be necessary to use more advanced tools of uncertainty assessment e.g. Monte Carlo Method [12]. It's not in the scope of this work, as not all the sources of uncertainty are evaluated and only the maximal and minimal varied parameters values are taken into account. Nonetheless, looking at the calculated values of introduced changes in stress, knowing that these are extreme cases and that there are more potential sources of uncertainties, it seems reasonable to conclude that the spread observed in measurements can be explained by variations of packaging parameters.

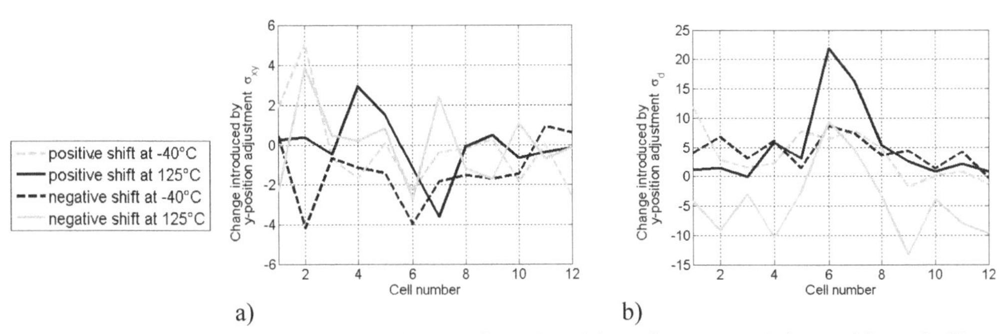

a) b)

Fig. 20 The differences in the values of stress introduced by adjustment of the position of silicon die in y-direction a) shear stress b) difference in normal in-plane stresses

Summary

In this paper the measurement uncertainties of a piezoresistive silicon-based stress sensor were evaluated. It can be observed that the variations between different samples are significant. The possible causes for this distribution were named. Furthermore, a FEA simulation was presented which is further used to perform a sensitivity study concerning the sources of uncertainties. Two potential sources of variations were considered – thickness of silicon die and position of the chip. The regions of the chip with highest standard deviations of results coincide with those from simulation. It is concluded, that the spread in parameters related to electronic packaging processes can explain the observed variations of the stress measurements by piezoresistive silicon based stress sensor. However, further investigation with a larger population is needed to make statistically significant conclusions. In the next steps, the methods to deal with the spread of measured values should be developed, in order to assess the health state of the system independent of manufacturing parameters of the sensor itself.

References

[1] T. Schreier-Alt; K. Unterhofer; F. Ansorge; K. Lang , "Stress analysis during assembly and packaging," Electronic Components and Technology Conference (ECTC), 2011 IEEE 61st , vol., no., pp.1684,1690, May 31 2011-June 3 2011

[2] Y. Zou; J.C. Suhling; R.W. Johnson; R.C. Jaeger; A.K.M. Mian, "In-situ stress state measurements during chip-on-board assembly," Electronics Packaging Manufacturing, IEEE Transactions on , vol.22, no.1, pp.38,52, Jan 1999

[3] P. Gromala; S. Fischer; T. Zoller; A. Andreescu; J. Duerr; M. Rapp; J. Wilde, "Internal stress state measurements of the large molded electronic control units," Thermal, Mechanical and Multi-Physics Simulation and Experiments in Microelectronics and Microsystems (EuroSimE), 2013 14th International Conference on , vol., no., pp.1,8, 14-17 April 2013

[4] F. Schindler-Saefkow; F. Rost; A. Schingale; D. Wolf; B. Wunderle; J. Keller; M. Michel; S. Rzepka, "Measurements of the mechanical stress induced in flip chip dies by the underfill and simulation of the underlying phenomena of thermal-mechanical and chemical reactions," Electronics System-Integration Technology Conference (ESTC), 2014 , vol., no., pp.1,6, 16-18 Sept. 2014

[5] G. Schlottig; F. Schindler-Saefkow; J. Zurcher; B. Michel; T. Brunschwiler, "Sequentially formed underfills: Thermo-mechanical properties of underfills at full filler percolation," Electronics Packaging Technology Conference (EPTC 2013), 2013 IEEE 15th , vol., no., pp.560,564, 11-13 Dec. 2013

[6] F. Schindler-Saefkow; F. Rost; A. Otto; W. Faust; B. Wunderle; B. Michel; S. Rzepka, "Stress chip measurements of the internal package stress for process characterization and health monitoring," Thermal, Mechanical and Multi-Physics Simulation and Experiments in Microelectronics and Microsystems (EuroSimE), 2012 13th International Conference on , vol., no., pp.1/10,10/10, 16-18 April 2012

[7] A. Palczynska; F. Pesth; P.J. Gromala; T. Melz; D. Mayer, "Acquisition unit for in-situ stress measurements in smart electronic systems," Thermal, mechanical and multi-physics simulation and experiments in microelectronics and microsystems (EuroSimE), 2014 15th international conference on , vol., no., pp.1,4, 7-9 April 2014

[8] Kittel H. , Endler S., Osterwinter H. , Oesterle S., Schindler-Saetkow F, "Novel Stress Measurement System for Evaluation of Package Induced Stress", Smart Systems Integration 2008 - 2nd European Conference & Exhibition on Integration Issues of Miniaturized Systems - MOMS, MOEMS, ICS and Electronic Components, Barcelona April 2008

[9] Robert Bosch GmbH, Abschlussbericht zum Verbundvorhaben iForceSens : Entwicklung eines integrierten Stressmesssystems zur Quantifizierung der 3D-Verformung von Sensorbauelementen in Abhängigkeit des Verpackungsprozesses, Technische Informationsbibliothek u. Universitätsbibliothek , Abstatt [u.a.] ; 2008

[10] R.C. Jaeger; J.C. Suhling; R. Ramani, "Errors associated with the design, calibration and application of piezoresistive stress sensors in (100) silicon," Components, Packaging, and Manufacturing Technology, Part B: Advanced Packaging, IEEE Transactions on , vol.17, no.1, pp.97,107, Feb 1994

[11] B. Wu; D.S. Kim; B. Han; A. Palczynska; P.J. Gromala, "Thermal Deformation Analysis of Automotive Electronic Control Units Subjected to Passive and Active Thermal Conditions," Thermal, mechanical and multi-physics simulation and experiments in microelectronics and microsystems (EuroSimE), 2015 16th international conference on , vol., no., pp.1,6, 20-22 April 2015

[12] BIPM, IEC, IFCC, ILAC, ISO, IUPAC, IUPAP, and OIML. Evaluation of measurement data - An introduction to the "Guide to the expression of uncertainty in measurement". Joint Committee for Guides in Metrology, JCGM 104, 2009.

CHAPTER 2:

Uncertainty of Structural Dynamic Improvements in Light Weight Design

Applied Mechanics and Materials Vol. 807 (2015) pp 57-66
© (2015) Trans Tech Publications, Switzerland
doi:10.4028/www.scientific.net/AMM.807.57
Submitted: 2015-04-15
Accepted: 2015-08-04

Comparison of a New Passive and Active Technology for Vibration Reduction of a Vehicle Under Uncertain Load.

Philipp Hedrich[1,a] , Ferdinand-J. Cloos[1,b] , Jan Wuertenberger[3,c], Peter F. Pelz[1,d*]

[1]Chair of Fluid Systems, Magdalenenstr. 4 64289 Darmstadt, Germany

[2]Institute for Product Development and Machine Elements, Magdalenenstr. 4 64289 Darmstadt, Germany

[a]philipp.hedrich@fst.tu-darmstadt.de, [b]ferdinand.cloos@fst.tu-darmstadt.de, [c]wuertenberger@pmd.tu-darmstadt.de, [d]peter.pelz@fst.tu-darmstadt.de

Keywords: vehicle, chassis, active air spring damper, fluid dynamic absorber, quarter car, vibration reduction, control of uncertainty

Abstract. This paper presents two new technologies in order to optimize the operation of a conventional spring-damper-system. Therefore, the function structure, such as the energy flow of a conventional system, is investigated and optimized. The first resulting technology is the fluid dynamic absorber (FDA), which is still a passive solution and improves the energy flow of the conventional spring-damper-system with the help of an absorber with a hydraulic transmission. The second technology is the active air spring damper (AASD), which is an active variant of a spring-damper-system and optimizes the energy flow by using electrical energy. We use a quarter car model to examine the performance of our technologies and compare them in the conflict diagram where driving comfort vs. driving safety is shown within the scope of uncertainty. The FDA improves the driving safety at almost the same comfort. The driving comfort is improved by using the AASD. We also examine the system behavior at uncertain loads. The results show that they are capable of handling this uncertainty.

Motivation

The requirements of a vehicle suspension system are to carry the load, to stabilize the body and to lead the wheel safely to reach optimal driving comfort and driving safety at minimal effort. The additional demands like minimal weight or constructed size need to be fulfilled. Generally, a vehicle consists of a body mass m_b, wheel mass m_w including the axle mass, wheel stiffness k_w and a passive spring-damper-system, i.e. stiffness k_b and damping constant d_b, connecting the the body mass and the wheel mass. The limitation of this spring-damper-system becomes obvious when investigating the vertical dynamics with a quarter car vehicle; see Fig. 1(a). This two degree-of-freedom-system (DOF) with base excitation is a sufficient approach to examine spring-damper-systems. The operating point of this conventional system is predefined by the spring stiffness and the damper constant. It is fixed if the damper is not adjustable. To fulfill the requirements, a trade-off between driving comfort and driving safety has to be made when tuning the spring and the damper. For simplification we use the standard deviation of the body acceleration $\sigma(\ddot{z}_\mathrm{b})$ to rate the driving comfort instead of the weighted vibration severity as defined in VDI 2057 [1]. The related wheel load fluctuation $\sigma(F_\mathrm{w})/F_\mathrm{w0}$, with the static wheel load F_w0 and the wheel load $F_\mathrm{w} = k_\mathrm{w}(z_0 - z_\mathrm{w})$, is equivalent to the driving safety. This is illustrated in the conflict diagram in Fig. 1(b). The Pareto front shows the limitation of the conventional topology of a spring-damper-system [2].

Our aim is to develop new and more flexible systems to connect the body mass with the wheel mass which is able to overstep the limitations of the conventional topology. The new topologies should control uncertainty, like different drivers or unknown loads, during the usage of a vehicle. That is why we develop and compare new topologies of spring-damper systems like Vergé et al. did with a hydrostatic transmission [18]. We show the capability of our new topologies to control uncertainty by simulations. By doing so, our new topologies are evaluated within the stress field of uncertain use,

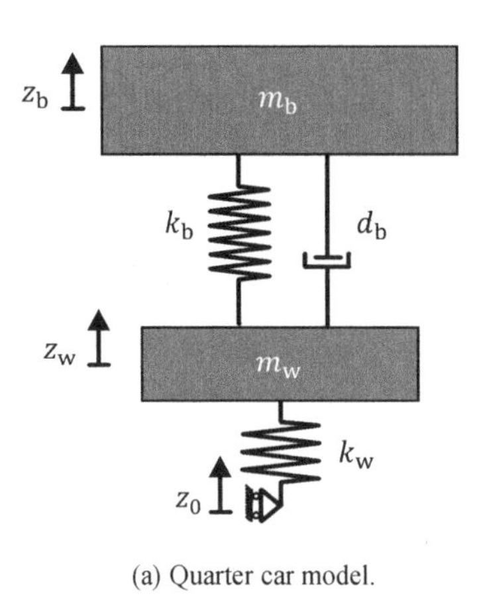

(a) Quarter car model.

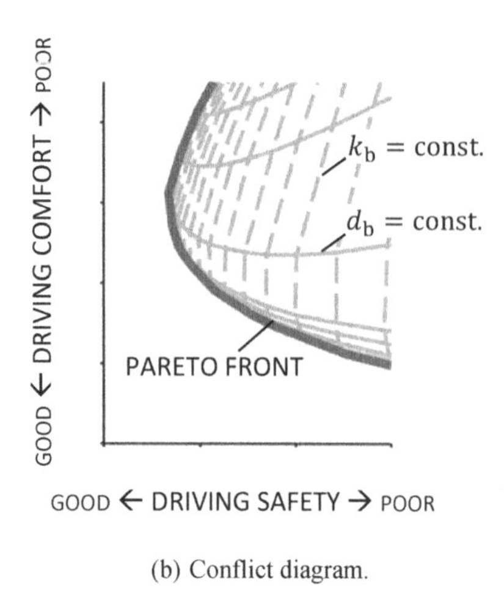

(b) Conflict diagram.

Fig. 1: Conventional topology of a vehicle suspension (a) and its limitation for vertical dynamics by a pareto front (b).

availability and equipment expenditure. Methods developed by the Collaborative Research Center (SFB) 805 are used to describe and analyze the uncertainty within use. After that, the prototypes need to be designed, constructed and investigated within the real SFB Demonstrator [3].

Introduction

Product development aims to identify an appropriate technical method to achieve a desired function. It is necessary to verify that the solution corresponds to the customer requirements considering cost aspects, quality, reliability and so on. The overall product development process used to identify an appropriate solution is provided in single working steps to reduce complexity. Starting with an abstract product idea, possible versions of a solution are identified on varying levels of abstraction. From that point a promising solution is chosen and all attributes of the product are gradually organized right up to an entire solution [4, 5]. All essential working steps are combined in a standardized procedure model. Therefore, a well-known and accepted formulation to structure the development process is guideline 2221 [6] elaborated by Verein Deutscher Ingenieure (VDI). It can be roughly assigned to a project definition (phase 1), a variation of abstract partial solutions and their combination in an overall concept (phase 2), an elaboration of layout (phase 3) and to a specification of the product documentation (phase 4). Chiefly the procedure model of VDI 2221 constitutes a first sector-independent guideline for a development project.

Function structure. In the context of product development, the function structure has a great importance. It ends up with a solution-neutral description of the development task. This avoids prefixing in order to get a wide range of possible solutions. The creation of a function structure can be applied for new and adaptive designs [4]. For new designs the function structure is created for further concretion based on a requirements list and advanced further into reality during the development process. For an adaptive design a known solution is constructed as a base for creating a function structure. It is selectively varied in order to generate additional potential solutions. A function structure itself depicts the conversion of energy, material and signals within a previously determined system boundary. The description of a function structure varies in literature, because it always gives a subjective point of view on a problem [7], so it is important that the practicality is constantly in focus [4]. For this approach the generally applicable sub-functions "channeling", "connecting", "changing", "storing" and "varying" by Pahl/Beitz are used [4].

Product models represent the product on a defined abstract level in the virtual product life cycle. All relevant properties and information which are necessary for the different working steps of product development are represented here [6]. In the context of the procedure model of VDI 2221 [6], product models are used to predict the product behavior or as a basis for decision-making in the development process, for example to define or verify product properties. On the one hand the creation of a product is based on ideas, assumptions, schemes or concepts on different levels of abstraction [8]. On the other hand there is incomplete information about the product and its product life-cycle, so there is a lack of information during the entire product development process. If relevant influences are ignored, only considered insufficiently or irrelevant influences are depicted during product modeling because of the lack of information, there might be a deviation between the realized and planned product behavior, so product modeling is subjected to uncertainty. According to Heinrich Hertz uncertainty in general is a lack of information [9], so relevant information for product models is partially incomplete or does not exist at all. This phenomenon can be illustrated by a map using set theory, see Fig. 2. At each step of concretion, from the relevant reality to the model and from the model to its parameters, uncertainty due to simplifications and assumptions occurs [10], so uncertainty is illustrated by a gap between the reality and the relevant reality.

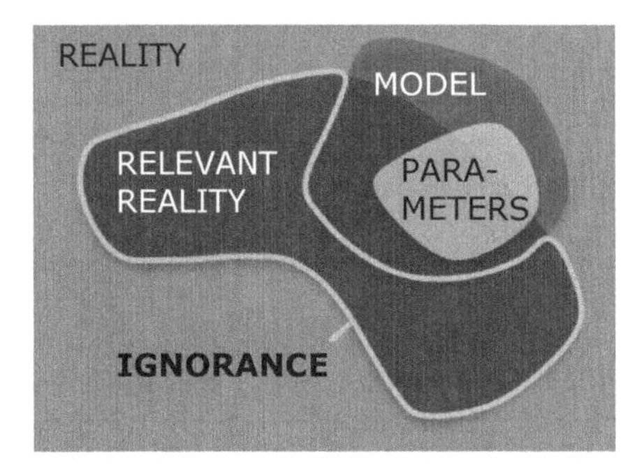

Fig. 2: Illustration of uncertainty by set theory [10].

SFB-805-uncertainty-model. The SFB 805 investigates uncertainty in order to be able to handle it [11]. In the context of SFB 805 uncertainty occurs in processes, so the SFB 805-process model has been developed in order to be capable of depicting and analyzing uncertainty [12]. It is based on different models like the SADT-model or the process model of Heidemann and contains a system boundary, in which a process with its initial and final state such as its influencing factors (resources, disturbances, user or information) is examined. By using semi-active or active systems there is a possibility to intervene in a process in order to react to occurring uncertainty. Based on this potential the process model has been extended in order to depict active systems and to differentiate them from semi-active and passive systems. For this purpose the appliance has been detached from the process which provides the working factor to realize them, so a fundamental aspect of the extension is the differentiation between the process itself and the product that the company produces for this purpose. Thus, the interaction between the appliance and the process can be examined. Active systems are differentiated from passive systems by the fact that they are able to provide an additional part to the existing working factor. Semi-active systems are only able to influence the appliance which has an indirect influence on the working factor.

Evaluation of three Solution Scenarios

This section is divided into two parts. The first part describes the development of two new technologies to overstep the limitation of driving safety vs. driving comfort. In the second part we investigate all three topologies and analyze their characteristics by simulation.

Applied Product Development. In order to uphold the driving comfort and driving safety of a spring-damper-system for an uncertain usage, the operating point has to be created flexibly. According to an adaptive design, a function structure is developed by using an existing spring-damper-system. The system boundary is drawn around the spring-damper-system and the adjusting energy flow is considered, whereby the sub-function "storing" represents the inertia and spring energy storage (Fig. 3). The mechanical energy exiting the system boundary should be preferably small and should approach zero quickly. An optimization of the existing spring-damper-system can be achieved by influencing the energy flow within the system boundary. On one hand, it is possible to extend the storing of energy in the system by adding another storage with a transmission, see Fig. 3(b). On the other hand, energy can be transferred to the system to influence the energy flow, see Fig. 3(c). Both possibilities are discussed hereafter.

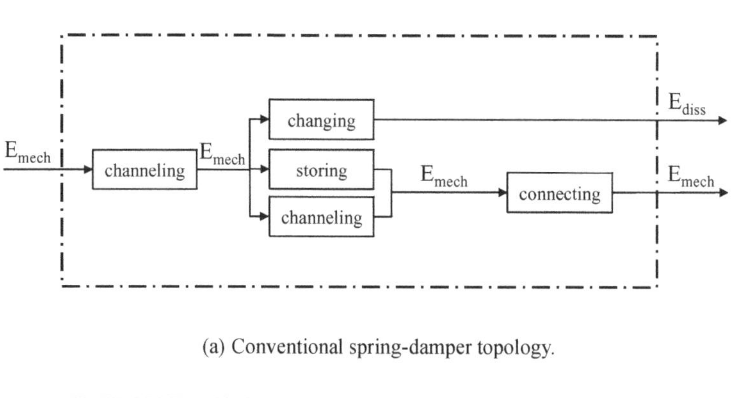

(a) Conventional spring-damper topology.

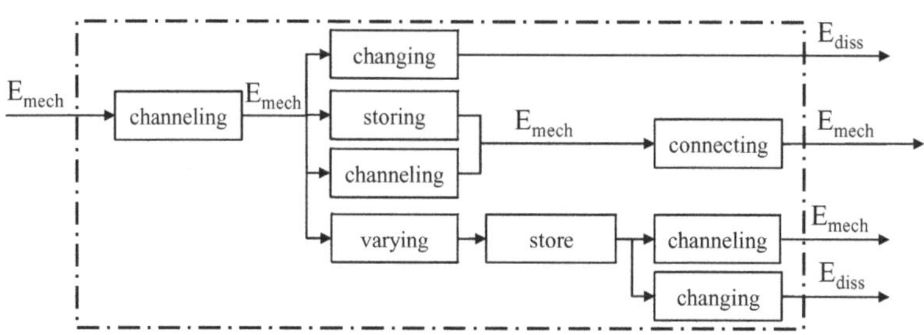

(b) New passive topology.

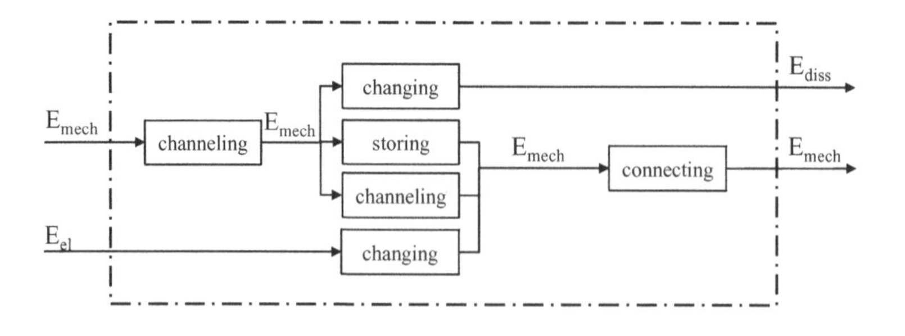

(c) New active topology.

Fig. 3: Function structures of the suspension.

Fluid Dynamic Absorber. This topology optimizes the energy flow within the system by displacing and partially dissipating the energy. The energy is displaced into a further degree of freedom and it is returned phased delayed to the main system [13]. The dissipation is realized by hydraulic losses and the tuning of the main system is not changed, i.e. m_b and k_b. Hence, the dynamic vibration absorber reduces the vibrations of the main degrees of freedom [13]. The inertia of the dynamic vibration absorber could increase by a transmission. Thus, the same performance is achieved by a smaller mass. For example, the transmission ratio can be realized hydraulically, electrically and mechanically. Advantages of the hydraulic transmission are the large power density and the simple implementation. Hence, the hydraulic transmission has better availability and lower equipment expenditure.

The result of the optimization is a fluid dynamic absorber with transmission (FDA) shown on Fig. 4(a) [14]. The FDA is connected with a spring of stiffness k_a to the wheel. The piston displaces the fluid in the reservoir when it is moving. The transmission is realized by alteration of the flow cross section, thus, the flow is accelerated in the channel. The transmission increases the inertia of the fluid and dissipates energy according to the inertia and friction pressure losses. The housing of the FDA is connected with the body and is sealed. The usage process of the FDA is shown in the process diagram; see Fig. 5(a).

The main advantage of the FDA is the very low ratio of weight to inertia and therefore it is attractive for mobile applications. The disadvantage of the FDA is that it needs to be supported by an initial system, here the body, to make the transmission work. Thus, the FDA stimulates the initial system. It is not suitable to absorb the body oscillation by the FDA because in this case it has to be supported by the wheel and the wheel mass is much smaller than the body mass. Hence, the wheel cannot be used as initial system.

Active Air Spring Damper. Another approach to overstep the conflict between driving comfort and driving safety is to use an active strut to apply forces during the usage. By using an active system one is able to influence the process and can thus respond to uncertainty in usage, such as unknown load or excitation. The system is adaptable and its operation range is more flexible.

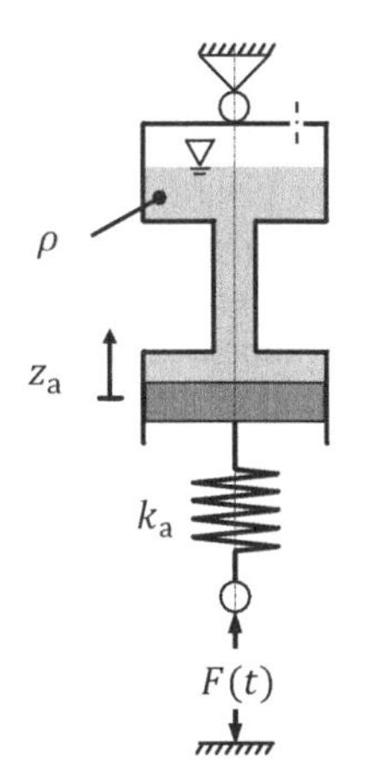

(a) Schematic diagram of the fluid dynamic absorber.

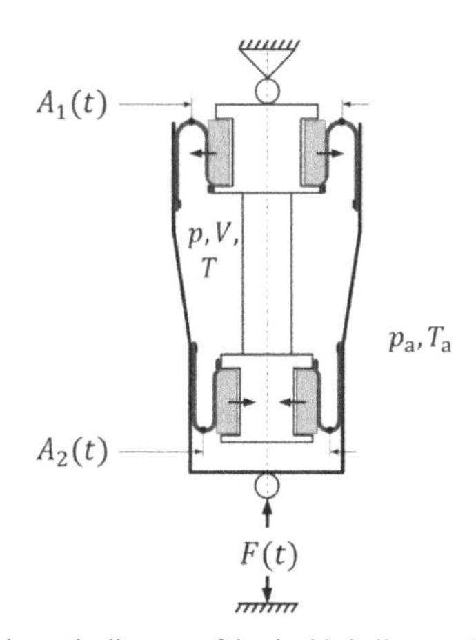

(b) Schematic diagram of the double bellows active air spring. The axial force is $F = A(t)(p(t) - p_a)$, with the load-carrying area $A(t) = A_1(t) - A_2(t)$.

Fig. 4: Schematics diagrams of the FDA (a) and the AASD (b).

The only fully active suspension system available on the market is the Magic Body Control by Daimler, the successor of the Active Body Control which was launched in 1999. A hydraulic piston in series with the steel spring applies forces with a frequency up to 5 Hz [15]. Due to the greater spring

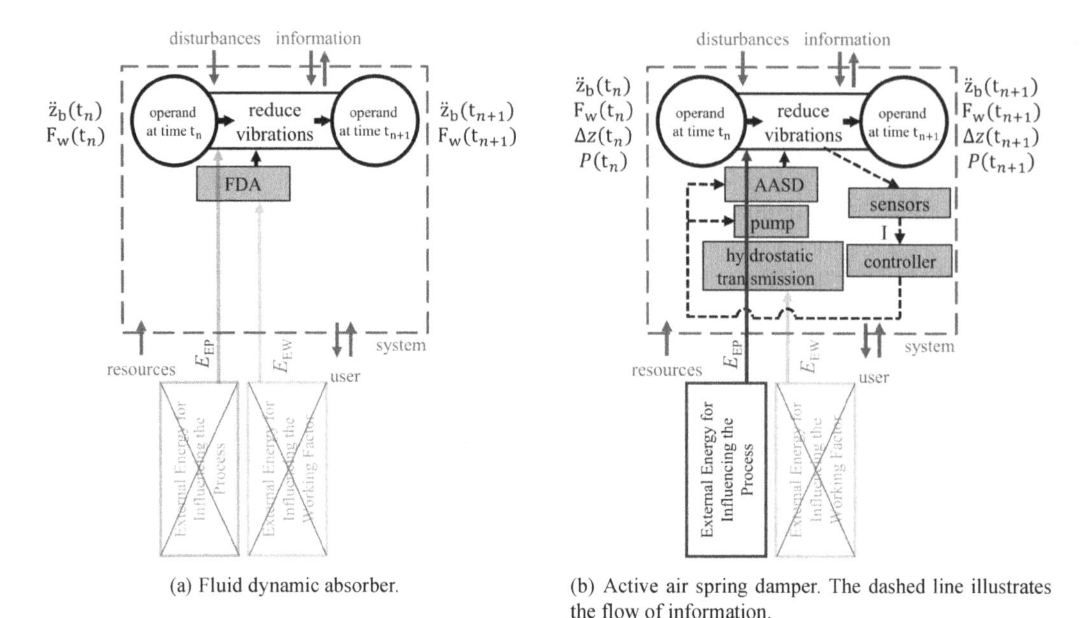

(a) Fluid dynamic absorber.

(b) Active air spring damper. The dashed line illustrates the flow of information.

Fig. 5: Process diagram of the topologies according to [16].

stiffness needed, the driving comfort at higher excitation frequencies is poor. Therefore we take a different approach and develop an active air spring damper (AASD). Motivation is to combine the advantages of an air spring damper, such as very good driving comfort and automatic level control, with those of an active system.

By altering the load-carrying area of the air spring during usage, axial tension and pressure forces are applied as shown in Fig. 4(b). The load-carrying area of an air spring with outside guiding of the bellows is a circle with the diameter $d_\mathrm{A} = \frac{1}{2}(d_\mathrm{p} + d_\mathrm{o})$, where d_p denotes the piston diameter and d_o the outside guide diameter. The piston diameter is changed with four radially positionable segments which are evenly distributed along the circumference of the piston. We use a double piston air spring (damper) shown in Fig. 4(b) with a ring circular load-carrying area. The advantage is that only small changes in the diameters d_{p1} and d_{p2}, always in opposite directions, are needed for large relative area changes. We already proved the operational capability of the active air spring damper experimentally [17]. An air spring damper combines the functions of a spring and a damper in one component. The oscillation energy is dissipated when compressed air flows through an orifice from one chamber into another. Due to its working principle the air damper is damping frequency specific. Normally the orifice is adjusted so that the damping energy reaches its maximum at the wheel natural frequency.

The usage process of the air spring damper is shown in the process diagram, see Fig. 5(b). The entire system consists of the air spring damper itself, the hydraulic pump, the hydrostatic transmission, the controller unit and displacement and acceleration sensors. Electric power enters the system and is transformed into hydraulic power in the pump. The hydrostatic transmission transmits the power to the air spring damper where it is used to alter the load-carrying area. Its wear can be minimized by an algorithmic structure synthesis as shown by Vergé et al. [18]. The controlled variables are the body acceleration, the wheel load fluctuation and the compression of the active strut Δz. The available electric power P is another important value. The increased complexity and effort of this active system compared to the passive solutions becomes obvious based on the numbers of components, but we will not address this issue in great detail.

Simulation. To identify the potential and performance of our new topologies, we use a quarter car model (two-mass system). Although this model is restricted due to the neglected effects of coupled masses [2], it is sufficient for our purpose. We simulate a ride on a very poor road (class E according to ISO 8608 [19]) at a velocity of 54 km/h. The parameters of the quarter car correspond to those of a typical middle class car with a conventional suspension consisting of a linear spring and damper (body mass $m_\mathrm{b} = 290$ kg, wheel mass $m_\mathrm{w} = 40$ kg, body damping rate $d_\mathrm{b} \approx 1100$ Ns/m, body stiffness $k_\mathrm{b} \approx 20$ kN/m and wheel stiffness $k_\mathrm{w} = 200$ kN/m). In a second step we examine the robustness of the suspension system with regard to uncertain loads (body masses of 100 and 290 kg). We use zero-d-models (lumped parameters) to model the FDA and the AASD. The model of the FDA consists of conservation equations of mass and vertical momentum. This set of equations is solved numerically in MatLab. The FDA is added in parallel to the conventional suspension system and therefore the damping rate is not changed. The object-oriented language Modelica is used to model the air spring damper [20]. The conservation equations of mass, vertical momentum and energy are solved. This model is integrated as a system function (s-function) in MatLab Simulink where the hydraulic actuator and the controller are modeled. The hydraulic actuator and transmission is modeled as a first-order lag element with a cutoff frequency of 5 Hz. To control the active system a simple PID-controller is used. The desired value is the body acceleration $\ddot{z}_b$.

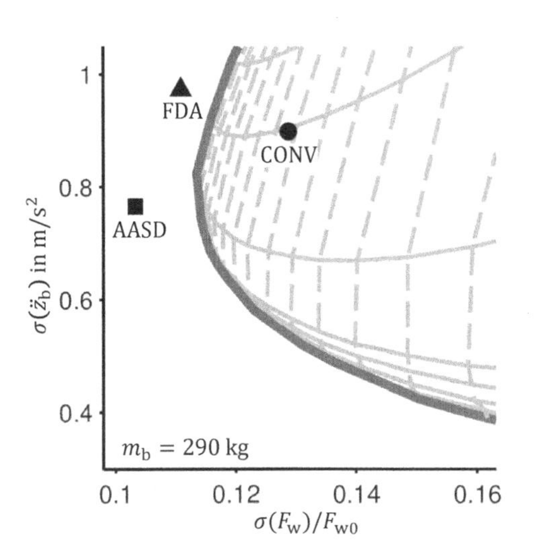

Fig. 6: Conflict diagram with all three topologies at the nominal body mass of 290 kg.

Each topology is simulated and analyzed. The results are shown in the conflict diagram, Fig. 6, which was introduced in the motivation. This diagram shows the conflict in finding the optimal setup for a spring damper system. It does not only depend on the spring stiffness and damping constant but also on the road and driving velocity [21]. The conflict diagram shows the body acceleration fluctuation $\sigma(\ddot{z}_b)$ over the wheel load fluctuation $\sigma(F_w)$ with respect to the static wheel load F_{w0}. The acceleration presents the intensity of vibrations. Therefore low body accelerations are equal to a good driving comfort. The driving safety is mainly dependent on the wheels remaining in constant contact with the road. This can be determined by the wheel load fluctuation. The driving safety is good if it is small.

The pareto line represents the limit of the convectional system. It can only work above this line. The new topologies meet our expectation and overstep the limitation by the pareto front of the conventional topology.

During the usage of the spring damper system many influences on the systems are uncertain but we only deal with an uncertain load for now. The change in load, which is equivalent to different body masses, has a large impact on the system's behavior and performance. The results are illustrated in the two conflict diagrams for a body mass of 100 kg and 350 kg, see Fig. 7.

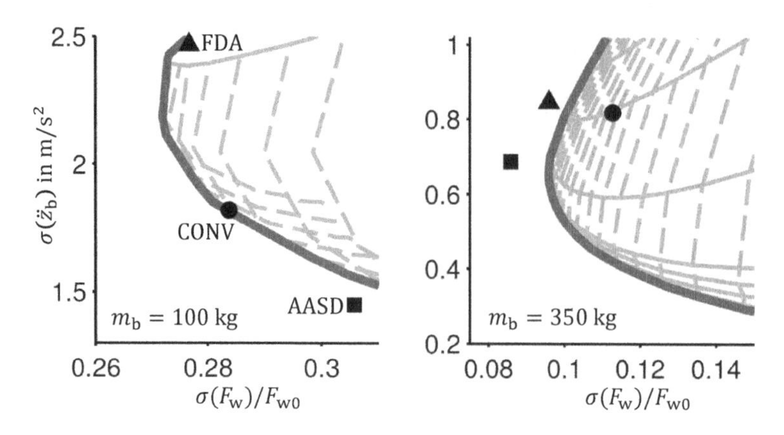

Fig. 7: The conflict diagrams for a body mass of 100 kg (a) and 350 kg (b).

Discussion

Analyzing the presented results, there are four facts to mention:

1. The new topologies overstep the limitation of the conventional topology.
2. The FDA increases the driving safety and reduces the comfort only a little bit. The latter is substantiated by the fact that the FDA is supported by the body. This becomes obvious for the lighter body mass. The stimulation of the body is reduced by scaling down the damper constant b_a, i.e. to 50 %. By doing so, the comfort is similar and the driving safety is 12 % better than using the conventional topology.
3. The advantages of the active system become obvious in the conflict diagram. The driving comfort is better than that of the passive ones for all three body masses. Even the driving safety is better for the body masses 290 kg and 350 kg but not for 100 kg. This is due to the great impact of the controller on the system performance. The simple PID-controller provides good results for great body masses, but not for the small one.
4. The equipment expenditure of AASD is higher than thats of the passive topologies. This is essentially because of the controller and the complex hydraulic actuator.

Conclusion

We presented two new topologies to overstep the limitation in driving comfort and safety of a conventional spring-damper-system by analyzing the function structure of the limited conventional system. By analyzing and improving its function structure we derived our new solutions. One of our topologies is passive, the so-called fluid dynamic absorber (FDA), an absorber with a hydraulic transmission. The other topology is active, which means external energy is used to reduce vibrations. The active solution is an active air spring damper (AASD).

To investigate the potential of our approaches we simulated the vertical dynamics of a car with a quarter car model, a two-mass-system, with parameters of a typical middle class car. We calculated the standard deviation of the body acceleration, i.e. the driving comfort and the wheel force deflection, i.e. the driving safety for a ride on a very poor measured road (road class E according to ISO8608 [19]) at a velocity of 54 km/h. We showed that we exceeded the limit of the conventional suspension system, which is represented by the pareto front in the conflict diagram where driving comfort vs. driving safety is illustrated. The FDA improves the driving safety at almost the same comfort. The driving comfort is improved by using the AASD. We also examined the system behavior at uncertain loads. The results show that they are capable of handling this uncertainty. Additionally, the air spring damper has the advantage that the full suspension travel is always available due to the adjustment of the initial pressure to the load.

In a next step we will build the FDA and a second prototype of the AASD to characterize them experimentally on our servo-hydraulic damper test system and to perform real time hardware-in-the-loop tests. Hardware-in-the-loop means that we have a real strut on the test rig interacting with the simulation of the virtual quarter car running on our dSpace system. After that we will test our technologies in our SFB-demonstrator. Based on these tests we will compare our two new topologies regarding performance and effort within the scope of “active vs. passive”. Besides this, the assumptions of the product models, on which the comparison between both solutions for a more flexible operating point is based on, has to be considered. So the uncertainty in modeling has to be analyzed. Therefore, a systematization of the modeling process such as the related assumptions has to be formalized in order to make a statement to the causes of modeling uncertainty. With the help of that, modeling uncertainty can be integrated into the comparison of both solutions. Finally, the question of whether the active or the passive solution is best for a suspension system can be answered.

Acknowledgment. We would like to thank Deutsche Forschungsgemeinschaft (DFG) for funding this project within the Collaborative Research Centre (SFB) 805.

References

[1] VDI, Human exposure to mechanical vibrations whole-body vibration, VDI 2057, part 1, Düsseldorf, 2002.

[2] M. Mitschke, H. Wallentowitz, Dynamik der Kraftfahrzeuge, 5. ed., Springer Vieweg, Berlin, 2014.

[3] C. Gehb, R. Platz and P. F. Pelz, Entwicklung des SFB-Demonstrators: Definition, Darmstadt, 2014.

[4] G. Pahl, W. Beitz, Konstruktionslehre, Heidelberg, 2005.

[5] K. Ehrlenspiel, H. Meerkamm, Integrierte Produktentwicklung, München, 2013.

[6] VDI 2221, Methodik zum Entwickeln und Konstruieren technischer Systeme und Produkte, Berlin, 1993.

[7] U. Lindemann, Methodische Entwicklung technischer Produkte, Berlin, 2007.

[8] N. Drémont, P. Graignic, N. Troussier, R. Whitfield, A. Duffy, A metric to represent the evolution of CAD/analysis models in collaborative design. In: International Conference in Engineering Design (ICED), Copenhagen, 2011.

[9] H. Hertz, Die Prinzipien der Mechanik in neuem Zusammenhange dargestellt, Gesammelte Werke Band 3, J.A. Barth, Leipzig, 1894.

[10] P. F. Pelz, P. Hedrich, Unsicherheitsklassifizierung anhand einer Unsicherheitskarte, interner Bericht des Instituts für Fluidsystemtechnik, Darmstadt, 2015.

[11] H. Hanselka, R. Platz, Ansätze und Maßnahmen zur Beherrschung von Unsicherheit in lasttragenden Systemen des Maschinenbaus, VDI Konstruktion (2010), Nr.11/12, pages 55-62.

[12] T. Eifler, G. Enss, M. Haydn, L. Mosch, R. Platz, H. Hanselka, Approach for a Consistent Description of Uncertainty in Process Chains of Load Carrying Mechanical Systems, Applied Mechanics and Materials (Volume 104), Vol. Uncertainty in Mechanical Engineering (2011), pages 133-144.

[13] J.P. Den Hartog, Mechanical Vibrations , Dover Publications, 1985, New York.

[14] T. Corneli, P. F. Pelz, Employing Hydraulic Transmission for Light Weight Dynamic Absorber, 9th IFK Proceedings Vol. 3, Aachen, 2014.

[15] M. Pyper, W. Schiffer and W. Schneider, ABC - Active Body Control, Verlag Moderne Industrie, Augsburg, 2003.

[16] A. Bretz, S. Calmano, T. Gally, B. Götz, R. Platz and J. Würtenberger, Darstellung passiver, semiaktiver und aktiver Maßnahmen im SFB 805-Prozessmodell, Preprint, SFB 805, TU Darmstadt, 2015.

[17] T. Bedarff, P. Hedrich and P. F. Pelz, Design of an Active Air Spring Damper, 9th IFK Proceedings Vol. 3, Aachen, 2014.

[18] A. Vergé, P. Pöttgen, T. Ederer, L. Altherr and P. F. Pelz, Lebensdauer als Optimierungsziel, OP-Journal, submitted, 2015.

[19] ISO, Mechanical Vibration - Road Surface Profiles - Reporting of Measured Data, ISO 8608, Geneva, 1995.

[20] T. Bedarff, P. F. Pelz, Modellbildung des aktiven Luftfederdämpfers und Modellierung mit Dymola, interner Bericht Institut für Fluidsystemtechnik, Darmstadt, 2015.

[21] B. Heißing, M. Ersoy, Chassis Handbook, Wiesbaden, 2011.

Applied Mechanics and Materials Vol. 807 (2015) pp 67-77
© (2015) Trans Tech Publications, Switzerland
doi:10.4028/www.scientific.net/AMM.807.67

Submitted: 2015-04-20
Accepted: 2015-08-04

Model verification and validation of a piezo-elastic support for passive and active structural control of beams with circular cross-section

Benedict Goetz[1,a*], Maximilian Schaeffner[1,b], Roland Platz[2,c] and Tobias Melz[1,d]

[1]System Reliability and Machine Acoustics SzM, Technische Universität Darmstadt, Magdalenenstraße 4, D-64289 Darmstadt, Germany

[2]Fraunhofer Institute for Structural Durability and System Reliability LBF, Bartningstraße 47, D-64289 Darmstadt, Germany

[a]goetz@szm.tu-darmstadt.de, [b]schaeffner@szm.tu-darmstadt.de, [c]roland.platz@lbf.fraunhofer.de, [d]melz@szm.tu-darmstadt.de

Keywords: beam, structural control, vibration attenuation, active buckling control, model verification, model validation

Abstract. Beams in lightweight truss structures are subject to axial and lateral loads that may lead to undesired structural vibration or failure by buckling. The axial and lateral forces may be transferred via the truss supports that offer possibilities for structural control of single beams and larger structures. In our earlier work, the concept of a piezo-elastic support for active buckling control and resonant shunt damping has been investigated. An elastic spring element is used to allow a rotation in the beam's bearing in any plane perpendicular to the beam's longitudinal axis. The rotation is laterally transferred to an axial displacement of piezoelectric stack transducers that are either used to generate lateral forces for active buckling control or to attenuate vibrations with a resonant shunt. In this paper, the model verification and validation of the elastic properties of the piezo-elastic support for passive and active structural control of beams with circular cross-section is presented. The rotational and lateral spring element stiffness is investigated numerically and experimentally and the existing models are updated in the verification process. The model is validated by comparing the numerical and experimental results for vibration attenuation.

Introduction

In mechanical and civil engineering, truss structures represent complex mechanical systems that comprise truss members, considered as beams if connected to each other via the relatively stiff truss supports. Truss structures bear and withstand static and kinetic loads in axial and lateral directions leading to structural vibrations and, in the worst case, buckling of slender truss members. Therefore, passive or active structural state control of truss structures or single beams subject to axial or lateral forces is desirable in certain applications.

Active buckling control of slender beams with rectangular and circular cross-section by active lateral forces or bending moments has been investigated in [1, 2, 3, 4, 5, 6]. Often, surface bonded piezoelectric patches were used to induce bending moments in the structure to counteract the deformation, [1, 2, 3, 4]. In our earlier work, lateral forces applied near the fixed support of a beam were used to control the lateral displacement of the structure and increase the maximum bearable load, [5, 6].

For vibration reduction in truss structures, piezoelectric shunt damping has been investigated in [7, 8, 9]. Generally, a piezoelectric transducer converts mechanical kinetic energy of a vibrating host structure into electrical energy due to the piezoelectric effect. Shunting the piezoelectric transducer with resistor and inductance, the resonant RL-shunt, an electrical oscillation circuit with the inherent capacitance of the transducer is created. This electromechanical system acts like a mechanical vibration absorber. Axial piezoelectric stack transducers were integrated, e. g. in one strut of a truss and investigations were focused on vibrations of global modes, resulting in compression and elongation of the transducer in axial direction of the strut, in [7, 8, 9]. In [8], a smart support with piezoelectric washers has been investigated and vibration attenuation in a truss substructure was achieved.

In [10, 11], a piezo-elastic support with integrated piezoelectric stack transducers for the structural control of beams by active buckling control and passive resonant shunt damping was investigated numerically and experimentally. In this paper, the rotational and lateral spring element stiffness used in the piezo-elastic supports is first verified by experimental data and the updated model is then validated by comparing numerical and experimental capability to attenuate vibration.

Description of beam with piezo-elastic supports

The investigated system is a slender beam of length l_b with circular solid cross-section of radius r_b with two elastic supports A at $x = 0$ and B at $x = l_\mathrm{b}$ with rotational stiffness k_φ and lateral stiffness k_l that are the same for both supports A and B and in both y- and z- direction, Fig. 1. The beam properties are radius r_b, bending stiffness EI_b and density ρ_b, they are assumed to be constant across the entire beam length l_b. In each support A and B at position $x = -l_\mathrm{ext}$ and $x = l_\mathrm{b} + l_\mathrm{ext}$, three piezoelectric stack transducers P_1, P_2, P_3 and P_4, P_5, P_6 are arranged in the support housing at an angle of $120\,^\circ$ to each other in one plane orthogonal to the beam's x-axis. They are connected to the beam via a relatively stiff axial extension of the beam with length l_ext, radius r_ext, bending stiffness EI_ext and density ρ_ext.

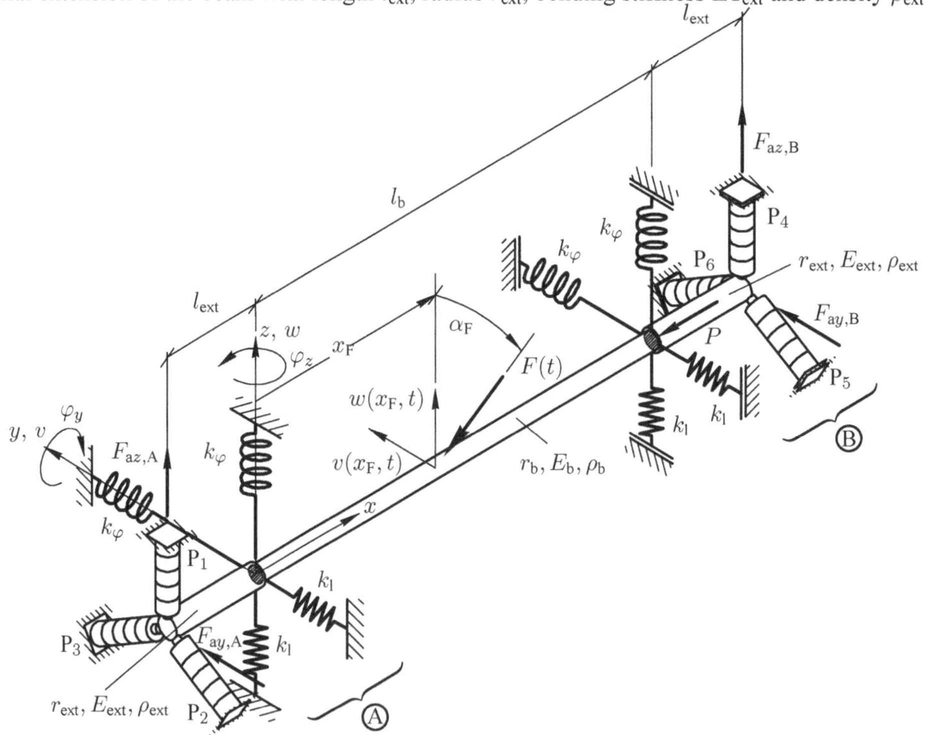

Fig. 1: Sketch of beam system

The beam may be loaded by a time-dependent lateral force $F(t)$ applied at $x = x_\mathrm{F}$ with variable angle $0\,^\circ \leq \alpha_\mathrm{F} < 360\,^\circ$ in y-z-plane and a constant axial load P acting at support B at position $x = l_\mathrm{b}$. The beam with circular cross-section has no preferred direction of deflection in case of buckling or vibration, so the beam may deflect in any plane lateral to the x-axis. The piezo-elastic support is used for two different applications of passive and active structural control, active buckling control and passive lateral vibration attenuation.

- Active buckling control: Lateral forces in arbitrary directions orthogonal to the beam's longitudinal x-axis are introduced in both supports A and B to control the beam against buckling when loaded with a constant axial load P that exceeds the first critical axial buckling load $P \geq P_{\mathrm{cr}}$.
- Passive resonant shunt damping: Lateral vibration in y-z-plane is transformed into the stack transducer's axial deformation. The transducer electrodes are shunted with resonant inductive resistive electrical networks for the purpose of vibration attenuation.

Fig. 2 illustrates active buckling control and passive vibration attenuation with RL-shunt within the SFB 805 uncertainty process model [12]. In case of active buckling control, a controller is connected to the transducer and the external energy E_{EP} is needed to induce control forces. In case of passive vibration attenuation, no external energy is needed. All four sources of uncertainty influence the structural state control, but are neglected in this paper and are subject of further investigations.

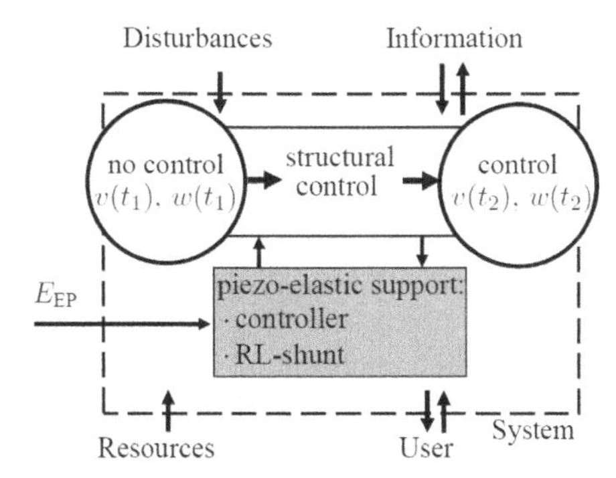

Fig. 2: SFB805 uncertainty process model for structural control with uncertainty sources

The experimental setup of the piezo-elastic support B, with three piezoelectric stack transducers P_4, P_5 and P_6 is shown in Fig. 3a. Fig. 3b shows a sectional view of the CAD-model with the piezoelectric stack transducers, the axial extension and the elastic spring element. The support housing and beam material is aluminum alloy EN AW-7075 and the extension is hardened steel 1.2312. The elastic spring element is made of spring steel 1.4310.

The model parameters describing the properties of the beam and the supports presented in Figures 1 and 3 are summarized in Table 1.

Table 1: Properties of the beam system

property	symbol	value	SI-units
beam length	l_{b}	0.4	m
beam radius	r_{b}	$[0.004,\ 0.005]$	m
Young's modulus aluminum EN AW-7075-T6	E_{b}	$70.0 \cdot 10^9$	$\mathrm{N/m^2}$
density aluminum EN AW-7075-T6	ρ_{b}	2710	$\mathrm{kg/m^3}$
axial extension length	l_{ext}	0.0075	m
axial extension radius	r_{ext}	0.006	m
Young's modulus steel 1.2312	E_{ext}	$210.0 \cdot 10^9$	$\mathrm{N/m^2}$
density steel 1.2312	ρ_{ext}	7810	$\mathrm{kg/m^3}$

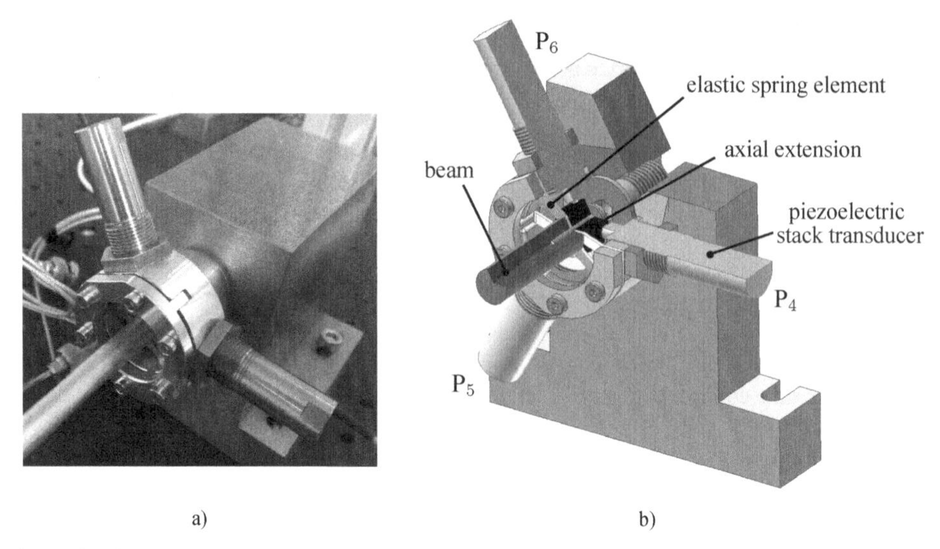

Fig. 3: Piezo-elastic beam support B with integrated piezoelectric stack transducers P_4, P_5 and P_6 a) experimental setup b) sectional view of CAD-model

Verification and validation of rotational and lateral spring element stiffness

The elastic spring element is an essential component of the piezo-elastic supports A and B, Fig. 1, as it bears the axial loads and allows rotations in any plane perpendicular to the beam's x-axis that is necessary for both applications active buckling control and passive resonant shunt damping. The characteristic of the elastic spring element in the model is reduced to a rotational stiffness k_φ and a lateral stiffness k_1. Ideally, the rotational stiffness is very low and, at the same time, the lateral stiffness is high, so that the spring element is similar to a pinned support. In real application, the spring element's rotational and lateral stiffness are not independent from each other. Therefore, a numerical and experimental investigation of the rotational and lateral spring element stiffness is performed.

Verification of spring element stiffness

Fig. 4a shows the CAD-model of the spring element. It has six spring arms that are bent with alternating bending angles of $\beta = \pm 22\,°$. The spring element's thickness t_s varies between 0.4 mm and 1 mm to achieve different rotational and lateral stiffness k_φ and k_1.

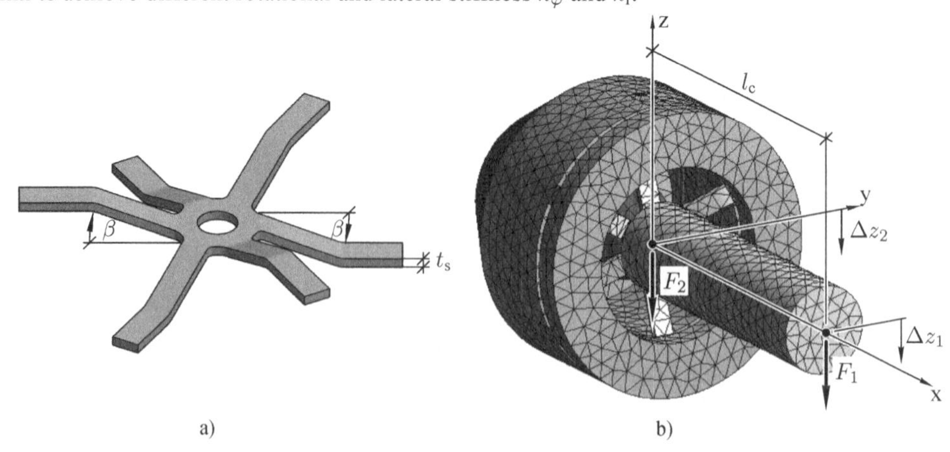

Fig. 4: a) CAD-model of spring element, b) elastic support with mesh in ANSYS Workbench

The spring element stiffness is calculated using a linear ANSYS Workbench FE-model of the elastic support without piezoelectric stack transducers. Fig. 4b shows the meshed model of one elastic support fixed at the rear end of the support housing and with a cantilever of length l_c as substitute for the actual beam of length l_b. The cantilever bending stiffness is much higher than the rotational stiffness of the spring element, so the cantilever is assumed to be rigid. Two different static simulations are performed to determine the rotational stiffness k_φ and lateral stiffness k_l. First, a static force F_1 is applied at the free end $x = l_c$ of the cantilever in negative z-direction. The cantilever rotates around the spring element center and with a small angle approximation using the displacement Δz_1 of the free end, the rotational stiffness is calculated by

$$k_\varphi = \frac{F_1 \, l_c^2}{\Delta z_1}. \tag{1}$$

Second, the lateral stiffness k_l is determined similarly with static force F_2 applied at the spring element center and lateral displacement in z-direction Δz_2 by

$$k_l = \frac{F_2}{\Delta z_2}. \tag{2}$$

The results of the numerically simulated spring element stiffness k_φ and k_l for varying spring element thickness $0.4\,\mathrm{mm} \leq t_s \leq 1\,\mathrm{mm}$ are shown in Fig. 5a. Both the rotational stiffness k_φ and the lateral stiffness k_l show an increase in the stiffness for increasing spring element thickness t_s. The rotational stiffness increases by a factor of 2.46 between the smallest and the highest spring element thickness and the lateral stiffness also increases by a factor of 1.5. The stiffness increase can be assumed to be linear for k_l for approx. $0.6\,\mathrm{mm} \leq t_s \leq 1\,\mathrm{mm}$, the stiffness increase for k_φ is nonlinear.

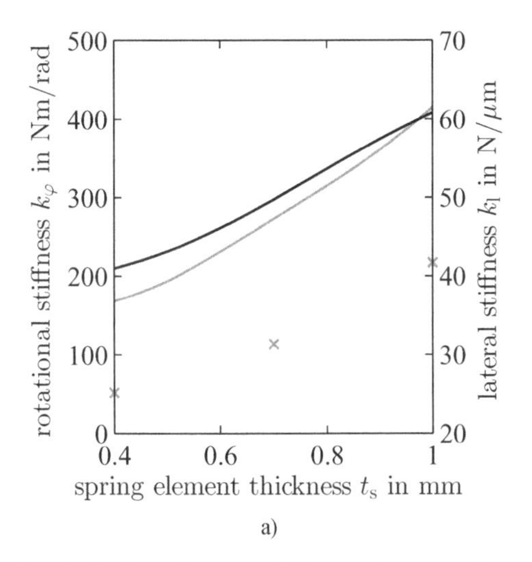

a)

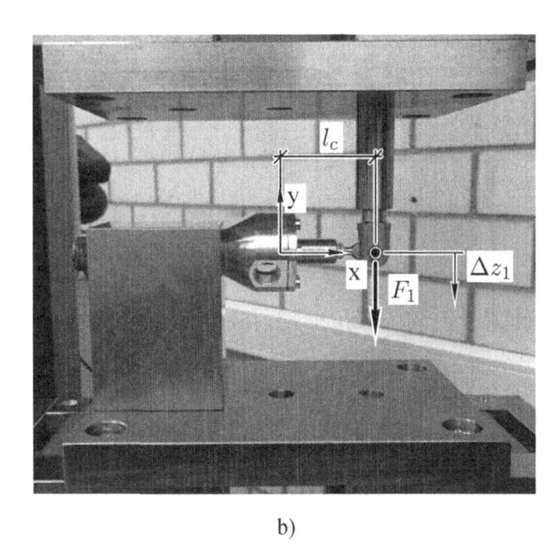

b)

Fig. 5: Spring element stiffness, a) comparison of rotational stiffness k_φ from simulation (—) and measurement (×) and simulated lateral stiffness k_l (—), b) experimental test of rotational stiffness k_φ

In an experimental test shown in Fig. 5b, the rotational stiffness k_φ is determined for spring element thickness $t_s = [0.4; 0.7; 1.0]$ mm for a maximum force $F_{1,\mathrm{max}} = 100\,\mathrm{N}$ at $x = l_c$. The experimental tests are performed on a static material testing machine Zwick Roell Allround-Line 100 kN. The spring element in the support housing is loaded with compression and tension forces and the force and displacement are measured for the load cycles. Measurements of the lateral stiffness k_l were not possible with the available test setup in Fig. 5b. The mean values of the rotational stiffness k_φ calculated according to Eq. (1) for two different spring elements for each spring element thickness t_s are shown in Fig. 5a. In the calculation, only the initial slope of the force-displacement curve up to 20 N is considered. The increasing rotational stiffness k_φ for increasing spring element thickness t_s is reproduced by the experimental tests. However, the experimentally determined values are smaller than the simulated values. For the smallest spring element thickness $t_\mathrm{s} = 0.4\,\mathrm{mm}$, the simulated value is three times higher than the measured value and for the largest thickness $t_\mathrm{s} = 1.0\,\mathrm{mm}$, the simulated rotational stiffness is overestimated by a factor of two. The lower stiffness in the experiment might result from a non-ideal clamping of the elastic spring element between cantilever and axial extension as well as in the support housing or from deviations in the material properties.

Validation of lateral dynamic transfer behavior

In a simplified model, the spring element stiffness in Fig. 1 is represented by discrete rotational and lateral support stiffness k_φ and k_l. k_φ was verified by a static test in the prior section. Now, the lateral dynamic behavior is validated and then updated by adjustments of the beam's elastic support properties. An adjustment of the beam's elastic properties, mass and bending stiffness, has also been conducted but is not included in this paper. An experimental modal analysis is conducted to determine the first lateral eigenfrequency f_1 of the beam supported by elastic spring elements with and without attached piezoelectric transducers.

The beam in Fig. 1 is assumed to be symmetric in y- and z-direction. Therefore, the finite element (FE) model is presented for vibration in the x-z-plane, only. The beam and axial extensions are discretized by $N-1=12$ one-dimensional EULER-BERNOULLI beam elements [13, 14], Fig. 6, with $N = 13$ nodes and one translational displacements in z-direction and one rotational displacement around the y-axis.

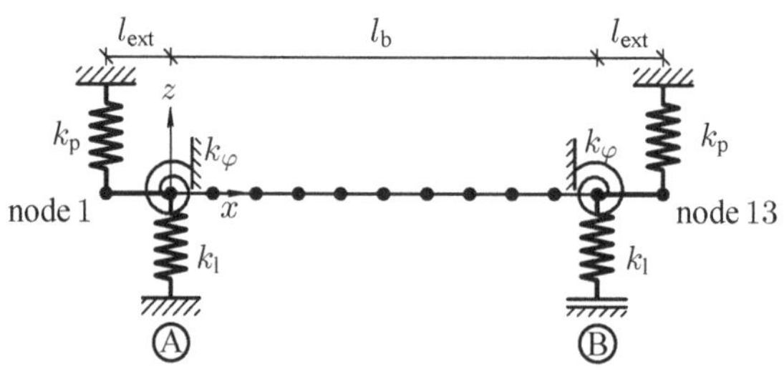

Fig. 6: FE-model of elastically supported beam in x-z-plane

The beam's n eigen angular frequencies $\omega_n = 2\,\pi f_n$ can be calculated by solving the eigenvalue problem

$$\det\left[\mathrm{K} - (\omega_n)^2\,\mathrm{M}\right] = 0. \tag{3}$$

M and K are the global mass and stiffness matrix with dimension $[2N \times 2N]$. The transducer stiffness k_p is added to the global stiffness matrix at node 1 and node 13. The lateral stiffness k_l and the rotational stiffness k_φ are added to the global stiffness matrix at node 2 and node 12.

The rotational and lateral stiffness k_φ and k_l are updated in the simulation to match the experimental eigenfrequency f_1 for spring element thicknesses $t_\mathrm{s} = [0.4; 0.7; 1]\,\mathrm{mm}$. In a first step, k_φ is updated to match the frequency f_1 of the elastically supported beam without transducers. In a second step, k_l is updated to match the eigenfrequency f_1 of the elastically supported beam with attached transducers. To eliminate influences of the transducer stiffness in the measurement, all transducers are replaced by stiff steel pins.

Fig. 7a shows the first eigenfrequency f_1 of the elastically supported beam without steel pins and with steel pins for spring element thickness $t_\mathrm{s} = [0.4; 0.7; 1]\,\mathrm{mm}$. f_1 without steel pins increases by a factor of 1.22 between the smallest and the highest spring element thickness. With steel pins, f_1 remains almost constant. The stiffness values for k_φ and k_l obtained from the frequency matching are presented in Fig. 7b. The rotational stiffness k_φ increases by a factor 4.25 between the smallest and highest spring element thickness and, therefore, shows a similar trend like the results from the static

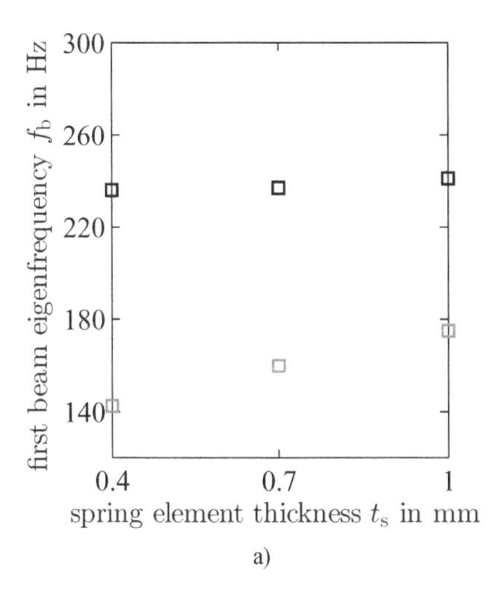

a)

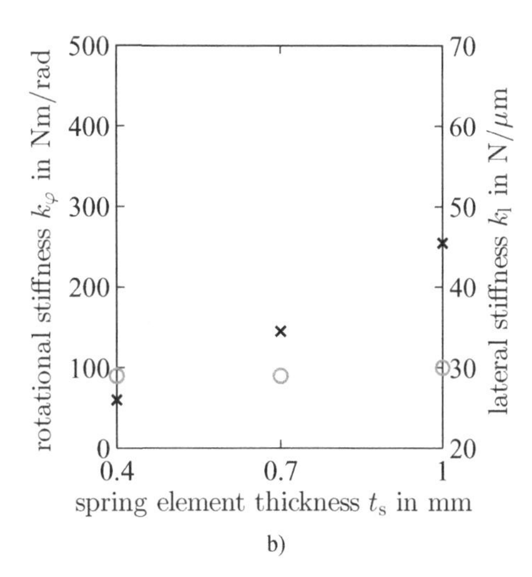

b)

Fig. 7: a) First eigenfrequency f_1 of elastically supported beam without stiff pins (□) and with stiff pins (□) as substitutes for the piezoelectric stack transducers, b) rotational and lateral spring element stiffness k_φ (×) and k_l (○) obtained from measured beam eigenfrequency f_1 and mathematical model update for spring element thickness $t_s = [0.4;\ 0.7;\ 1]$ mm

measurement in Fig. 5b, see Table 2. The lateral stiffness k_l remains almost constant, indicating that the eigenfrequency is not dominated by the lateral stiffness but rather by the clamping of the steel pins. Therefore, k_l may not be validated finally. Nevertheless, the values of k_l in Fig. 7b are used to update the simulation models in order to account for a lower experimental spring stiffness compared to the simulation.

Table 2: Rotational stiffness from verification $k_{\varphi,\mathrm{veri}}$, Fig. 5b, and validation $k_{\varphi,\mathrm{vali}}$, Fig. 7b

	$t_s = 0.4$ mm	$t_s = 0.7$ mm	$t_s = 1$ mm
$k_{\varphi,\mathrm{veri}}$ in Nm/rad	51	113	217
$k_{\varphi,\mathrm{vali}}$ in Nm/rad	60	145	255

Application of piezo-elastic support for structural control

The piezo-elastic support is used for two different applications of passive and active structural control. In the application of active buckling control, the stack transducers induce active lateral forces. A detailed numerical investigation of the concept for active buckling control is given in [11]. In the application of passive resonant shunt damping for vibration attenuation, bending of the beam is transformed into the transducer's axial deformation and vibration reduction is achieved in combination with resonant shunts. A detailed numerical and experimental investigation of the concept for vibration attenuation is given in [10]. In the earlier studies, the numerically simulated values for the rotational and lateral stiffness k_φ and k_l, Fig. 5a, were used to represent the stiffness of the elastic supports. In this section, the experimentally validated values, Fig. 7b, are used to update the beam model and investigate the effect on active buckling control and passive resonant shunt damping.

Active buckling control

For the application of active buckling control of a beam with circular cross-section loaded with a supercritical constant axial load $P > P_{cr}$, the elastic supports A and B with integrated piezoelectric stack transducers in Fig. 1 induce forces $F_{\mathrm{a}y/z,\mathrm{A/B}}$ that act on the axial extensions below the elastic spring element, Fig. 1. The beam has a radius of $r_\mathrm{b} = 4\,\mathrm{mm}$ and the spring element thickness is $t_\mathrm{s} = 1.0\,\mathrm{mm}$. The maximum theoretical critical buckling load of a beam with properties given in Table 1 is achieved for a beam with infinitely stiff supports to

$$P_\mathrm{e} = \frac{\pi^2 E I_\mathrm{b}}{(0.5\,l)^2} = 3472.7\,\mathrm{N}, \tag{4}$$

representing EULER case IV, [15]. With the simulated values for rotational and lateral stiffness k_φ and k_l, Fig. 5a, the critical buckling load of the uncontrolled elastically supported beam was calculated to $P_\mathrm{cr} = 3122.5\,\mathrm{N} \approx 0.90\,P_\mathrm{e}$ being slightly lower than the maximum theoretical critical buckling load P_e, [11]. With the updated values for the rotational and lateral stiffness k_φ and k_l, Fig. 7b, that are lower than the earlier simulated values, the critical buckling load of the beam system is reduced to $P_\mathrm{cr} = 3013.4\,\mathrm{N} \approx 0.87\,P_\mathrm{e}$. Thus, the updated stiffness parameter for the elastic supports with piezoelectric stack transducers results in a smaller critical buckling load for the passive beam system.

For the active buckling control, the supercritically loaded beam is disturbed by a lateral impulse force $F(t)$, Fig. 1. The maximum axial load P for which the beam can be stabilized is $P = 7700\,\mathrm{N} \approx 2.22\,P_\mathrm{e}$. This limitation is based on the assumption of a maximum controller force of 750 N for the chosen piezoelectric stack transducers. Fig. 8 shows the beam displacement $w(x_\mathrm{F})$ in z-direction of the controlled and uncontrolled beam, Fig. 1, and the control forces $F_{\mathrm{a}z,\mathrm{A/B}}$ in z-direction. The controlled beam is initially deflected by the impulse force, but then quickly returns to its initial straight form due to the control forces in supports A and B.

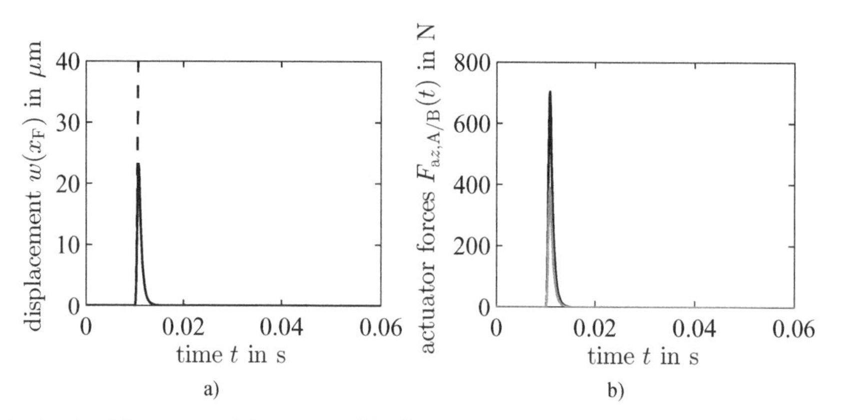

Fig. 8: Active buckling control for supercritically loaded beam, a) beam displacement $w(x_\mathrm{F})$ controlled (—) and uncontrolled (– –), b) controller forces $F_{\mathrm{a}z,\mathrm{A}}$ (—) and $F_{\mathrm{a}z,\mathrm{B}}$ (—)

In [11] with the stiffer passive rotational and lateral stiffness k_φ and k_l, the same maximum bearable axial load of $P = 7700\,\mathrm{N}$ was determined for the active buckling control. Therefore, the smaller stiffness of the spring element does influence the passive critical buckling load P_cr, but not the maximum bearable load for the active buckling control. This shows that the active buckling control is robust against changes in the spring element stiffness k_φ and k_l.

Passive resonant shunt damping

For the application of passive resonant shunt damping, a beam with radius $r_\mathrm{b} = 5\,\mathrm{mm}$ and spring element thickness $t_\mathrm{s} = 0.7\,\mathrm{mm}$ is investigated. Only the piezoelectric transducers P_1, P_2 and P_3 in support A, Fig. 1, are taken into account. In support B, effects of piezoelectric transducers are neglected, [10]. A harmonic force with amplitude F excites the beam, Fig. 1. Frequency response functions

$$H_y(\Omega) = \frac{v(x_F, \Omega)}{F(\Omega)} \quad \text{and} \quad H_z(\Omega) = \frac{w(x_F, \Omega)}{F(\Omega)} \tag{5}$$

for excitation angular frequency $\Omega = 2\pi f$ of the lateral displacements $v(x_F)$ and $w(x_F)$ in y- and z-direction, Fig. 1, are investigated with and without resistive inductive shunts, RL-shunts, connected to the piezoelectric transducers.

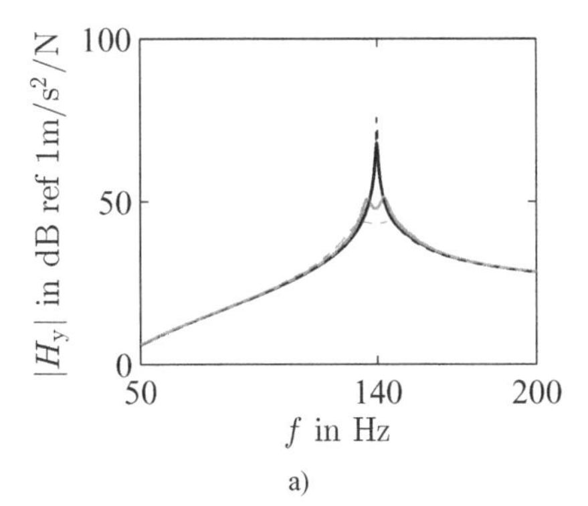

a)

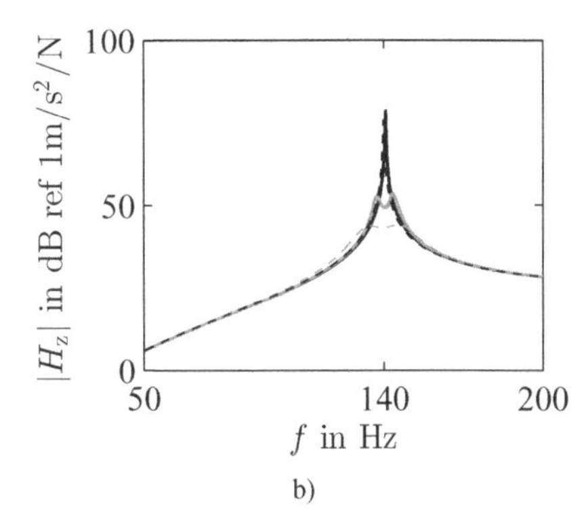

b)

Fig. 9: Frequency transfer function of the beam with elastic supports and three integrated resonantly shunted piezoelectric transducers in support A without RL-shunts: experiment (—) and simulation (- -), with RL-shunt: experiment (—) and simulation (- -) in a) y-direction and b) z-direction

Fig. 9a and Fig. 9b show the amplitudes of the frequency response functions $|H_y|$ and $|H_z|$ of the beam in y- and z-direction with and without connected RL-shunts for the experimental setup and for the numerical simulation. With the simulated values for rotational and lateral stiffness k_φ and k_l, Fig. 5a, the vibration reduction in y-direction and z-direction is 31 dB, [10]. With the updated values for the rotational and lateral stiffness k_φ and k_l, Fig .7b, that are lower than the earlier simulated values, the vibration reduction in both y-direction and z-direction is reduced by 10 %. Furthermore, the eigenfrequency of the beam without connected RL-shunts is reduced by 8 %. For the experimental setup, the vibration reduction in y-direction is 16 dB and in z-direction is 26 dB. Therefore, the updated numerical simulation still overestimates the possible vibration reduction compared to experimental results, Fig. 9. This might result from deviations in the piezoelectric properties or an influence of the support housing stiffness that was neglected in the simulations.

Conclusion

A model verification and validation of the elastic properties of a piezo-elastic support for passive and active structural control of beams with circular cross-section is performed. The piezo-elastic support comprises an elastic spring element characterized by rotational and lateral stiffness. Simulated values for the rotational and lateral stiffness are overestimated in comparison to the experimentally determined values that are used to update the simulation models for active buckling control and passive resonant shunt damping. For the active buckling control of supercritically loaded beams, the reduced support stiffness results in lower critical buckling loads of the passive beam. However, it does not significantly affect the maximum bearable axial load for the active stabilization. For the passive resonant shunt damping, the achieved vibration attenuation as well as the resonance frequency of the beam without connected shunts is reduced by the softer supports. However, the updated numerical simulation still overestimates the possible vibration reduction compared to experimental results. Hence, further investigations on the different aspects reducing the piezo-elastic support's performance in structural control such as the piezoelectric properties, influence of the support housing stiffness, non-ideal clamping and deviations in the material properties as well as a static measurement of the lateral spring element stiffness are necessary.

Acknowledgements

The authors like to thank the German Research Foundation (DFG) for funding this project within the Collaborative Research Center (SFB) 805.

References

[1] Meressi, T. ; Paden, B.: Buckling control of a flexible beam using piezoelectric actuators. *Journal of Guidance, Control, and Dynamics*, 16 (5): 977 – 980, 1993.

[2] Wang, Q. S.: Active buckling control of beams using piezoelectric actuators and strain gauge sensors. *Smart Materials and Structures*, 19: 1 – 8, 2010.

[3] Thompson, S. P. ; Loughlan, J.: The active buckling control of some composite column strips using piezoceramic actuators. *Composite Structures*, 32: 59 – 67, 1995.

[4] Berlin, A. A. ; Chase, J. G. ; Yim, M. ; Maclean, J. B. ; Olivier, M. ; Jacobsen, S. C.: Mems-based control of structural dynamic instability. *Journal of Intelligent Material Systems and Structures*, 9: 574 – 586, 1998.

[5] Enss, G. C. ; Platz, R.: Statistical approach for active buckling control with uncertainty. In *Proc. IMAC XXXII*, 209, 2014.

[6] Schaeffner, M. ; Enss, G. C. ; Platz, R.: Mathematical modeling and numerical simulation of an actively stabilized beam-column with circular cross-section. In *Proc. SPIE*, 9057, 2014.

[7] Hagood, N. W. ; Crawley, E. F.: Experimental investigation of passive enhancement of damping for space structures. *Journal of Guidance, Control, and Dynamics*, 14 (6): 1100 – 1109, 1991.

[8] Hagood, N. W. ; Aldrich, J. B. ; von Flotow, A. H.: Design of passive piezoelectric damping for space structures. *NASA Contractor Report*, 4625, 1994.

[9] Preumont, A. ; de Marneffe, B. ; Deraemaeker, A. ; Bossens, F.: The damping of a truss structure with a piezoelectric transducer. *Computers & Structures*, 86:227 – 239, 2008.

[10] Götz, B. ; Platz, R. ; Melz, T.: Lateral vibration attenuation of a beam with circular cross-section by supports with integrated resonantly shunted piezoelectric transducers. In *Proc. SMART2015, 7th ECCOMAS Conference on Smart Structures and Materials*, 2015

[11] Schaeffner, M. ; Platz, R. ; Melz, T.: Active buckling control of an axially loaded beam-column with circular cross-section by active supports with integrated piezoelectric actuators. In *Proc. SMART2015, 7th ECCOMAS Conference on Smart Structures and Materials*, 2015

[12] Bretz, A. ; Calmano, S. ; Gally, T. ; Götz, B. ; Platz, R. ; Würtenberger, J.: Darstellung passiver, semi-aktiver und aktiver Systeme auf Basis eines Prozessmodells (Representation of Passive, Semi-Active and Active Systems Based on a Prozess Model). unreleased paper by the SFB 805 on www.sfb805.tu-darmstadt.de/media/sfb805/f_downloads/150310_AKIII_Definitionen_aktiv-passiv.pdf

[13] Klein, B.: *FEM*. Springer Vieweg, Wiesbaden, 2012.

[14] Przemieniecki, J. S.: *Theory of Matrix Structural Analysis*. McGraw-Hill, New York, 1968.

[15] Timoshenko, S. P. ; Gere, J. M.: *Theory of Elastic Stability*. McGraw-Hill, New York, 1961.

Applied Mechanics and Materials Vol. 807 (2015) pp 78-86
© (2015) Trans Tech Publications, Switzerland
doi:10.4028/www.scientific.net/AMM.807.78

Submitted: 2015-04-30
Accepted: 2015-08-04

Opportunities and Limitations of Structural Intensity Calculation regarding Uncertainties in the NVH design of Complex Vehicle Body Structures

Torsten Stoewer[1,a *], Johannes Ebert[2,b] and Tobias Melz[3,c]

[1,2]Knorrstr. 147, 80788 München, Germany

[3] Magdalenenstraße 4, 64289 Darmstadt, Germany

[a]Torsten.Stoewer@bmw.de, [b]Johannes.Ebert@bmw.de, [c]melz@szm.tu-darmstadt.de

Keywords: structural intensity, energy flow, NVH, FEM, uncertainty, acoustics

Abstract. The calculation of the structural intensity allows for a better understanding of the NVH behavior of complex structures as it shows vibratory energy flows between an excitation and radiating areas. However, the information gathered is underlying aleatory and epistemic uncertainties and needs to be dealt with carefully. In this paper two aspects are discussed: Firstly how the structural intensity calculation helps to reduce uncertainty in NVH design and secondly what currently existing uncertainties need to be considered and how they can be further reduced. This does not only include an improvement of the current calculation process itself but also an extension towards an integrated, holistic calculation of vibratory energetic quantities for structure-borne and air-borne sound.

Introduction

As premium vehicles costumers' expectations towards their cars NVH comfort grows, manufacturers are constantly faced with the challenge of improving the demanded properties. This improvement has to go along with the enhancement of other characteristics like crashworthiness, operational stability and lightweight design. At the same time the product development process is tightened and the number of costly hardware prototypes is reduced, which requires a strong and robust virtual design that allows for a deep knowledge of the future vehicles' body-in-white vibro-acoustic behavior.

In order to enhance this knowledge, the method of structural intensity (STI) has been proposed in numerous publications (e.g. [1, 2, 3]). Analogous to the well established quantity of (air-borne) sound intensity it enables the calculation of structure-borne sound transfer paths in detail and by that closes the gap between excitation and radiation of a structure as schematically shown in Fig. 1.

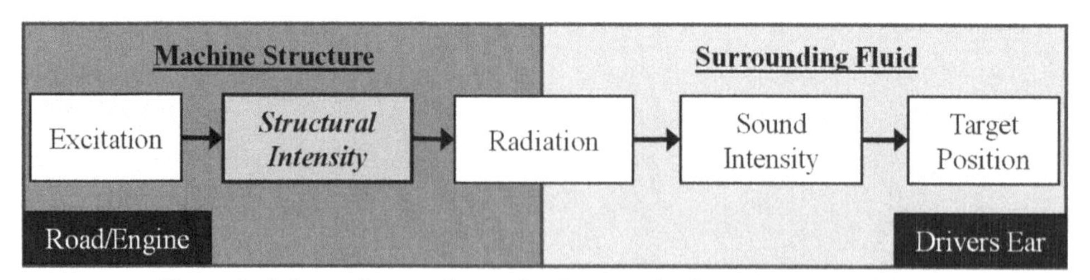

Figure 1: Schematic sound transfer from excitation to target position.

However, structural intensity results are underlying aleatory and epistemic uncertainties that have to be taken into account. While aleatory uncertainties usually arise from the variation of structural characteristics due to variation in the production process like body shell or assembly, epistemic uncertainties result from the vaguely known properties of the construction like material anisotropy, joint stiffness or structural damping that are typically only roughly included in standard NVH calculations and unknown boundary conditions as well as excitation spectra. In addition there are

deviations in the calculation process between the analytical calculation approach and an implementation for discrete finite element models.
Both types of uncertainty can have a major impact on structural intensity results as the quantity provides very detailed results for every point of the structure. It is therefore necessary to know the parameters influencing the structural intensity and understand their effect on the results in order to be able to interpret the results of complex structures correctly.

Basics of structural intensity calculation

The quantity of the structural intensity $\boldsymbol{I}_s$ describes the amount of energy flowing through a plain element of a solid body. It is the product of the multiplication of the stress tensor $\boldsymbol{\sigma}$ and the velocity vector $\boldsymbol{v}$:

$$\mathbf{I}_s = -\boldsymbol{\sigma} \cdot \mathbf{v} = -\begin{bmatrix} \sigma_{xx} v_x + \tau_{xy} v_y + \tau_{xz} v_z \\ \tau_{yx} v_x + \sigma_{yy} v_y + \tau_{yz} v_z \\ \tau_{zx} v_x + \tau_{zy} v_y + \sigma_{zz} v_z \end{bmatrix}. \tag{1}$$

The structural intensity is an energetic quantity that is based on the first law of thermodynamics, where, according to [4], it is defined that the sum of in-coming and out-coming energies is always zero for a defined control volume of an elastic medium

$$\iiint\limits_V \frac{\partial e}{\partial t}\,\mathrm{d}V = -\iiint\limits_V \nabla \cdot \mathbf{I}_s\,\mathrm{d}V + \iiint\limits_V (\pi_{\mathrm{in}} - \pi_{\mathrm{diss}})\,\mathrm{d}V, \tag{2}$$

with e as energy density, π_{in} as the input power density, and π_{diss} as the dissipated power density. Calculated in frequency domain the structural intensity becomes complex and consists of a real $\mathbf{I}_a(\omega)$ and an imaginary $\mathbf{I}_r(\omega)$ part. In the further context the parts are referred to as active and reactive structural intensity with the main focus on the active, "real" energy flow.

$$\boldsymbol{I}_{s,a}(\omega) = \frac{1}{2}\Re\left(\underline{\boldsymbol{I}_s}(\omega)\right) \tag{3}$$

$$\boldsymbol{I}_{s,r}(\omega) = \frac{1}{2}\Im\left(\underline{\boldsymbol{I}_s}(\omega)\right) \tag{4}$$

In case of the vehicle body development most structures are thin walled and can be approximated by shell elements in finite element calculation. With the presupposition of small displacements Eq. 1 can be transformed to the equation for the structural intensity of shells $\mathbf{I}_s'$

$$\mathbf{I}_s' = \begin{bmatrix} I_{s,x}' \\ I_{s,y}' \end{bmatrix} = -\begin{bmatrix} N_x v_x + N_{xy} v_y + M_x \dot{\varphi}_y - M_{xy} \dot{\varphi}_x + Q_x v_z \\ \underbrace{N_y v_y + N_{yx} v_x}_{\text{in-plane}} \underbrace{- M_y \dot{\varphi}_x + M_{yx} \dot{\varphi}_y + Q_y v_z}_{\text{out-of-plane}} \end{bmatrix} \tag{5}$$

which includes the section forces $N_{x,y}$ and moments for bending $M_{x,y}$, torsion $M_{xy,yx}$ and shear $Q_{x,y}$ of one shell element and allows for a differentiation between in-plane (ip) and out-of-plane (oop) wave components like shown in Fig. 2. The advantage of the separated evaluation of in-plane and out-of-plane waves is discussed in [3, 5]. According to [6] the energy transport in local z-direction is negligible for shell elements.

Figure 2: Basic wave forms

As a result of the structural intensity calculation vectorial information about vibratory energy flows of operational vibrations is obtained either in the time domain or via transformation by the cross-spectral density in the frequency domain [7]. Exemplary for the results in the frequency domain the active energy flow $\mathbf{I}_{s,a}$ at the fourth resonance frequency of an orthogonally excited, simply supported plate is shown in Fig. 3. Looking at the structural intensity on the right side of the figure the reason for the shape of the mode can be observed. Emanating from the point of excitation on the lower right side the vibratory energy flows to the left side where it is transported circularly in a vortex and by that forming three modes in cross direction.

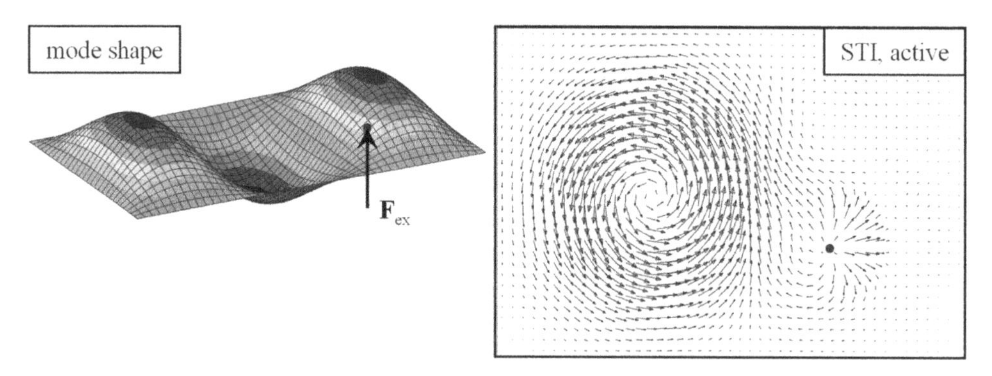

Figure 3: Mode and structural intensity of an orthogonally excited plate

This additional detailed information can be used in various ways to deepen the understanding for vibroacoustic occurrences in structures and develop effective and efficient countermeasures. By that the calculation of the structural intensity helps to reduce uncertainties in NVH engineering.

Comparison between the analytical and the discrete numerical solution of the structural intensity

In order to verify the mathematical correctness of the analytical structural intensity results in the frequency domain for shell structures, an option is to look at its convergence over the included number of modes when calculating the modal superposition of the simply supported plate. Like [3] and [5] have shown for the discrete numerical solution, the structural intensity results become more defective when including not enough modes in the modal superposition. However, it is shown that the error is vanishing with including rising numbers of modes.
In contrast to that differences in the result can be seen for the analytical solution even for very high numbers of included modes (Fig. 4). The question is whether the results would eventually converge.

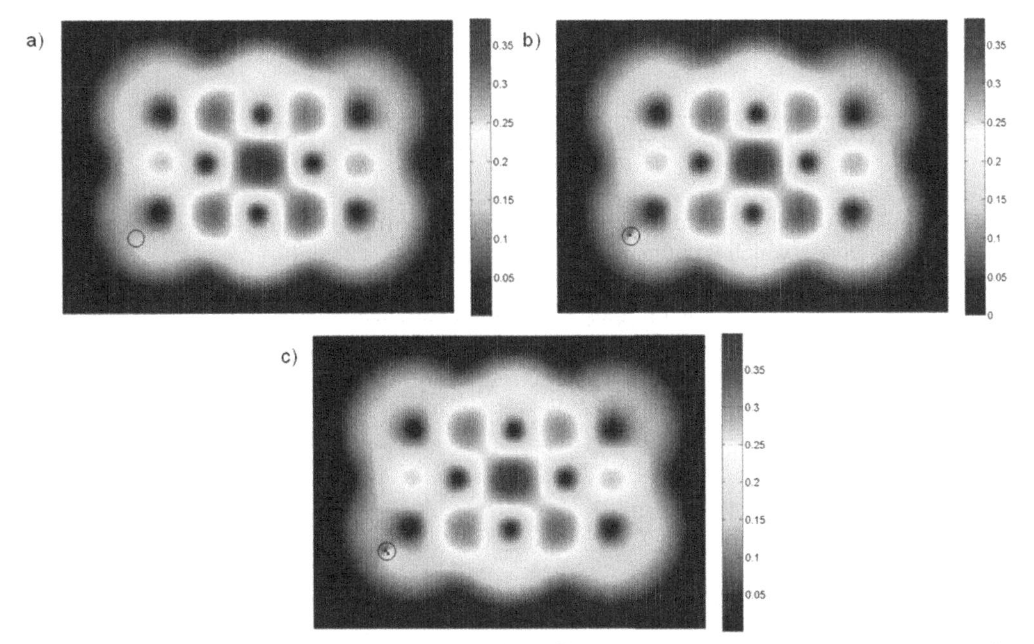

Figure 4: Example of the plates structural intensity with a) 100, b) 10.000 and c) 1.000.000 modes included. Note the change at the excitation position marked with a black circle

Starting point of the derivation is the out-of-plane share of Eq. 5. Only the shear force part is extracted as its rising share in the result in Fig.4 is causing the deviation at the excitation point. For this part it applies that

$$Q_x v_z = -\underline{B}' \left(\underbrace{\frac{\partial^3 \underline{u}_z}{\partial x^3}}_{C} + \frac{\partial^3 \underline{u}_z}{\partial x \partial y^2} \right) \left(\frac{\partial \underline{u}_z}{\partial t} \right)^* . \tag{6}$$

Exemplary, the convergence is analyzed for the term indicated with C.
According to [3] this term can be written as

$$C = \sum_{m=1}^{\infty} \sum_{n=1}^{\infty} - \frac{4\underline{F}_0}{\rho h l_x l_y} \cdot \frac{\sin(\alpha_m x_0) \sin(\beta_n y_0)}{\frac{B'}{\rho h} ((\alpha_m)^2 + (\beta_n)^2)^2 (1 + i\eta) - \omega^2} \alpha_m^3 \cos(\alpha_m x) \sin(\beta_n y) \tag{7}$$

with

$$\alpha_m = \frac{m\pi}{l_x} \text{ and } \beta_n = \frac{n\pi}{l_y}, \tag{8}$$

and $l_{x,y}$ being the length and width and h the thickness, ρ the density, η the loss factor and B' the bending stiffness of the plate.
To deduce the convergence of Eq. 7 a comparison to the harmonic series in Eq. 9 is performed:

$$\sum_{n=1}^{\infty} \frac{1}{n} = \infty . \tag{9}$$

With the negligence of constant factors and the trigonometric functions Eq. 7 can be transformed to

$$\sum_{m=1}^{\infty}\sum_{n=1}^{\infty}\frac{\alpha_m^3}{\alpha_m^4+2\alpha_m^2\beta_n^2+\beta_n^4}=\frac{1}{\alpha_m+2\frac{\beta_n^2}{\alpha_m}+\frac{\beta_n^4}{\alpha_m^3}}\approx\frac{1}{m+2\frac{n^2}{m}+\frac{n^4}{m^3}}\,. \tag{10}$$

This is the relevant term for the deduction of the series property. The next steps are based on the direct comparison test, which is defined as follows:

"If the infinite series $\sum b_k = \infty$ and $0 \leq b_k \leq a_k$ for all sufficiently large k, then the infinite series $\sum a_k = \infty$".

The direct comparison test is applied to the series

$$\sum_{m=1}^{\infty}\sum_{n=1}^{\infty}\frac{1}{m+2\frac{n^2}{m}+\frac{n^4}{m^3}}\,. \tag{11}$$

With $m = n$ an estimation towards the lower limit can be performed based on Eq. 12

$$\sum_{m=1}^{\infty}\sum_{n=1}^{\infty}x_{nm}\geq\sum_{m=1}^{\infty}\sum_{n=m}x_{nm}=\sum_{m=1}^{\infty}x_{mm}\,. \tag{11}$$

For this case it leads to

$$\sum_{m=1}^{\infty}\sum_{n=1}^{\infty}C(mn)=\sum_{m=1}^{\infty}\sum_{n=1}^{\infty}\frac{1}{m+2\frac{n^2}{m}+\frac{n^4}{m^3}}$$

$$\rightarrow\sum_{m=1}^{\infty}C(mn)=\sum_{m=1}^{\infty}\frac{1}{m+2\frac{m^2}{m}+\frac{m^4}{m^3}}=\sum_{m=1}^{\infty}\frac{1}{m+2m+m}=\sum_{m=1}^{\infty}\frac{1}{4m}=\frac{1}{4}\sum_{m=1}^{\infty}\frac{1}{m}=\infty\,. \tag{11}$$

As the low estimation diverges, the full term will diverge. This means, that the shear force part of the structural intensity diverges and also the structural intensity diverges.
However, due to limited computational precision, the numerical solution based on modal superposition will eventually give similar results to the direct solution. Still, the direct solution offers more precise results.

Uncertainties in numerical structural intensity calculation of complex models

As stated in the preceding chapter the quantity of structural intensity provides very detailed information about the NVH behavior of structures. Still, or hence, the results have to be dealt with very carefully in order not to draw wrong conclusions. This is explicitly true for the influence of locally changing damping. Adding a highly damping patch to the plate already shown in Fig. 3 changes the resulting structural intensity for the fifth resonance frequency considerably (Fig. 5) as the vibratory energy is channeled towards the damping patch where it is absorbed. At the same time quantities like the eigenfrequency or the mode shape will stay pretty constant though the change of the energy flow suggests something different.

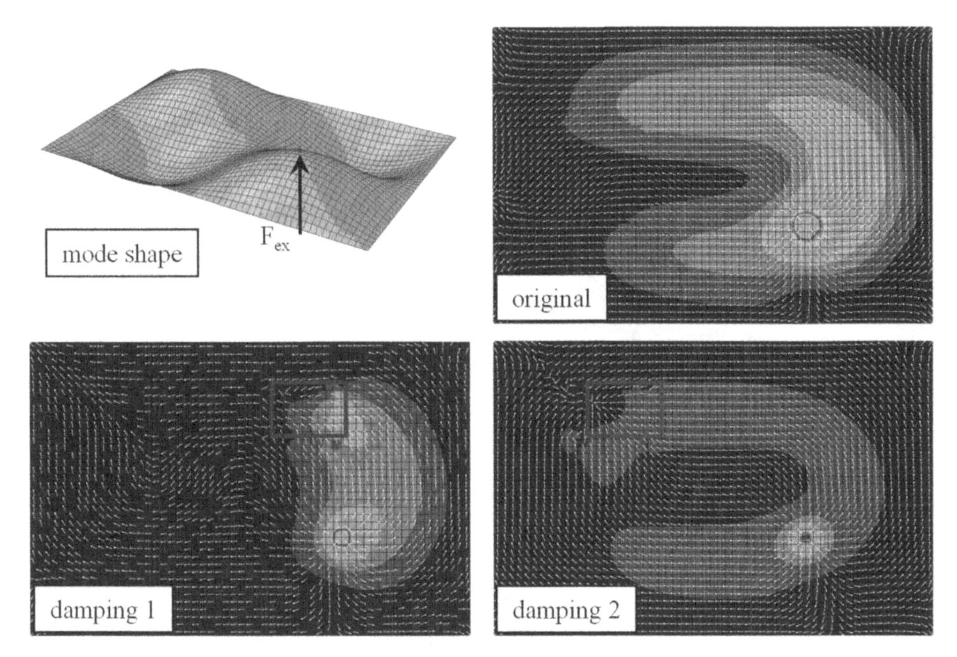

Figure 5: Simply supported plate with damping patch at two different positions

For a simple plate this causality is easy to deal with, but for more complex, jointed structures uncertainty arises from inconstant structural damping resulting out of differing material properties, metal forming and various forms of joining such as welding, gluing, screwing, etc. All these points influence the local vibratory energy flow but are typically not adequately included in standard finite element models as their complexity and resulting calculation time would be to high to efficiently support a vehicle series development. Figure 6 gives an impression of a typical structural intensity calculation result for a front vehicle. It is clear to see, that the energy flux changes considerably at local structure changes. Yet, no direct information about the customer relevant acoustic behavior or necessary structural changes of the vehicle can be derived.

Two conclusions can be drawn out of this example:

1. It is necessary to have very exact finite element models in order to obtain exact results for a sensitive physical quantity such as the structural intensity. As this level of accuracy is almost impossible to reach for complex structures like vehicle bodies without being at odds with the development time line and computational costs, it is important to find a good compromise between these objectives. This includes dealing with uncertainty in the information of the structural intensity.
2. In order to be able to easier interpret the structural intensity results additional quantities can be calculated that link the vibratory energy flow to the costumer relevant quantity of sound pressure level. This point will be discussed in the following chapter.

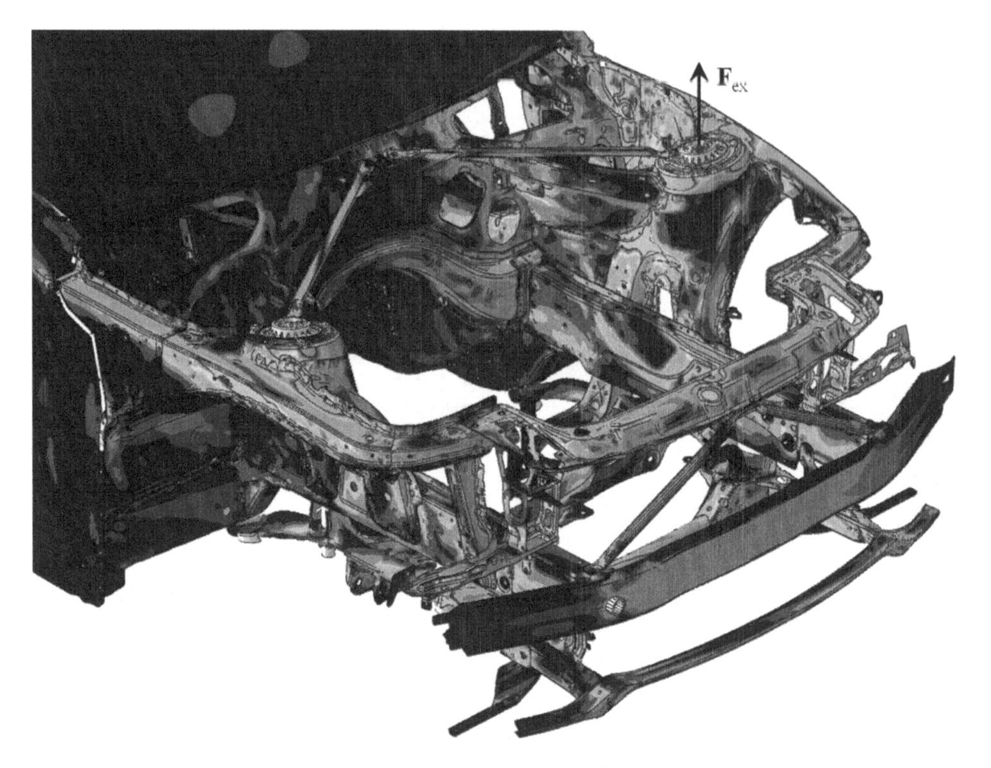

Figure 6: Scalar structural intensity $|\mathbf{I}'_{s,a}|$ for a front vehicle

Approaches to reduce uncertainty in the structural intensity calculation

As the numerical simulation of the structural intensity is based on finite element models it seems obvious to improve the modeling in order to reduce deviations and uncertainty. However, the improvement of accuracy in the finite element models goes along with a rising modeling effort and higher computational costs. Therefore two approaches are presented briefly that have the potential to further improve the evidence of the structural intensity calculation and by that reduce the uncertainty in NVH design.

A way to link the structural intensity directly to recommendations for constructional improvements is to calculate its divergence. According to [8] the divergence is the derivative of all real structural intensity components and is connected to the potential energy density e_p:

$$\Re(\nabla \cdot \underline{\mathbf{I}}_s) = \Re\left(\frac{\partial \underline{I}_{s,x}}{\partial x} + \frac{\partial \underline{I}_{s,y}}{\partial y} + \frac{\partial \underline{I}_{s,z}}{\partial z}\right) = -2\eta\omega e_p \,. \tag{12}$$

The calculation of the divergence allows for accurate predictions of energy sources and, more importantly, sinks. These sinks mark areas of high dissipation where the application of damping layers is very effective. In [9] chances and limitations of the calculation of the divergence are discussed.

In order to gain more knowledge about the energy flow beyond the quantity of structural intensity a promising option is to simulate the interaction of the structure with the surrounding fluid by the quantity of sound intensity as it is shown in Fig. 1. The simultaneous simulation of both energetic

quantities allows for an integrated view on the occurrences happening between the excitation of a structure and the resulting air-borne sound at certain positions, for example the drivers ear.
An example for the interaction between an excited structure and its coupled fluid is shown at the example of a hard-walled box with a simply supported plate as cover in Fig. 7, a detailed study of this calculation approach can be found at [10].

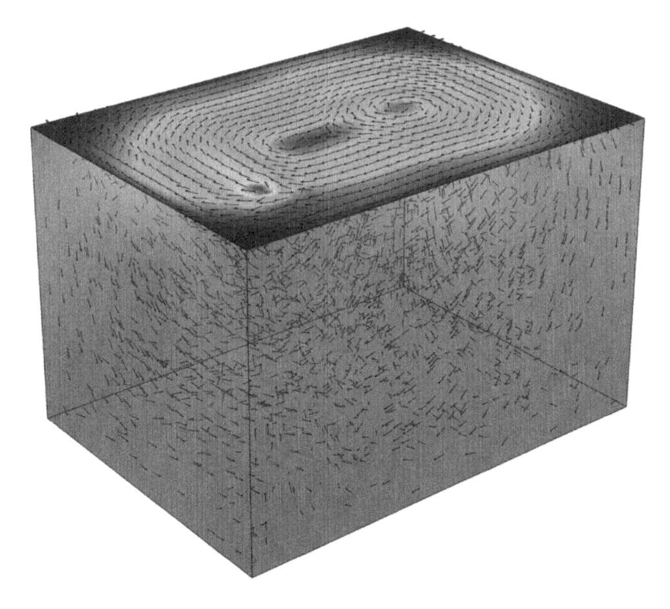

Figure 7: Structural intensity and sound intensity of a box at 148 Hz

Summary

The quantity of structural intensity is a very powerful tool in order to enhance NVH design of complex structures like vehicle bodies. It helps to reduce epistemic uncertainty through very detailed information of the vibratory energy flow within structures. However, this detailed information goes along with a high sensitivity towards inaccuracies in finite element modeling. Special care has to be taken with the modeling of local damping that is inherent to every type of joining technique. As it is very costly to model complex structures in this detail two approaches are presented that further increase the value to the existing calculation through additional information: Including fluid-structure interaction by the quantity of sound intensity as well as calculating the divergence of the structural intensity. The resulting integral view on the NVH behavior of a complex system helps to efficiently tackle occurring problems in every design phase.

Acknowledgements

The author would like to thank Peter Groba for his contribution to this paper. His works in the field of the convergence of the analytical solution of the structural intensity have considerably supported the in-depth understanding of resulting uncertainties.

References

[1] D. U. Noiseux, Measurement of Power Flow in Uniform Beams and Plates. in: Journal of the Acoustical Society of America 47, 1970, pp. 238-247.

[2] L. Gavric, G. Pavic, Computation of structural intensity in beam-plate structures by numerical modal analysis using FEM. in: Proceedings of the Third International Conference on Intensity Techniques, Senlis, 1991, pp. 207-214.

[3] T. Hering, Strukturintensitätsanalyse als Werkzeug der Maschinenakustik, Darmstadt, TU: Diss., 2012.

[4] O. M. Bouthier, R. J. Bernhard, Simple Models of Energy Flow in Vibrating Membranes, in: Journal of Sound and Vibration 182(1), 1995, pp. 129-147

[5] T. Stoewer, et al., Mit Schwingungsenergieflussberechnungen zur effizienten akustischen Auslegung von Fahrzeugstrukturen, in: Aachener Akustik Kolloquium, Aachen, 2011, pp. 215-221.

[6] W. Maysenhölder, Körperschallenergie, Hirzel, Stuttgart/Leipzig, 1994.

[7] J. W. Verheij, Cross Sprectral Density Methods for Measuring Structure Borne Power Flow on Beams and Pipes. in: Journal ofSound and Vibration 70, 1980, pp. 133-139

[8] G. Pavic, The role of damping on energy and power in vibrating systems, in: Journal of Sound and Vibration 281, 2005, pp. 45-71.

[9] B. Lamarsaude, Y. Bousseau, A. Jund, Structural intensity analysis for car body design: going beyond interpretation issues through vector field processing, in: Proceedings of the ISMA 2014, Leuven, 2014.

[10] J. Ebert, et al., NVH development of car structures by means of numerical simulation of energetic quantities, in: Aachener Akustik Kolloquium, Aachen, 2014, pp. 117-126.

CHAPTER 3:

Modular Design and Scaling for Reduced Uncertainties in the Design Process

Applied Mechanics and Materials Vol. 807 (2015) pp 89-98
© (2015) Trans Tech Publications, Switzerland
doi:10.4028/www.scientific.net/AMM.807.89

Submitted: 2015-04-15
Accepted: 2015-08-04

Uncertainty in product modelling within the development process

Jan Würtenberger [1,a *], Sebastian Gramlich [1,b], Tillmann Freund [1,c], Julian Lotz [1,d], Maximilian Zocholl [2,e], Hermann Kloberdanz [1,f]

[1] Institute for Product Development and Machine Elements, Magdalenenstraße 4, 64289 Darmstadt, Germany

[2] Institute for Computer Integrated Design, Otto-Berndt-Straße 2, 64287 Darmstadt, Germany

[a]wuertenberger@pmd.tu-darmstadt.de, [b]gramlich@pmd.tu-darmstadt.de, [c]freund@pmd.tu-darmstadt.de, [d]lotz@pmd.tu-darmstadt.de, [e]zocholl@dik.tu-darmstadt.de, [f]kloberdanz@pmd.tu-darmstadt.de

Keywords: uncertainty, product modelling, product development process, pragmatism, representation, shortening.

Abstract. This paper gives an overview of how to locate uncertainty in product modelling within the development process. The process of product modelling is systematized with the help of the characteristics of product models and typical working steps to develop a product model. Based on this, a conceptual framework is created and it is possible to distinguish between product modelling uncertainty, mathematic modelling uncertainty, parameter uncertainty, simulation uncertainty and product model uncertainty.

Introduction

Product models are used during the product development process to predict product behavior in the product lifecycle or as a basis for decision making, for example, to define or verify product properties where product behavior is influenced. Therefore, anticipation of information from the product lifecycle is necessary, for example, in integrated design approaches (Fig. 1).

Product modelling is based on assumptions, concepts, schemata or designer ideas, all of which have different levels of abstraction [1].

In product modelling, there is always a lack of information, especially in the early stages of product development. Information relevant to the product model, e.g. about the product lifecycle, is partially incomplete or does not exist at all, so the modelling process is subject to uncertainty [2].

If relevant information is neglected or regarded as insufficient, it is possible that there are variations between planned and realized product behavior, creating a need to handle uncertainty. To be able to handle uncertainty in product modelling, sources of uncertainty have to be identified. Therefore, this paper gives an overview of how to systematize the process of product modelling and how to allocate uncertainty with the help of a conceptual framework.

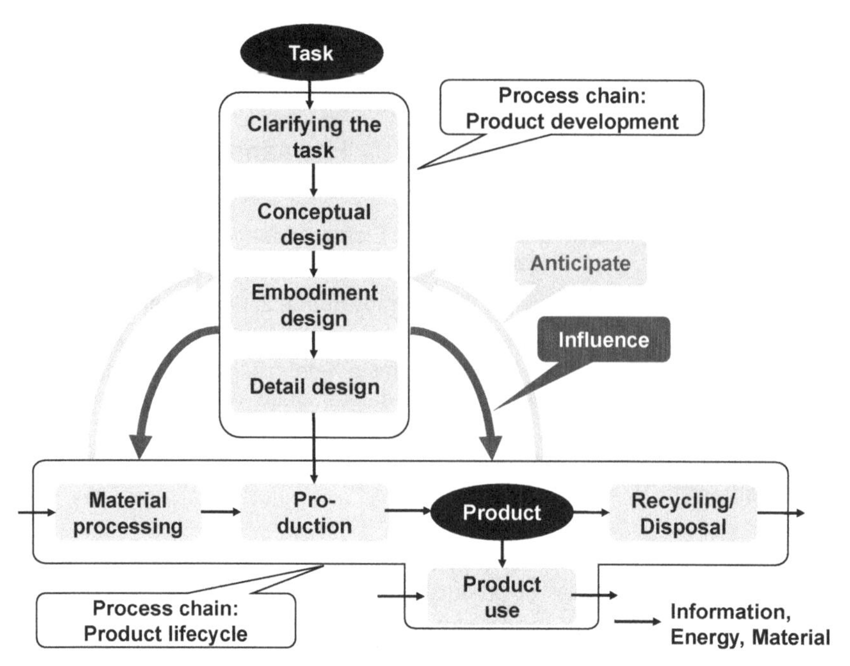

Figure 1: Integrated product and process development [3]

Definitional basics

Model and Modelling. A model is a simplified, abstracted depiction of the complex reality with a certain purpose [4, 5]. It separates between the essential and the non-essential depending on the point of view [2].
Stachowiak defines characteristics in order to describe models in general [6]:

- *Pragmatism*: Pragmatism describes for whom the model is created for. Furthermore, the point of time is set in which the model is used and it is specified for what it is used for.
- *Representation*: Representation describes in which way the original attributes are represented by the model attributes. Attributes in this context are properties which describe what individuals perceive. This characteristic can be compared to the understanding of attribution from mathematics.
- *Shortening*: Only the attributes of the original which are relevant for the model designer are investigated in the model. The selection of attributes made here can be partly pragmatic, so shortening can be made random up to a certain purpose.

Based on that, modelling is the application of the mentioned characteristics according to a certain topic, so modelling is a process in which the model is created [6].

Product model and product modelling. Product models represent the product at a defined abstract level in the virtual product lifecycle. All relevant properties and information that are necessary for the working steps of product development are represented [4], so each product model has a certain purpose. Product modelling is a process during which the characteristics of models have to be set.

Pragmatism defines for whom the product model is developed. Primarily, product models are a base for the designer to practice synthesis and analysis, two of the main tasks in the product development process [5]. The point of time is set, enabling the product model to be allocated to a working step during the development process. At the least, the purpose of the model has to be defined as well as the way in which it is to be achieved. The purpose can be deduced from requirements of the product or from tasks resulting from synthesis and analysis processes. Representation describes the way in which the product is represented by a product model, so the model room is defined in which the allocation from product properties to the product model takes place. Shortening selects product properties and known influences that are probably necessary for the product model.

Pyramid of product models. In order to represent the working results of each step within the product development process, the pyramid of product models is used in this paper (Fig. 2). It consists of four levels, where the pyramid's shape indicates the possible range of variants and the increasing number of defined product properties. Alignment with the requirements is carried out at each level. The results are verified with the requirements and new requirements are defined on the basis of the results [7].

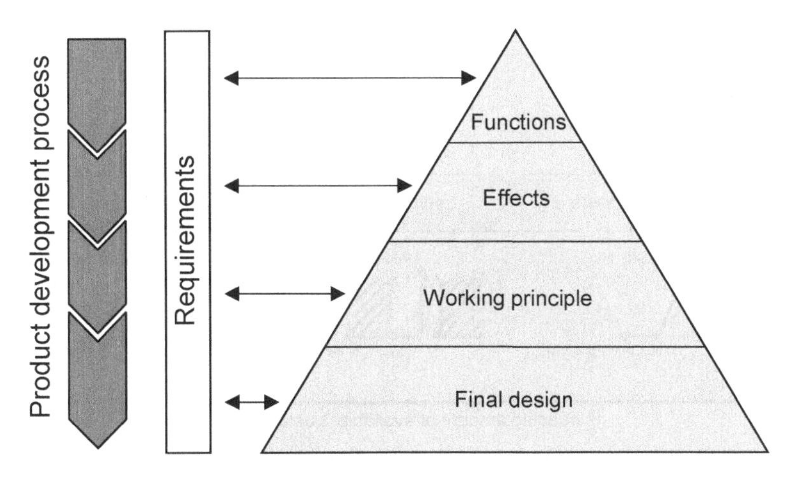

Figure 2: Pyramid of the product models (partly adapted from [7])

The functional level describes the product in a solution-neutral manner with the help of sub functions that have to be fulfilled [4]. A function in this context is a general intended connection between the input and output of a system that aims for a certain task [5]. The definition of sub-functions varies in literature. Pahl et al., for example, define the generally applicable subfunctions channeling, connecting, changing, storing and varying to describe the product [5].

After this, the sub-functions are concretized using physical, biological or chemical effects, for example, the leverage effect, piezo effect or stick-slip effect. They are allocated to each sub-function in order to realize them. By setting first geometrical and material properties, the product is further detailed and the working principle becomes visible [5]. The geometrical and material properties are then concretized in more detail until the final design of the product is set [8].

Uncertainty. Collaborative Research Centre SFB 805 investigates uncertainty to inform its handling [9]. In SFB 805, uncertainty occurs in processes [10]. Process always stays in interaction with a product, so uncertainty leads to divergences, for example, in product properties or process parameters.

Uncertainty is a situation that results from completely or partially missing information, understanding or knowledge [11]. This definition shows that the level of information is an indicator that can be used to describe uncertainty. Depending on the available level of information, the literature distinguishes between aleatoric and epistemic uncertainty. In this case, epistemic uncertainty is a situation of imprecise knowledge about the given task; aleatoric uncertainty describes the random variation of influencing factors from nature [12,13]. With increasing levels of information, epistemic uncertainty is reduced until aleatoric uncertainty exists [12].

Keeping the distinction between aleatoric and epistemic uncertainty in mind, SFB 805 developed a model of uncertainty that focuses on describability. Uncertainty is differentiated into three categories: unknown uncertainty, estimated uncertainty and stochastic uncertainty (Fig. 3).

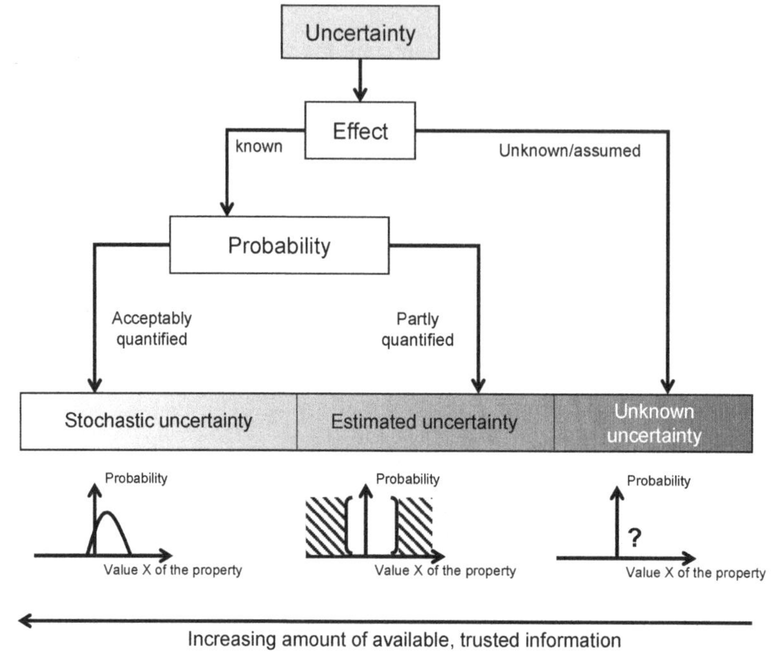

Figure 3: Model of uncertainty [14]

- *Unknown uncertainty* describes a situation in which relations and dependencies in a process are completely unknown: the level of information is low. Based on this, no decisions to handle uncertainty can be made.
- *Estimated uncertainty* describes a situation in which the effects of uncertainty are known. The resulting divergences of product properties, process parameter etc. can be partially described.
- *Stochastic uncertainty* is a situation in which all effects of uncertainty are known and the resulting divergences in properties or process parameters are sufficiently or, ideally, entirely defined.

The level of available, trusted information grows, and occurring uncertainty changes from epistemic to aleatoric. Using the model, strategies or recommendations have to be defined in order to handle uncertainty.

Systematization of product modelling

After giving some definitional basics, the process of product modelling is analyzed and systematized. The definitional basics are combined in order to create a conceptual framework to analyze uncertainty.

The framework is mainly based on an approach in the case of product modelling by Negele. He developed an approach to create a model to solve complex problems in the area of systems engineering, which is used as the basis for the systematization of product modelling (Fig. 4) [15]. He defines three essential levels, which are based on Computer Integrated Manufacturing Open System Architecture (CIM-OSA) [15]. First, the problem or task is defined, including the purpose for which the model is needed. Requirements or influences of the relevant part have to be considered, for example, economical or ecological factors.

Based on this, the generic level defines generic model constructs that are used as a formal description to create the model. Rules are set in the style of a syntax to describe how the model constructs can stay in interaction.

At the partial level, a suitable model structure for the problem or task is developed. Keeping that in mind, the generic constructs are then adapted and integrated into the structure, considering the given syntax.

The individual level specifies the partial model to an individual model by considering, for example, boundary conditions for the aspirated solution of the problem or specific information that is necessary to generate a solution. This makes the model usable only for the specific problem or task; it cannot be transferred to similar ones. At the minimum, the individual model is used as a basis for solving the given problem or task.

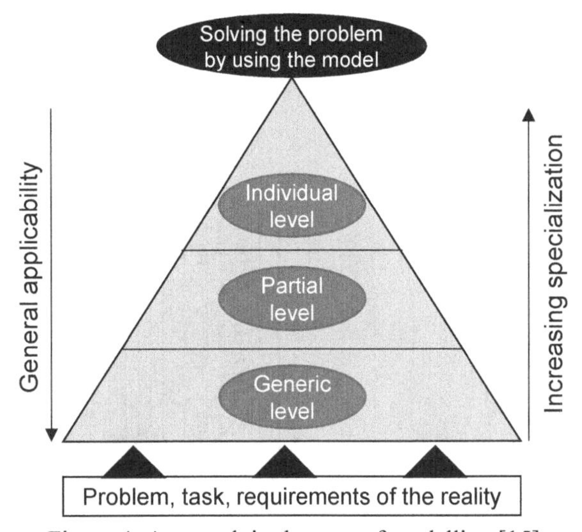

Figure 4: Approach in the case of modelling [15]

The approach in modelling is combined with the characteristics of product models to systematize the process of product modelling and to create a conceptual framework (Fig. 5). The initial point is a problem or a task for which the product model is needed, The problem or task is mainly deduced from synthesis and analysis processes within the product development process. The characteristic pragmatism is set by the designer, so as the purpose and the related approach are defined, a solution strategy arises.

Based on this, generic constructs are defined. As mentioned above, the pyramid of product models represents the working results of each product development step, such as increasing concretization of the product. Each level of the pyramid describes the product at a defined degree of abstraction, so a product can always be described by the elements functions, effects, and working principles, such as geometric and material properties. These elements are used as generic constructs for product modelling and constitute a model kit.

By choosing elements relevant to the solution strategy, the first part of characteristic representation is set and product model components are defined. The elements can be chosen from one or more levels of the pyramid. A model structure is then defined by describing how the chosen elements are connected in principle. From this, a model room is visible and the principle product model can be deduced, so the second part of the characteristic representation is set.

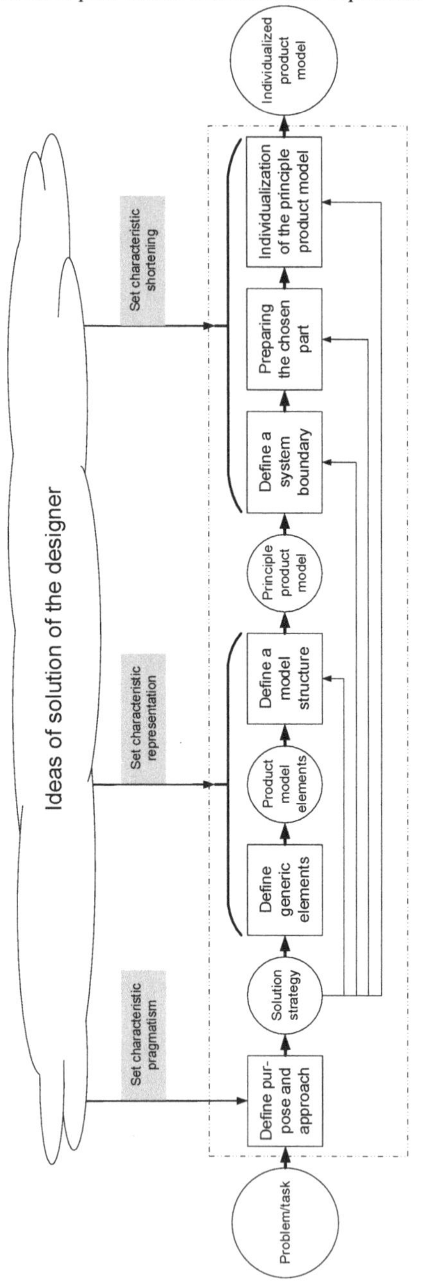

Figure 5: Conceptual framework to systematize the process of product modelling

Characteristic shortening adapts the principle product model to an individualized product model, so the idea of how the product model should look is matched with the investigated product. The investigated product in the development process is a work status that can have different degrees of concretization. The system boundary defines the part of the product that is relevant to the solution strategy. The chosen part of the product is then prepared.

The preparation picks up elements from inside the system boundary and adapts them in a way that the principle model can be performed. Relevant influences are identified that are also necessary for the individualization. Assumptions are made, for example, whether two elements should be summed up or whether identified influences should be investigated.

Finally, individualization attributes the relevant influences, such as the prepared elements, from the system boundary to the principle model, where the individualized product model is created. The individualized product model can then be used to solve the problem or task. The process of product modelling shows that pragmatism influences representation and shortening setting, so pragmatism guides the way to the individualized product model.

The conceptual framework to systematize the process of product modelling is now explained using a simple example of a tetrahedron out of the SFB 805 demonstrator (Fig. 6). The task is to make a statement on how a tetrahedron behaves if it is loaded with a force for which a product model is needed. Pragmatism is set here to analyze the stability of one of the bars. The chosen generic constructs are, for example, the effects 'elastic bending', and 'compression force', such as the properties 'length of the bar' and 'modulus of elasticity'. How the chosen constructs work together is explained using a combination of a balanced condition between the force and the lateral deflection of the bar, such as the law of elasticity. The combination of both describes the theory of buckling by Euler, whereby the principle product model is set [16]. The tetrahedron then has to be prepared to investigate the theory of buckling by Euler. Therefore, different assumptions have to be made. For example, it is assumed that the first case of Euler is applicable here, where the bar is shorted to a fixed connection on the one side and to a free end on the other side. With this, the theory of buckling can be adapted to the bar and an individualized product model is created that can be used to determine the critical buckling load.

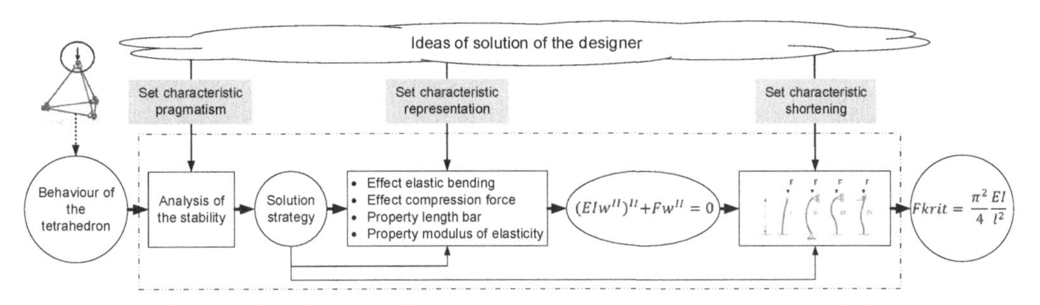

Figure 6: Creating an individual product model using a tetrahedron [16]

Depending on the task, problem or defined pragmatism, changes in a system can also be relevant, which means that there is a need for simulation. In this context, simulation means running experiments on the defined abstraction level of a product model [17]. Therefore, the individualized product model has to be adapted into a mathematical model, so an extension of the framework is necessary (Fig. 7). The relevant literature divides simulation models into static and dynamic, qualitative and quantitative, and continuous and discrete, such as deterministic and stochastic [15, 17]. The process of mathematic modelling sets a simulation model type, where the individualized product model is adapted. After that, values of parameters, which are defined during the process of product and mathematic modelling, have to be set. In the tetrahedron, for example, the parameters of the properties 'modulus of elasticity' and 'length of the bar' are set. After that, the process of simulation is performed, where specific simulation settings, like the simulation time, are set. The simulation results can then be used to find a solution for the given problem or task.

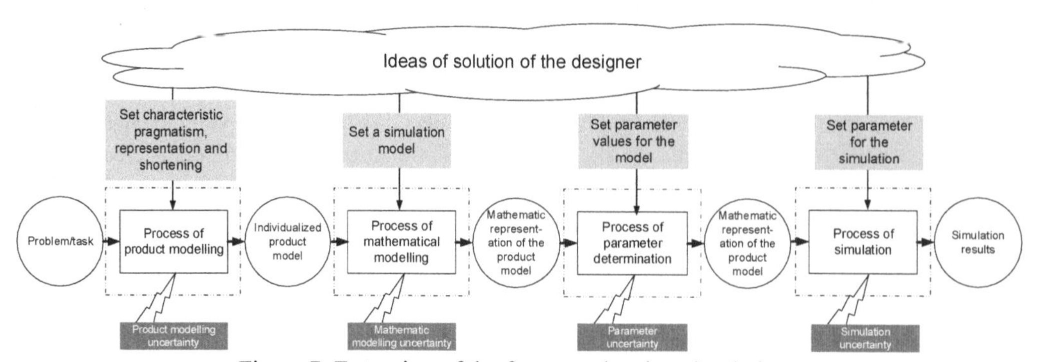

Figure 7: Extension of the framework using simulations

The process of product modelling mentioned in the framework is based on assumptions that have to be set at each sub-process. They depend on the available level of information and the solution ideas of the designer. As mentioned above, the combination of incomplete or missing information and assumptions may lead to uncertainty. With the help of the formalized process of product modelling, uncertainty can be allocated to each of the sub-processes. The following definitions can be deduced and are all based on the solution ideas of the designer, where the definition of mathematic modelling uncertainty and simulation uncertainty leans towards the understanding of uncertainty quantification [18]:

- *Product modelling uncertainty* occurs if it is not clear that all relevant influences and dependencies are considered during the creation of the product model. Product modelling uncertainty is made up of pragmatism uncertainty, representation uncertainty and shortening uncertainty.
- *Mathematic modelling uncertainty* occurs if it is not clear that the mathematical model represents the product model sufficiently.
- *Parameter uncertainty* occurs if all relevant influences made by defining values of the parameter are not considered or estimated correctly.
- *Simulation uncertainty* occurs if it is not clear that all relevant influences and dependencies are considered sufficiently by defining simulation settings.

All types of uncertainty defined here can be described with the help of the model of uncertainty, by describing their effects as acceptably or partly quantified, or by assuming them. Product modelling uncertainty is described here with the help of the short example above. According to pragmatism in the example, the assumption is made to investigate bar stability only. It is not currently clear whether an investigation of the strength of the bars is also relevant to the given task from which pragmatism uncertainty occurs. Representation in the example is based on the theory of buckling by Euler, which includes some simplifications, for example, the bars are rigid in the direction of shearing. Depending on the direction and position of the force, the simplification could ignore a critical bar deformation, not investigated here and where representation uncertainty results. Shortening uncertainty occurs when choosing the Euler case that is applicable to the investigated bar. In the example, Case One is chosen, which is a conservative shortening of the bar. Depending on the assumptions about how the connection of the bars is shortened, Case 2 can also be considered here, where the individualized product model is influenced, such as the solution to the problem or task based on it. This example shows that the assumptions behind the characteristic of a product model lead to occurring uncertainty.

As mentioned above, there is a lack of information during the product development process. The process of product modelling always depends on the solution ideas of the designer, such as the level of information, so it is not clear that the resulting individualized product model sufficiently represents reality. In a hypothetic situation in which all information about the product and its lifecycle is known, it is possible to create an ideal individualized product model, depending on the given problem or task. Comparing the ideal with the created individualized product model, it is possible to define product model uncertainty:

- *Product model uncertainty* occurs when it is not clear that the created individualized product model sufficiently describes the reality represented by an ideal model.

Understanding of product model uncertainty can be compared to the understanding of verification in uncertainty quantification [18]. In this context, verification is the process of determining that model implementation accurately represents the developer's conceptual description of the model and the solution to the model. In order to make a verification statement, numerical analysis and tests are necessary.

Applying this to the definition of product model uncertainty, experiments are also necessary to carry out verification. For example, prediction of product behaviour based on a product model has to be compared to the prediction based on the product prototype to estimate model implementation success. However, the estimation cannot be completed because the prototype is also subject to uncertainty.

Product model uncertainty contains the parts of uncertainty that can be allocated to the process of product modelling, such as an unknown part. The unknown part results from deviation between the designer's solution ideas on how to solve a problem and reality.

Conclusions and Outlook

This paper gives an overview of how to allocate uncertainty in product modelling. Based on the characteristics of a model, the approach of modelling and the pyramid of product models, the process of product modelling is formalized and a conceptual framework is created. Depending on a given task or problem, pragmatism defines the purpose and the way that it could be solved. Generic constructs are chosen using the elements of the pyramid, and are then used to create a principle product model. These two working steps represent the characteristic representation. Shortening prepares the investigated product, adapting the principle product model into an individualized product model. Depending on the task, simulation of the individualized product model may be necessary, so a mathematic representation is created and the framework is extended. The parameters of the individualized product model have to be set to perform the simulation. With this help, the given task or problem can be solved.

Along the process of product modelling, different types of uncertainty are defined that could cause discrepancies between planned and realized product behavior. To avoid this, any occurring uncertainty has to be handled and reduced. The example above shows that uncertainty results from the assumptions that are necessary to set the characteristic of the product model, therefore the assumptions have to be verified using information from the product lifecycle.

Verification can be carried out using the Uncertainty Mode and Effects Analysis (UMEA) methodology. It connects information from the product lifecycle with the product development process in a 5-step approach that determines uncertainties in technical systems [14]. The UMEA approach has to be adapted to the process of product modelling. Each step in product modelling has to be verified using the adapted UMEA and its methodological support. With this, occurring uncertainty can be reduced and variation between planned and realized product behavior can be minimized.

Acknowledgement

Collaborative Research Centre SFB 805 "Control of uncertainty in load-carrying mechanical systems" achieved the results presented here in subproject "A1: Development of Models, Methods and Instruments for the Acquisition, Description and Evaluation of Uncertainties". Thank you to the Deutsche Forschungsgemeinschaft (DFG) for funding and supporting SFB 805.

References

[1] N. Drémont, P. Graignic, N. Troussier, R. Whitfield, A. Duffy, A metric to represent the evolution of CAD/analysis models in collaborative design. In: International Conference in Engineering Design (ICED), Copenhagen, 2011.

[2] K. Ehrlenspiel, H. Meerkamm, Integrierte Produktentwicklung, München, 2013.

[3] H. Birkhofer, K. Rath, S. Zhao, Umweltgerechtes Konstruieren. In: F. Rieg, R. Steinhilper, Handbuch Konstruktion, München, 2012, pp.563-581.

[4] VDI 2221, Methodik zum Entwickeln und Konstruieren technischer Systeme und Produkte, Berlin, 1993.

[5] G. Pahl, W. Beitz, J. Feldhusen, K.H. Grote, Engineering Design, London, 2007.

[6] H. Stachowiak, Allgemeine Modelltheorie, Wien, 1973.

[7] T. Sauer, Ein Konzept zur Nutzung von Lösungsobjekten für die Produktentwicklung in Lern- und Anwendungssystemen, Düsseldorf, 2006.

[8] M. Wäldele, Erarbeitung einer Theorie der Eigenschaften technischer Produkte – Ein Beitrag für die konventionelle und algorithmenbasierte Produktentwicklung, Düsseldorf, 2012.

[9] H. Hanselka, R. Platz, Ansätze und Maßnahmen zur Beherrschung von Unsicherheit in lasttragenden Systemen des Maschinenbaus, VDI Konstruktion, Nr.11/12, S.55-62, 2010.

[10] T. Eifler, G. Enss, M. Haydn, L. Mosch, R. Platz, H. Hanselka, Approach for a Consistent Description of Uncertainty in Process Chains of Load Carrying Mechanical Systems, Applied Mechanics and Materials (Volume 104), Vol. Uncertainty in Mechanical Engineering, pp. 133-144, 2011.

[11] DIN EN ISO 31000, Risikomanagement - Grundsätze und Leitlinien, Berlin, 2009.

[12] T. Knetsch, Unsicherheiten in Ingenieurberechnungen, Magdeburg, 2003.

[13] D. P. Thunnissen. Propagating and Mitigating Uncertainty in the Design of Complex Multidisciplinary Systems, Pasadena, 2005.

[14] R. Engelhardt et al., A model to categorise uncertainty in load-carrying systems, In: Proceedings of the 1st MMEP International Conference on Modelling and Management Engineering Processes, Cambridge, 2010, pp.53-64.

[15] H. Negele, Systemtechnische Methodik zur ganzheitlichen Modellierung am Beispiel der integrierten Produktentwicklung, München, 2006.

[16] D. Gross et. al, Technische Mechanik Band 2 – Elastostatik, Heidelberg, 2007.

[17] G. Patzak, Systemtechnik – Planung komplexer innovativer Systeme, Berlin, 1982.

[18] W. L. Oberkampf, T. G. Trucano, Validation Methodology in Computational Fluid Dynamics, In. Proceedings of Fluids 2000 Conference and Exhibit, Denver, 2000.

Applied Mechanics and Materials Vol. 807 (2015) pp 99-108
© (2015) Trans Tech Publications, Switzerland
doi:10.4028/www.scientific.net/AMM.807.99
Submitted: 2015-04-15
Accepted: 2015-08-04

Uncertainty Scaling – Motivation, Method and Example Application to a Load Carrying Structure

Angela Vergé[1,a*], Julian Lotz[2,b], Hermann Kloberdanz[2,c], Peter F. Pelz[1,d]

[1]Technische Universität Darmstadt, Chair of Fluid Systems

Magdalenenstr. 4, 64289 Darmstadt, Germany

[2]Technische Universität Darmstadt, Product Development and Machine Elements

Magdalenenstr. 4, 64289 Darmstadt, Germany

[a]angela.verge@fst.tu-darmstadt.de, [b]lotz@pmd.tu-darmstadt.de, [c]kloberdanz@pmd.tu-darmstadt.de, [d]peter.pelz@fst.tu-darmstadt.de

Keywords: Scaling, Size Range, Uncertainty, Uncertainty Scaling, Dimensional Analysis, Laws of Growth, Product Development Process

Abstract. Scaling methods allow the estimation of the impact of changes in individual parameters on system performance. In the technical context, physical similarity is the focus. This paper demonstrates the extension of scaling methods to include uncertainty scaling. The advantages of using scaling uncertainty for the development of scaled products and the contribution of extended scaling methods to the analysis and assessment of uncertainty are illustrated. Uncertainty scaling based on dimensional analysis and complete similarity is derived. The potential of this method is demonstrated using a load carrying structure - a buckling beam.

Introduction

An engineer developing a product that comes in different sizes (size range) or transferring information from a model prototype to a life-sized product has to predict the properties of the scaled draft correctly. A more efficient method than designing the product with different numerical parameter values for each iteration exists: scaling using dimensional analysis or laws of growth, both of which refer to similarity laws. Being based on a basic design that is fully worked out, the designer can create new product sizes or predict the consequence of altering design parameter efficiently, leading to faster design processes for scaled products.

These methods are described well in literature. At the one hand, the best known methodology for size range development is that of Pahl and Beitz [1], which introduced the use of laws of growth derived from physical relations with respect to similarity laws. They also provide a brief overview of literature relevant to size range development. Simpson et al. [2] present an approach that is more strongly related to computational optimization methods, though they do not employ similarity relations to reduce complexity and design effort. One major concern in size range development is size-related uncertainty. Having variances in product properties that do not scale equally to the product property value or having disturbances that stay constant no matter the size of the product (e.g. the intensity of sunlight, temperature of the environment, etc.), critical design areas often have to be adapted and reworked. To prevent the designer from getting in too much trouble while mitigating the impact of size-related uncertainty, Lotz et al. developed laws of growth for product properties influenced by uncertainty [3,4]. The existing approaches to scaling uncertainty in product properties using static and dynamic laws of growth do not scale the uncertainty itself; they only address product property, including uncertainty, which is a less effective way of scaling uncertainty than being able to scale the variance caused by uncertainty directly.

On the other hand scaling is based on similarity laws derived from dimensional analysis. This powerful method can be traced back to the forefather of the modern world, Galilei Galileo. Joseph Fourier, in 1822, was the first to write of it in "Analytical Theory of Heat", where he reflected on a model: "...This relation depends in no respect on the units of length, which from its very nature is contingent, that is to say, if we took a different unit to measure linear dimensions, the equation ... would still be the same." Today, this is called the Bridgman postulate [5].

The technical focus is physical similarity, including geometrical similarity. The most prominent example of a complete physical similarity is the explosion by Taylor, von Neumann and Sedvo [6, 7, 8, 9]. Incomplete similarity is treated by Pelz et. al. [10, 11], particularly in the context of turbomachinery. The strength of scaling, not only in a geometric sense, became obvious recently in the paper of Pelz and Vergé [12]: Type and discrete numbers of proper formula units on vehicle speed was precisely predicted through physical and allometric scaling.

However, as found with laws of growth, none of the mentioned methodologies for scaling with dimensional analysis investigates the effects of uncertainty. Within this paper, scaling is expanded to the contemplation of uncertainty, with the focus being on parameter uncertainty. The strength of the uncertainty scaling is demonstrated by analyzing a load carrying structure.

Motivation for Uncertainty Scaling

A product passes through several phases in its lifecycle, including the production of material and then the product itself, as well as its usage and later disposal or recycling. These phases are characterized by processes in which uncertainty occurs [13]. As well as the product lifecycle phases, there is the product development (PD) process of the product itself. The uncertainty occurring in product lifecycle processes can be addressed within the development of the product by anticipation and analysis of the processes through which the product goes or that it enables. This can be done using product-process models, such as the process model of Collaborative Research Centre SFB 805 [14].

As well as the development of singular products, there is a class of products that face another type of uncertainty: scaled products. Scaling, as mentioned above, uses methods based on similarity to predict the product properties that the product will have when its size or pressure, material, etc., are changed. This prediction has the big advantage that the complexity of physical relations is reduced by monitoring only the relative change of parameters (Eq. 3). This makes the dimensional analysis a valuable and efficient scaling method. The reduction of scaling to similarity laws as it is performed in literature has a downside: In being able to determine the values of scaled parameters, the possibility of their deviation when the product is scaled is often overlooked. For example, production tolerances, as in the example later in this paper, grow less than proportionally compared to the product's size because the relative precision of production processes increases with the size of the manufactured product. This shows that the parameters describing the product's properties have to be scaled, and that there is also a need to scale the deviation from the nominal value (scaling uncertainty).

The scaling of uncertainty is not only a necessity, the methods existing in literature [3, 4] (that only target laws of growth) as well as the method developed in this paper using dimensional analysis can be a valuable contribution to the PD process for scaled products. They can be integrated into the existing methodologies for the analysis, assessment and control of uncertainty, which currently lack methods that address scaling uncertainty.

Uncertainty scaling methods need information to contribute to the analysis and assessment of uncertainty, which is their main contribution to the PD process. The information needed can be prepared systematically, using the existing models and methods. The scaling of products, while simultaneously using uncertainty scaling methods, can be displayed in a procedural model (Fig. 1). The advantage of integrating uncertainty analysis into the scaling of products is that iterations, including expensive ones that become necessary in the latter stages of the development process, can be reduced by anticipating critical processes and parameters in early design stages.

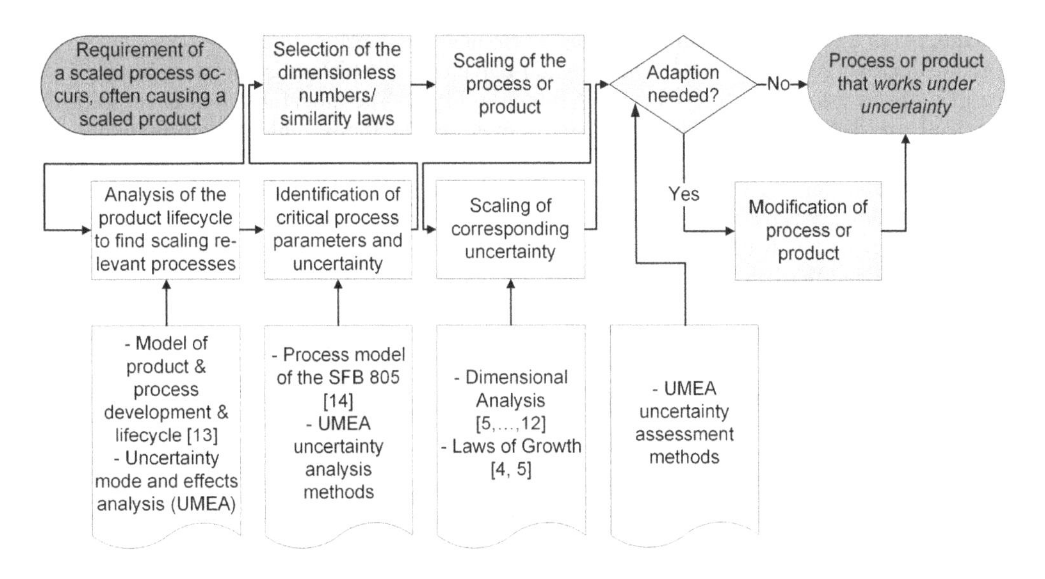

Figure 1: A procedural model of the scaling process, including the analysis and assessment of uncertainty. The light grey processes are also part of the scaling process without considering uncertainty. Supporting methods are displayed in the third row of the diagram.

A few additional details about uncertainty scaling are relevant: The scaling process can be performed at different stages of the process or product concretization. An estimation of basic product or process properties can be carried out in the early stages of the development process. This also allows an initial estimate of the occurring uncertainty, which is often based on expert estimates of the uncertainty of certain parameters. If the process or product that is to be scaled is known in detail (which is often the case when the need to scale arises), a brief analysis and assessment of uncertainty can be achieved as concretization occurs during the development process at a higher level, where more information is available. Scaling uncertainty can be used as a method for uncertainty analysis within the UMEA framework created by Collaborative Research Centre SFB 805, and therefore expands the methodical framework for controlling uncertainty to include the control of scaling uncertainty.

Method and Example Application

Fig. 2 shows the example application of a load carrying structure - a buckling beam.

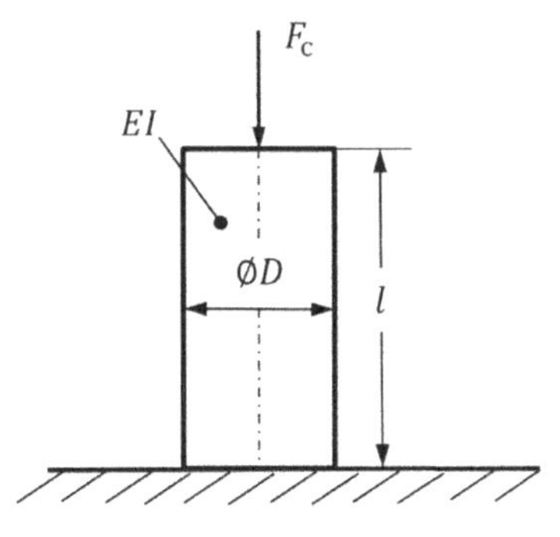

Figure 2: Load carrying structure.

In the first place we make a picture representing the reality. The model of the load carrying structure in our example is a cylindrical beam of circular cross-section (diameter D), of length l and constant flexural rigidity $EI \sim ED^4$. The question remains if this picture is complete or incomplete. The incompleteness of a picture = model is shown in Fig. 3. The model does not cover everything of the relevant reality to the buckling problem example shown, for example predeformation.

This discussion, summarized in Fig. 3, demonstrates that the model is a source of uncertainty. The model parameters are the second source of uncertainty. The third and, in most cases, dominant source of uncertainty is those that are critical or transferred through the boundary conditions: In this example, one boundary is the fixing of the beam on the ground. The second boundary is the applied load. The first and third sources of uncertainty are not considered here; this paper concentrates on parameter uncertainty only.

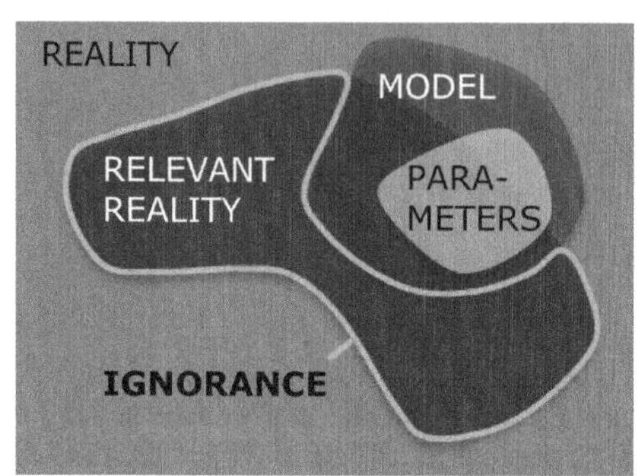

Figure 3: The uncertainty map according to Pelz and Hedrich [15].

Dimensional Analysis. Using the example of a buckling beam, the dimensional analysis content is extended to account for uncertainty. To do this, a brief review of the basic concept of scaling itself is required. To describe the physical problem with $j = 1 \dots n$ dimensioned values p_j, generally the value of interest, here the critical load F_c, is calculated from other measurable values: the flexural rigidity EI and the beam length l. As well as the measured values, $i = 1 \dots m$ fundamental units P_i are also required. The buckling of a beam is a static problem, thus, the fundamental units are given with force and length: $[F, L]$. Accordingly, it is possible to describe the relation of the physical values with

$$p_1 = fn(p_2 \dots p_n), \tag{1}$$

where the critical load F_c is a function of the flexural rigidity EI and the length of the beam l

$$F_c = fn(EI, l). \tag{2}$$

Corresponding to the Bridgman postulate [5], only relative values have absolute significance and the fundamental units have no relevance. This reduction follows an equivalent relation to Eq. 1

$$\Pi_1 = fn(\Pi_2 \dots \Pi_d), \tag{3}$$

with $d = n - r$ dimensionless products Π_i. r is the minimum number of fundamental units needed to describe the physical values and the rank of the dimension matrix $\left(a_{ij}\right)_{m,n}$. In general, the number of the fundamental units corresponds to the rank of the dimension matrix $m = \mathrm{rg}\left(\left(a_{ij}\right)_{m,n}\right) = r$.

For $n = 3$ and $r = 2$ Eq. 3 simplifies to $\Pi = \text{const.}$ and the dimensionless products are products of the physical values, i.e. parameters and quantities

$$\Pi = \prod_{j=1}^{n} p_j^{k_j}, \tag{4}$$

as well as

$$[\Pi] = 1 = \prod_{j=1}^{n} [p_j]^{k_j}. \tag{5}$$

The square brackets are used here as an operator to give the dimensions of any operand. For the dimension quantities apply

$$[p_j] = \prod_{i=1}^{m} P_i^{a_{ij}}. \tag{6}$$

From Eq. 5 and Eq. 6 follows

$$[\Pi] = 1 = \prod_{j=1}^{n} \prod_{i=1}^{m} P^{a_{ij}k_j}. \tag{7}$$

In the special and most simple case $n = m + 1$, which is sufficient for our example, Eq. 7 is satisfied, as long as

$$\sum_{j=1}^{n} a_{ij}k_j = 0, \qquad i = 1 \dots m, \tag{8}$$

has no trivial solutions for the n unknown k_j. The dimension matrix $(a_{ij})_{m,n}$ is represented with

	p_1, ..., p_n
P_1	
.	$(a_{ij})_{m,n}$
.	
P_m	

for the buckling beam

	F_c	EI	l
F	1	1	0
L	0	2	1

Eq. 8 has r linearly independent equations with $d = n - r$ linearly independent solutions, thus the values of k_j can be determined. In the general case $n > m$ which has more unknown values than independent equations, Eq. 8 change to an inhomogeneous linear equation from which $d = n - r$ dimensionless products Π_i are derived [16].

The result of Eq. 8 in our example is given $k_{F_C} = 1$, $k_l = 2$, $k_{EI} = -1$, i.e.

$$\Pi = \frac{F_C l^2}{EI}. \tag{9}$$

Using model $EI \sim ED^4$ yields

$$\Pi = \frac{F_C l^2}{ED^4}. \tag{10}$$

Scaling. In the technical context the basic design and behavior of size ranges and scaling of physical quantities is of interest. It is important to mention that not only geometric scaling can be treated but also Young's modulus can be scaled. However, looking at every sequential design is not efficient: using scaling methods enables the prediction of the behavior of a scaled design. The physical properties of the basic design correlate with the full-scale by

$$p_j := p_{0j} M_j, \qquad j = 1 \dots n, \tag{11}$$

which defines the scale factors M_j. The basic and full-scale designs behave in a completely physically similar manner under the condition that all dimensionless products are similar. This means

$$\Pi_i = \Pi_{0i}, \qquad i = 1 \dots d, \tag{12}$$

and is equal to

$$1 = \prod_{j=1}^{n} M_j^{k(i),j}, \quad i = 1 \dots d. \tag{13}$$

Usually, scaling in the context of heat and mass transfer, turbomachinery, etc. is based on Eq. 13. If there is complete similarity between basic design and scaled design, is given as long as the systems of equations (Eq. 13) are fulfilled. The requirements in Eq. 12 and 13 are the basis of model theory [16].

In our example the scale factor of the critical buckling load is given by

$$M_{F_c} = \frac{M_E M_D^4}{M_l^2}. \tag{14}$$

Uncertainty Scaling. The concept of scaling based on Eq. 13 is ready to be extended to include uncertainty scaling. The assumption of deterministic values does not correspond to the reality that any value is uncertain (Fig. 4).

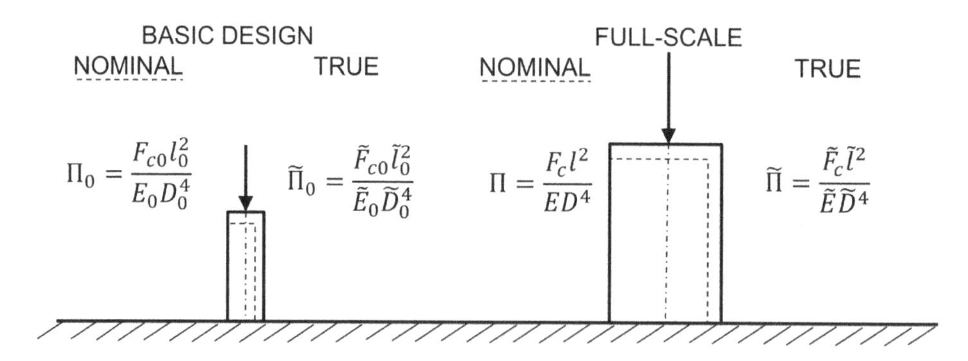

Figure 4: Nominal and true values of the basic design and a geometric scale of a buckling beam.

To account for uncertainty it is necessary to differentiate between the nominal value p_j and the tolerance range Δp_j. The true value $\tilde{p}_j$ is between

$$\tilde{p}_j = p_j \pm \Delta p_j, \qquad j = 1 \dots n. \tag{15a}$$

With the definition of uncertainty $U_j := \Delta p_j / p_j$ the true value is

$$\tilde{p}_j = p_j (1 \pm U_j), \qquad j = 1 \dots n. \tag{15b}$$

In the same manner, applies a nominal value for the dimensionless products, Eq. 4, and a true value

$$\widetilde{\Pi}_i = \prod_{j=1}^{n} \tilde{p}_j^{k(i),j}, \qquad i = 1 \dots d, \tag{16}$$

as well as

$$\widetilde{\Pi}_i = \Pi_i \prod_{j=1}^{n} (1 \pm U_j)^{k(i),j}, \qquad i = 1 \dots d. \tag{17}$$

The same description holds for the basic design

$$\widetilde{\Pi}_{0i} = \Pi_{0i} \prod_{j=1}^{n} (1 \pm U_{0j})^{k(i),j}, \qquad i = 1 \dots d. \tag{18}$$

When considering uncertainty the product of the scaling factors reads

$$1 = \prod_{j=1}^{n} \widetilde{M}_j^{k(i),j} = \prod_{j=1}^{n} M_j^{k(i),j} \prod_{j=1}^{n} (1 \pm U_j)^{k(i),j}, \quad i = 1 \dots d. \tag{19}$$

Since $\prod_{j=1}^{n} M_j^{k(i),j} = 1$ the desired result for the uncertainty scaling follows:

$$1 = \prod_{j=1}^{n} (1 \pm U_j)^{k(i),j}, \qquad i = 1 \dots d. \tag{20}$$

Eq. 20 allows the calculation of the uncertainty of the interesting, most critical, values of a physical system in terms of complete similarity. In this paper we focuses on the production uncertainty that affects the geometric properties of the beam. The nominal geometric value, where p represents either length l or diameter D, is linked to the related tolerance factor i through the empirical relationship

$$i = L_1 \left(\frac{p}{L_{10}}\right)^{1/3}. \tag{21}$$

The constants L_1 and L_{10} in Eq. 21 are defined by DIN ISO 286 to $L_1 := 10^{-3}$ mm and $L_{10} := 10$ mm [17]. The specified tolerance range Δp is a multiple of the tolerance factor i

$$\Delta p = fi. \tag{22}$$

The constant f is set according to DIN ISO 286 [17]. For a diameter with the fundamental tolerance IT7 $f = 16$, for IT9 $f = 40$. Larger components can be manufactured more precisely than smaller components [18].

From the condition for complete geometrical similarity and similarity of the material $M_E = 1$ the scaling factor of the critical buckling load is

$$\widetilde{M}_{F_c} = \frac{\widetilde{M}_D^4}{\widetilde{M}_l^2}, \tag{23}$$

and the uncertainty of the buckling load is given by

$$\pm U_{F_c} = \frac{(1 \pm U_D)^4}{(1 \pm U_l)^2} - 1. \tag{24}$$

In relation to the basic design with a scaling factor, $\widetilde{M}_D = \widetilde{M}_l = 1$, the uncertainty of the buckling load decreases with higher scaling factors, up scaling, and increases considerably more strongly with down-scaling, as seen in Eq. 24 and Fig. 5. Fig. 5 also shows that the uncertainty of the critical load varies in a higher range than the production uncertainties. Non-symmetrical behavior of the positive and negative curves of the worst case buckling load uncertainty results from Eq. 24. In this case, the effect is not recognizable in Fig. 5.

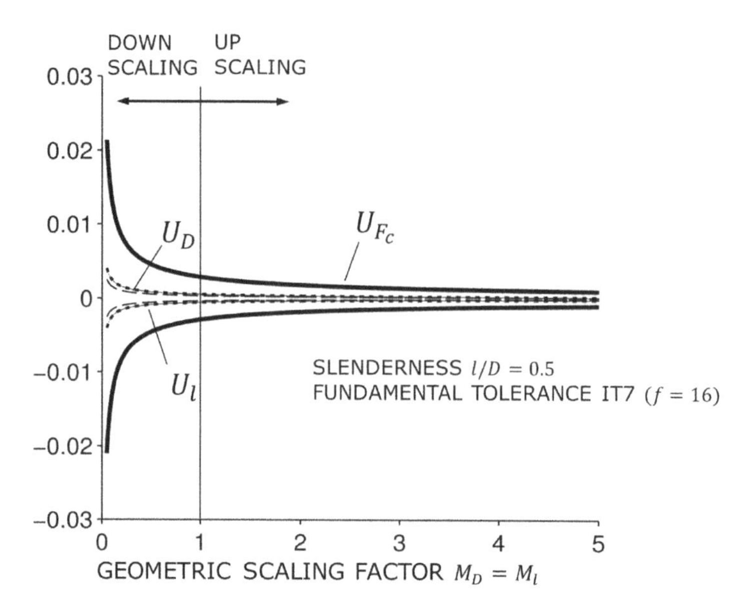

Figure 5: Uncertainty of the geometric values and the critical buckling load of the beam.

Fig. 5 reflects the need for uncertainty scaling. The uncertainty of the critical load due to the manufacturing tolerances increases sharply in down-scaling. If this is not considered in the scaled design, it can lead to failure due to high load. The method of uncertainty scaling introduced here allows the direct capture of the critical impact of scaling so that necessary actions can be performed. The method further provides the advantage of considering uncertainty directly during the scaling process for the entire region, so that it does not need to be considered anew for each execution.

Summary and Outlook

The main motivation for uncertainty scaling, ensuring the proper function of scaled products, leads to the advantage of integrating uncertainty scaling into product or process scaling: deviations that prevent proper working of process or product are not monitored with conventional scaling but can be analyzed with the expanded scaling methods, i.e. uncertainty scaling, derived in this paper. The expansion of the scaling methods is done by creating additional similarity relations for uncertainty, leading to a product model that is more detailed and therefore represents reality in a more precise way. This allows existing uncertainty control methodologies like UMEA to address not only solitary but also scaled products. Information about the class of size-dependent uncertainty influences uncertainty scaling (Fig. 5) and can be fed into the product development process to determine the true behavior of products and processes.

Within this paper, uncertainty scaling was derived, based on a load carrying structure – a buckling beam. Our focus was on complete geometric similarity and demonstrating the influence of production uncertainty on a size range. It is important to keep in mind that not only geometric scaling can be treated.

The advantage of uncertainty scaling is being able to show the uncertainty of critical correlations directly. It was also demonstrated that it is insufficient to only care about individual sources of uncertainty – in the example, production uncertainty – it is also necessary to consider the scaled uncertainty of the critical value, which is often calculated through simulation or measurement and can be a multiple higher, as seen in the example of the buckling beam. The uncertainty of the buckling load varies in a higher range than the production uncertainties: Our example shows the nonlinear amplification of geometric manufactured uncertainty by the power of two (Eq. 24 $U_D \approx U_l$). Generally, the product development process for size ranges and scaled products can be shortened by anticipating the effects of uncertainty in a size-related manner, reducing the risk of creating scaled products that exceed the limits of scaling, which would lead to reduced functionality or even failure. Iterations in the development process can be reduced.

Future work treats uncertainty scaling in products with more than one dimension and uncertainty scaling for incomplete similarity in dimensional analysis, if Eq. 13 is not fully satisfied.

Acknowledgment

We would like to thank Deutsche Forschungsgemeinschaft (DFG) for funding this project within the Collaborative Research Centre (CRC) 805.

References

[1] G. Pahl, W. Beitz, J. Feldhusen, K.H. Grote, Pahl Engineering Design. A Systematic Approach, third ed., Springer-Verlag, London, 2007.

[2] T.J. Simpson, J.R.A. Maier, F. Mistree, Product platform design: method and application, Research in Engineering Design ,Vol. 13 (2001), pp. 2-22.

[3] J. Lotz, H. Kloberdanz, T. Freund, K. Rath, Estimating Uncertainty of Scaled Products Using Similarity Relations and Laws of Growth, Design Conference 2014, 19.-22.05.2014, Dubrovnik. Proceedings of the 13th International Design Conference DESIGN 2014 Dubrovnik.

[4] J. Lotz, H. Kloberdanz, in T.J. Howard, T. Eifler (eds.), Scaling under Dynamic Uncertainty using Laws of Growth, Proceedings of the International Symposium on Robust Design – IsoRD14, Copenhagen (2014), pp. 17-27.

[5] P.W. Bridgman, Dimensional Analysis, Yale University Press, New Haven, 1922.

[6] G.I. Taylor, The formation of a blast wave by a very-intense explosion, Proceedings of the Royal Society, London, Vol. A201 (1950), pp. 175–186.

[7] J. von Neumann, The point source solution, NDRC Division B Rept AM-9, 1941, Reprinted in Taub AH (ed) John von Neumann collected works, Pergamon, Oxford, (1963), pp 219-237.

[8] L.I. Sedov ,The movement of air in a strong explosion, Journal of Applied Mathematics and Mechanics, Vol. 10 (1946), pp. 241 – 250.

[9] L.I. Sedov, Similarity and dimensional methods in mechanics, Academic Press: New York, (1959), Chap. 4.

[10] P.F. Pelz, S. Karstadt, Tip Clearance Losses – A Physical Based Scaling Method, In: International Journal of Fluid Machinery and Systems, Vol. 3 (2010), pp. 279-84.

[11] P.F. Pelz, S. Stonjek, A Second Order Exact Scaling Method for Turbomachinery Performance Prediction, In: International Journal of Fluid Machinery and Systems (2013).

[12] P.F. Pelz, A. Vergé, Validated biomechanical model for efficiency and speed of rowing, Journal of Biomechanics, Vol.47 (2014), pp. 3415-3422.

[13] H. Hanselka, R. Platz, Ansätze und Maßnahmen zur Beherrschung von Unsicherheit in lasttragenden Systemen des Maschinenbaus, Konstruktion, November/December (2010), pp. 55-62.

[14] A. Bretz, S. Calmano, T. Gally, B. Götz, R. Platz, J. Würtenberger, Darstellung passiver, semi-aktiver und aktiver Maßnahmen im SFB 805-Prozessmodell, Preprint 2015, SFB 805, TU Darmstadt.

[15] P.F. Pelz, P. Hedrich, Unsicherheitsklassifizierung anhand einer Unsicherheitskarte, technical Report of Instituts for Fluid Systems, Darmstadt, 2015.

[16] J.H. Spurk, Dimensionsanalyse in der Strömungslehre, Springer Verlag, Berlin Heidelberg, 1992.

[17] Deutsches Institut für Normung, DIN ISO 286

[18] W. Steinhilper, B. Sauer, Konstruktionselemente des Maschinenbaus 1, Springer-Lehrbuch, Springer Verlag, Berlin Heidelberg, 2012.

Applied Mechanics and Materials Vol. 807 (2015) pp 109-117
© (2015) Trans Tech Publications, Switzerland
doi:10.4028/www.scientific.net/AMM.807.109

Submitted: 2015-04-15
Accepted: 2015-08-04

An approach to using elemental interfaces to assess design clarity

Tillmann Freund[1, a *], Jan Würtenberger[1,b], Hermann Kloberdanz[1,c] and Petrit Blakaj[1,d]

[1]TU Darmstadt, Institute for Machine Elements and Product Development (pmd), Magdalenenstraße 4, 64289 Darmstadt, Germany

[a]freund@pmd.tu-darmstadt.de, [b]wuertenbeger@pmd.tu-darmstadt.de, [c]kloberdanz@pmd.tu-darmstadt.de, [d]p.blakaj@gmx.de

Keywords: Robust Design, Design Clarity, Uncertainty

Abstract. A method is proposed to assess the clarity of a design already at the working principle level. To do this, the Contact and Channel Model is used in an adapted form to differentiate elemental interfaces based on their geometry and function. This classification is used to derive a generic catalogue of elemental interfaces. A product can be represented through a combination of these interfaces and the working structures in between. The design clarity can then be assessed through generic information stored in the catalogue. A procedure model is proposed to implement this method and demonstrate its applicability. Additionally, the main design parameters that greatly affect product behavior can be identified with the help of the generic catalogue for every elemental interface. Finally, further steps for variability in design parameters are discussed.

Introduction

Although a lot of Robust Design Methodology is available, for an overview see *Göhler* [1], there is a lack of methods that are applicable in early design phases [2,3,4]. The design principle of *Clarity* proposed by *Pahl* and *Beitz* [5] can be seen as a precondition [3,4]. As long as the relation between input and output parameters (Transferfunction [6]) cannot be determined, the evaluation of a design for robustness is not possible [4].

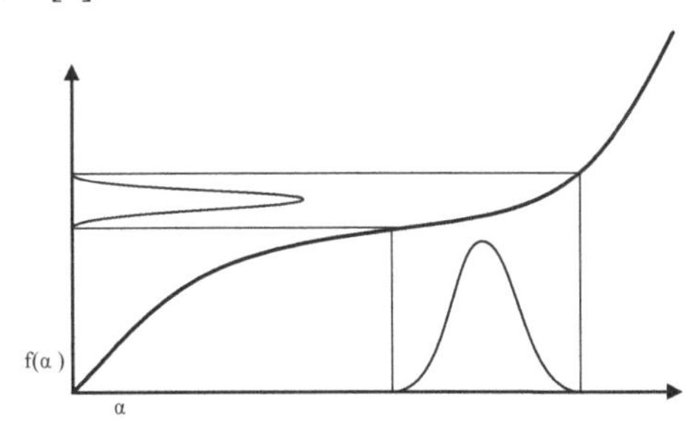

Figure 1 Relation between variant input and output parameters, according to *Ulrich* and *Eppinger* [7]. Robust transfer behaviour occurs if variety in input parameters results in non-variant output.

Ebro et al. contribute to filling this gap, presenting a method to assess design clarity at a system level, using the *Kutzbach Grübler Equation* (kinematic design), and a method at the interface level that uses matrix-based representations containing information about actual degrees of freedom enriched with information about possible design failures due to ambiguity [3].

It is in the earliest development phase where, in the case of a green field approach, Robust Design principles can be applied. In *Freund et al.* [8], elemental interfaces were derived that have a generic character. Assuming that a technical system at the working principle level can be described sufficiently by properties and functions that can be addressed to these interfaces, the systematic investigation of the nominal design clarity can be processed. The following sections present theoretical basics, necessary models and measures to discuss the value of the idea.

The adapted Contact and Channel Model

The Contact and Channel Model (C&CM), originally developed by *Albers* and *Matthiesen* [9,10], is a model that combines functions with geometric data in a visually reduced way. The structure and the principle of how the function is implemented can be seen at a glance. Table 1 shows an overview of the different model elements. *Freund et al.* propose that the C&CM is used for analysing interfaces (Fig. 2, a)) [8]. The Working Surface Pairs can be classified according to contact surface geometry (Fig. 2, b)). Based on this, the basic C&CM is specifically enriched with information on interfaces. *Elemental interfaces* have a generic character and every technical system at the Working Principle Level can be represented through a combination of these *elemental interfaces* and Working Structures. This idea is the starting point for the approach presented in this paper. If generic properties can be addressed in the elemental interfaces, designers can be supported through additional information, derived from elemental interfaces and provided via a catalogue or database.

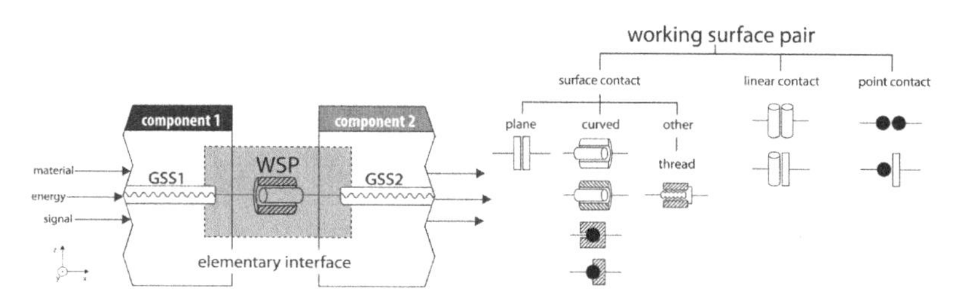

Figure 2 a) Depiction of Contact and Channel Model at the elemental interface level.
b) Classification of Working Surface Pairs [7].

Table 1 Model elements of the Contact and Channel Model [6].

Working Surface	Surfaces of bodies, liquids, gases or fields that have intermittent or continuous contact with another working surface and contribute to an intended exchange of energy and information in a technical system.
Residual Surface	Surfaces that are not intended to be working surfaces
Working Surface Pair	Two Working Surfaces that have at least intermittent or partial contact and exchange energy and information.
Working Structure	Volumes of bodies, liquids, gases or field-flushed spaces that allow at least intermittent transmission of energy or information between working surfaces.
Residual Structures	Volumes that are not intended to transmit energy.

Elemental Interfaces

To fulfil functions, parts and assemblies must be connected with each other. The connection allows transmission of energy between these parts. The connection type and its properties characterise the behaviour of the connection. *Roth* provides a matrix-based approach to describe this behaviour [11]. *Eifler*, *Söderberg* and *Ebro* choose similar approaches [6,4,3]. *Roth* differentiates the degrees of freedom in positive and negative directions, therefore the freedom of a part can be represented using a matrix with 12 values. This matrix is called *Contact Matrix*. *Roth* uses a binal code for each value, where 1 means a contact and 0 not.

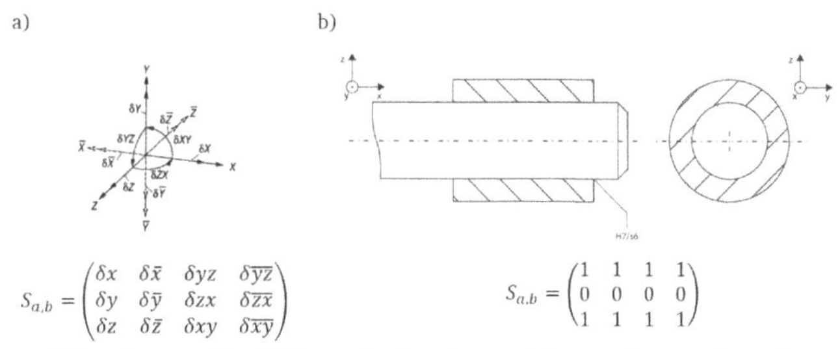

$$S_{a,b} = \begin{pmatrix} \delta x & \delta \bar{x} & \delta yz & \delta \overline{yz} \\ \delta y & \delta \bar{y} & \delta zx & \delta \overline{zx} \\ \delta z & \delta \bar{z} & \delta xy & \delta \overline{xy} \end{pmatrix} \qquad S_{a,b} = \begin{pmatrix} 1 & 1 & 1 & 1 \\ 0 & 0 & 0 & 0 \\ 1 & 1 & 1 & 1 \end{pmatrix}$$

Figure 3 Degrees of freedom with differentiation in positive and negative directions. a) general notation. b) Example of a cylindrical Working Surface Pair.

The *Contact Matrix* does not contain forces, therefore *Roth* introduces the *Closing Type Matrix*. It can be derived from the *Contact Matrix*. A 1 in the *Contact Matrix* without a force in normal direction is a form fit (f). A force in normal direction leads to an elastic fit (E). Frictional forces can occur tangential to normal forces. These forces lead to a frictional closing if the operational load is lower than the possible frictional force determined through frictional coefficient and normal force. Considering these conditions, the *Digitalized Closing type matrix* can finally be derived. It contains information about the movability under assumed conditions. Fig. 4 contains the closing types according to *Roth*.

example	name	orientation	occur on	loose fit	label
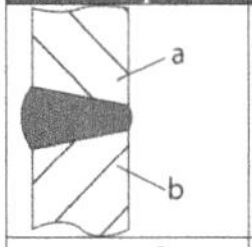 a, b	**adhesive bond**	any	welding, soldering, gluing	no	s
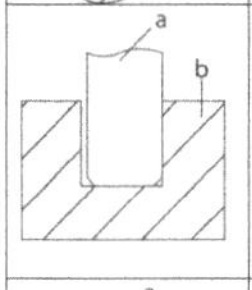 a, b	**form locking without surface pressure**	normal	geometric shape fit without surface pressure	yes	f
		tangential	insignificant friction due to loose fit		r
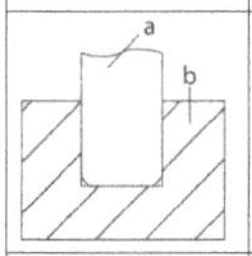 a, b	**form locking with surface pressure**	normal	geometric shape fit with surface pressure	no	E_f
		tangential	friction due to surface pressure		r
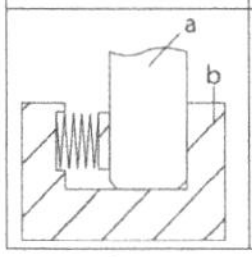 a, b	**elastic form locking with surface pressure**	normal	elastic shape fit with surface pressure		E
		tangential	friction due to surface pressure		r

Figure 4 Closing type notations according to *Roth*.

Table 2 illustrates the relations between *Contact Matrix*, *Closing Type Matrix* and *Digitalised Closing Type Matrix* for two uses of the cylindrical working surface pair in Fig. 3. In the transverse press fit, elastically pretensioned bodies in normal directions are necessary (E) to transmit a rotational moment and prevent movement. In linear guidance, movability of the system is necessary to ensure its function.

Table 2 Contact Matrix, Closing Type Matrix and Digitalised Closing Type Matrix for a cylindrical Working Surface Pair with two different functions (Transverse Press fit and Linear Guidance).

Type of coupling	Contact Matrix	Closing Type Matrix	Digitalised Closing Type Matrix
Transverse pressfit	$S_{a,b} = \begin{pmatrix} 1 & 1 & 1 & 1 \\ 0 & 0 & 0 & 0 \\ 1 & 1 & 1 & 1 \end{pmatrix}$	$S_{a,b}^{A} = \begin{pmatrix} E & E & E & E \\ r & r & r & r \\ E & E & E & E \end{pmatrix}$	$S_{a,b}^{D} = \begin{pmatrix} 1 & 1 & 1 & 1 \\ 1 & 1 & 1 & 1 \\ 1 & 1 & 1 & 1 \end{pmatrix}$
Linear guidance	$S_{a,b} = \begin{pmatrix} 1 & 1 & 1 & 1 \\ 0 & 0 & 0 & 0 \\ 1 & 1 & 1 & 1 \end{pmatrix}$	$S_{a,b}^{A} = \begin{pmatrix} f & f & f & f \\ r & r & r & r \\ f & f & f & f \end{pmatrix}$	$S_{a,b}^{D} = \begin{pmatrix} 1 & 1 & 1 & 1 \\ 0 & 0 & 0 & 0 \\ 1 & 1 & 1 & 1 \end{pmatrix}$

Linking elemental interfaces at the Contact matrix and Closing Type matrix level.

To generate information about overall system behaviour, the elemental interfaces must be combined. Depending on the part structure, serial and parallel relations can be differentiated. The difference affects the underlying rules that are necessary to combine Working Surface characteristics. To combine elemental interface properties at the *Closing Type Matrix* level it is necessary to assess which locking type dominates the function. Related to the interface layout, there are 2 different procedures.

1. If two connections lock one movability independently, the stronger one has to be considered.

$$E \vee r = E \,, if \; E > r \tag{1}$$

$$E \vee r = r \,, if \; E < r \tag{2}$$

2. If two connections lock one movability dependently, the weaker one has to be considered.

$$E \wedge r = r \,, if \; E > r \tag{3}$$

$$E \wedge r = E \,, if \; E < r \tag{4}$$

working surface pair	plane working surface pair				
icon			closing matrix		
			$\begin{pmatrix} 0 & 1 & 0 & 0 \\ 0 & 0 & 0 & 0 \\ 0 & 0 & 0 & 0 \end{pmatrix}$		
function	**positioning**				
orientation	physical effect	interference type	digitalization	characteristics	parameters
x-axis	contact	$\begin{pmatrix} 0 & f & - & - \\ - & - & - & - \\ - & - & - & - \end{pmatrix}$	$\begin{pmatrix} 0 & 1 & - & - \\ - & - & - & - \\ - & - & - & - \end{pmatrix}$	• positioning require contact of two surfaces	• $\sigma_p = \frac{F_n}{A} \leq \sigma_{p,zul}$ • $F_N = \frac{\Delta l}{l_0} EA \ with \ \Delta l \ll l_0$
y-axis	no contact	$\begin{pmatrix} - & - & - & - \\ 0 & 0 & - & - \\ - & - & - & - \end{pmatrix}$	$\begin{pmatrix} - & - & - & - \\ 0 & 0 & - & - \\ - & - & - & - \end{pmatrix}$		
z-axis	no contact	$\begin{pmatrix} - & - & - & - \\ - & - & - & - \\ 0 & 0 & - & - \end{pmatrix}$	$\begin{pmatrix} - & - & - & - \\ - & - & - & - \\ 0 & 0 & - & - \end{pmatrix}$		
force transmission					
orientation	physical effect	closing type	digitalization	characteristics	parameters
normal	hook's law	$\begin{pmatrix} 0 & E_f & - & - \\ - & - & - & - \\ - & - & - & - \end{pmatrix}$	$\begin{pmatrix} 0 & 1 & - & - \\ - & - & - & - \\ - & - & - & - \end{pmatrix}$	• E-modulus • surface area	• $\sigma_p = \frac{F_n}{A} \leq \sigma_{p,zul}$ • $F_N = \frac{\Delta l}{l_0} EA \ with \ \Delta l \ll l_0$
tangential	coulomb's law	$\begin{pmatrix} 0 & E_f & - & - \\ r & r & - & - \\ r & r & - & - \end{pmatrix}$	$\begin{pmatrix} - & - & - & - \\ 1 & 1 & - & - \\ 1 & 1 & - & - \end{pmatrix}$	• friction, attrition • dependency on jacking force and coefficient of friction • potential relative movement cause slippage	• $F_N = \sigma_p A$ • $F_R = \mu F_N$
torque transmission					
orientation	physical effect	interference type	digitalization	characteristics	parameters
normal	coulomb's law	$\begin{pmatrix} - & E_f & r & r \\ - & - & - & - \\ - & - & - & - \end{pmatrix}$	$\begin{pmatrix} - & 1 & 1 & 1 \\ - & - & - & - \\ - & - & - & - \end{pmatrix}$	• friction, attrition • dependency on jacking force and coefficient of friction • potential relative movement cause slippage	• $F_N = \sigma_p A = \sigma_p \left(\frac{\pi}{4} d^2\right)$ • $M_R = \mu r F_N$
tangential	normal force	$\begin{pmatrix} - & E_f & - & - \\ - & - & g & g \\ - & - & g & g \end{pmatrix}$	$\begin{pmatrix} - & - & - & - \\ - & - & 1 & 1 \\ - & - & 1 & 1 \end{pmatrix}$	• depending on the geometrical measurements of the surface as well as the point of the applied normal force, the working surface pair will be resistant against rotation in the tangential orientation	$\begin{pmatrix} - & - & - & - \\ - & - & \overline{z_s}F_N & z_s F_N \\ - & - & \overline{y_s}F_N & y_s F_N \end{pmatrix}$

working surface pair	plane working surface pair				
icon			closing matrix		
			$\begin{pmatrix} 0 & 0 & 0 & 0 \\ 1 & 1 & 1 & 1 \\ 1 & 1 & 1 & 1 \end{pmatrix}$		
function	**positioning**				
orientation	physical effect	closing type	digitalization	characteristics	parameters
x-axis	no contact	$\begin{pmatrix} 0 & 0 & - & - \\ - & - & - & - \\ - & - & - & - \end{pmatrix}$	$\begin{pmatrix} 0 & 0 & - & - \\ - & - & - & - \\ - & - & - & - \end{pmatrix}$	• positioning require contact of two surfaces	• $\sigma_p = \frac{F_n}{A} \leq \sigma_{p,zul}$ • $F_N = \frac{\Delta l}{l_0} EA \ with \ \Delta l \ll l_0$
y-axis	contact	$\begin{pmatrix} - & - & - & - \\ f & f & - & - \\ - & - & - & - \end{pmatrix}$	$\begin{pmatrix} - & - & - & - \\ 1 & 1 & - & - \\ - & - & - & - \end{pmatrix}$		
z-axis	contact	$\begin{pmatrix} - & - & - & - \\ - & - & - & - \\ f & f & - & - \end{pmatrix}$	$\begin{pmatrix} - & - & - & - \\ - & - & - & - \\ 1 & 1 & - & - \end{pmatrix}$		
force transmission					
orientation	physical effect	closing type	digitalization	characteristics	parameters
normal	hook's law	$\begin{pmatrix} - & - & - & - \\ E_f & E_f & - & - \\ E_f & E_f & - & - \end{pmatrix}$	$\begin{pmatrix} - & - & - & - \\ 1 & 1 & - & - \\ 1 & 1 & - & - \end{pmatrix}$	• E-modulus • surface area	• $\sigma_p = \frac{F_n}{A} \leq \sigma_{p,zul}$ • $F_N = \frac{\Delta l}{l_0} EA \ with \ \Delta l \ll l_0$
tangential	coulomb's law	$\begin{pmatrix} r & r & - & - \\ E_f & E_f & - & - \\ E_f & E_f & - & - \end{pmatrix}$	$\begin{pmatrix} 1 & 1 & - & - \\ - & - & - & - \\ - & - & - & - \end{pmatrix}$	• friction, attrition • dependency on jacking force and coefficient of friction • potential relative movement cause slippage	• $F_N = \sigma_p A$ • $F_R = \mu F_N$
torque transmission					
orientation	physical effect	closing type	digitalization	characteristics	parameters
normal	hook's law	$\begin{pmatrix} - & - & - & - \\ - & - & f & f \\ - & - & f & f \end{pmatrix}$	$\begin{pmatrix} - & - & - & - \\ - & - & 1 & 1 \\ - & - & 1 & 1 \end{pmatrix}$	• E-modulus • surface area	• $\sigma_p = \frac{F_n}{A} \leq \sigma_{p,zul}$ • $F_N = \frac{\Delta l}{l_0} EA \ with \ \Delta l \ll l_0$
tangential	coulomb's law	$\begin{pmatrix} - & - & r & r \\ E_f & E_f & - & - \\ E_f & E_f & - & - \end{pmatrix}$	$\begin{pmatrix} - & - & 1 & 1 \\ - & - & - & - \\ - & - & - & - \end{pmatrix}$	• friction, attrition • dependency on jacking force and coefficient of friction • potential relative movement cause slippage	• $F_N = \sigma_p A = \sigma_p \left(\frac{\pi}{4} d^2\right)$ • $M_R = \mu r F_N$

Figure 5 Catalogue for elemental interfaces.

Design Clarity Assessment

As shown in the introduction, clarity in literature is seen as a precondition for robustness. The approach presented here contributes to design clarity assessment using generic elemental interfaces. General Procedure.

Based on the VDI 2221 [12] procedure model for the development process and related to the procedure that was proposed by *Ebro* [3], the following steps are proposed to include generic elemental interfaces in the procedure. Fig. 6 shows the steps together with results after certain steps and the data required. The grey boxes are steps that are affected by the proposed procedure.
Determine physical effects: *Choose a physical effect that is not influenced by disturbances that occur during product life.* Mathias [13] proposes a catalogue of physical effects, and their sensitivity to disturbances, that can be applied to satisfy the rule.

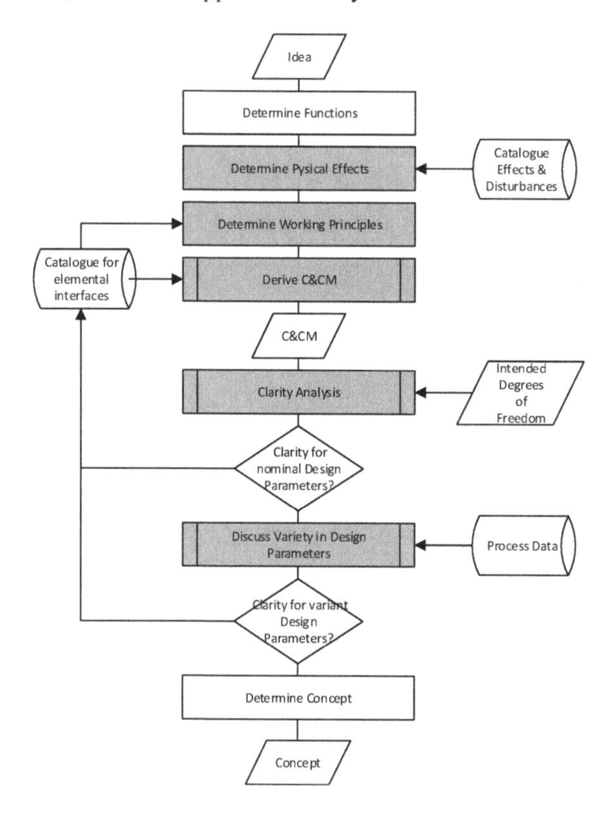

Figure 6 Procedure Model for design clarity assessment using elemental interfaces.

Determine Working Principle:
The database of elemental interfaces contains information about functions and effects. Therefore, the database can be used to choose sufficient elemental interfaces. Referring to *Suh's* axioms [14], the following rule can be derived: *Choose as few interfaces as possible.*

Derive Contact and Channel Model:
To obtain the C&CM, elemental interfaces and working bodies have to be identified. The *Contact Matrix* can be derived based on the geometry of the WSP. Taking the intended function and loads into account, the *Contact Matrix* can be transferred into the *Digitalised Closing Type Matrix*. The corresponding data are stored in the catalogue of elemental interfaces.

Clarity analysis:
This step is closely related to *Ebro's* Kinematic Design Step [3] and aims for the detection of design failures caused by to incorrect constraints, though it does not take malfunction caused by varying design parameters into account. Once the elemental interfaces are identified (geometry, function, and effect), the actual design clarity can be directly derived, applying the combination rules according to *Roth* introduced above.

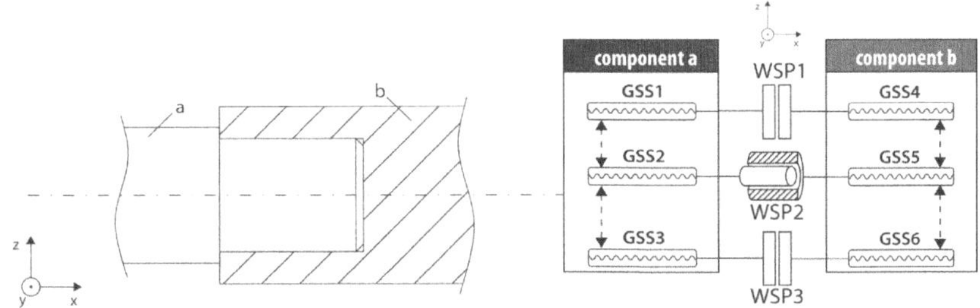

Figure 7 Example of clarity assessment with elemental interfaces for a shaft (a) to shaft (b) connection.

Fig. 7 illustrates the applicability of the approach. A simple shaft to shaft connection should be analysed. The right-hand side of Fig. 7 shows the corresponding structure of the C&CM. Three WSP can be identified. Two of them are plain WSP, the third is curved and cylindrical. The catalogue of elemental interfaces provides all necessary data. The *Contact Matrix* is shown in Table 3. It shows a double contact in positive x direction, and none in negative x or rotational x directions.

Table 3 Contact Matrix, Closing Type Matrix and Digitalised Closing Type Matrix for a cylindrical Working Surface Pair with two different functions (Transverse Press fit and Linear Guidance).

		components		translation						rotation					
	WSP	moved	fixed	x	$\bar{x}$	y	$\bar{y}$	z	$\bar{z}$	R_x	$\overline{R_x}$	R_y	$\overline{R_y}$	R_z	$\overline{R_z}$
contact matrix	1	a	b	1	0	0	0	0	0	0	0	0	0	0	0
	2	a	b	0	0	1	1	1	1	0	0	1	1	1	1
	3	a	b	1	0	0	0	0	0	0	0	0	0	0	0
	T	a	b	1^2	0	1	1	1	1	0	0	1	1	1	1
closing type matrix	1	a	b	E_f	g	r	r	r	r	r	r	g	g	g	g
	2	a	b	r	r	f	f	f	f	r	r	f	f	f	f
	3	a	b	E_f	g	r	r	r	r	r	r	g	g	g	g
	T	a	b	E_f^2	g	f	f	f	f	r	r	f	f	f	f
digitalised closing typ	1	a	b	1	0	0	0	0	0	0	0	0	0	0	0
	2	a	b	0	0	1	1	1	1	0	0	1	1	1	1
	3	a	b	1	0	0	0	0	0	0	0	0	0	0	0
	T	a	b	1^2	0	1	1	1	1	0	0	1	1	1	1

WSP 1 transmits forces in x direction via form closure, therefore a surface pressure occurs (Ef). The rest of the degrees of movability can be locked through friction (r) or additional field forces (g) (Table 3). The cylindrical WSP 2 has the function of centring the shafts relative to each other. Therefore, there is just form closure (f) in y and z directions. In x directions, friction may occur to lock the movability. In combination, the closing types that lock each degree of movability must now

be assessed to find the dominant type. For example, the form closure realised through WSP 2 in y direction dominates the (unintended) frictional locks through WSP 1 and WSP 3.
The *Digitalised Closing Type Matrix* (Table 3) contains the decision of whether a degree of movability is locked or not. In the example, WSP 2 does not constrain (0) the x directions; because of the form closure in y and z transversal forces cannot realise a frictional lock. Finally, there are three remaining degrees of movability (x negative and both rotations around x). Additionally, the system is over-constrained in positive x direction.
Following the procedure model, the next step would be adaption of the design at the working principle level, using the catalogue of elemental interfaces for support. In the example, a solution would be to delete one of the plain WSP.

Discuss variability in design parameters:
Although often assumed to be deterministic, design parameters differ in their values due to uncertainty in processes. These deviations can affect the performance of a system. To obtain robust solutions, designers have to be supported by purposeful methods. The UMEA [15], for example, contributes to the systematic analysis and assessment of uncertainty; *Eifler* [6] provides sensitivity indices for the life phases of a product to assess how much a particular process parameter affects a particular property, i.e. a temperature change during production influences the length property of a shaft. *Mathias* [16] introduces three robust design strategies to create designs that are more robust against disturbances. *Ebro* [3] use a catalogue with experience-based depictions of typical variety during manufacturing that can affect a design and provide exemplary solutions.

The catalogue of elemental interfaces can contribute to deriving robust products. It contains information about the design parameters that are necessary to realise a particular function. These parameters are affected in uncertain processes, so it is the design parameters that finally realise uncertain functions. Knowledge of relevant design parameters combined with sensitivity indices can help to specifically investigate life phase processes and discuss their impacts. Either variability due to processes exceeds the acceptable parameter values and the design has to be reconsidered, i.e. tensile stress is too high → enlarge working surface, or processes have to be restricted → manufacturing tolerance not higher than H7, Do not use your watch in the water.
In the example of Fig. 5, WSP 1 transmits forces. The catalogue directly refers to the admissible surface pressure and therefore provides the force to be transmitted, the surface area and material as parameters for the designer. These parameters now can be discussed related to life phases to assess sources and impact of variability. Life phases then have to be modelled with adequate process models [8, 17, 9].

Conclusions

The catalogue of elemental interfaces provides a possible way to investigate technical systems for their clarity in nominal design parameters based on generic interface characteristics. The presented procedure can be applied by hand, but there is potential for a computer-based implementation, which would reduce time. The catalogue also provides the ability to identify design parameters that are of functional relevance. To consider uncertainty in design parameters, processes and design parameters have to be combined in integrated measures. Further work has to focus on this step.

Acknowledgement

The results of this paper were achieved by the Collaborative Research Centre SFB 805 "Control of uncertainty in load-carrying mechanical systems" in subproject "A2: Robust Design – Methodology for the Design of Systems with Optimum Uncertainty Solutions". The authors wish to thank the Deutsche Forschungsgemeinschaft (DFG) for funding and supporting the SFB 805.

References

[1] S. M. Göhler, T. J. Howard, A Framework for the Application of Robust Design Methods and Tools, Proceedings of the First International Symposium of Robust Design 2014, pp.123-133, (2014).

[2] P. Andersson, A Process Approach to Robust Design In Early Engineering Design Phases, Department of Machine Design Lund Institute of Technology Lund, Lund, (1996).

[3] M. Ebro, T. J. Howard, J. J. Rasmussen, The foundation for Robust Design: Enabling Robustness through Kinematic Design and Design Clarity, INTERNATIONAL DESIGN CONFERENCE - DESIGN 2012, volume 2, pp. 817, (2012).

[4] R. Söderberg, L. Lindkvist and J. S. Carlson, Managing physical dependencies through location system design, Journal of Engineering Design, Vol. 17, No. 4, August 2006, 325–346

[5] G. Pahl, W. Beitz, J. Feldhusen, K.H. Grote, Engineering Design, London, (2007).

[6] T. Eifler, Modellgestützte Methodik zur systematischen Analyse von Unsicherheit im Lebenslauf technischer Systeme, Düsseldorf, (2015).

[7] K.T. Ulrich, S.D. Eppinger, Product Design and Development, New York, (2008).

[8] T. Freund et al., An Approach to analysing Interface Uncertainty using the Contact and Channel Model, International Conference on Engineering Design 2015, (2015).

[9] A. Albers, M. Ohmer, Engineering Design in a different way: Cognitive Perspective on the Contact & Channel Model Approach, Visual and Spatial Reasoning in Design III, pp. (3-21), (2006).

[10] S. Matthiesen, A contribution to the basis definition of the element model "Working Surface Pairs & Channel and Support Structures" about the correlation between layout and function of technical systems, (2002).

[11] K. Roth, Konstruieren mit Konstruktionskatalogen, Springer Verlag, Heidelberg, (1982).

[12] VDI2221 – Systematic Approach to the Design of Technical Systems and Products, Düsseldorf, (1987).

[13] J. Mathias, Selection of Physical Effects based on Disturbances and Robustness Ratios in the early Phases of Robust Design, International Conference on Engineering Design 2011, pp. 324-335, (2011).

[14] N.P. Suh, The Principles of Design, New York, (1990).

[15] R. A. Engelhardt, Uncertainty Mode and Effects Analysis – heuristische Methodik zur Analyse und Beurteilung von Unsicherheiten in technischen Systemen des Maschinenbaus, Darmstadt/ Technische Universität Darmstadt, (2013).

[16] J. Mathias, Strategies and Principles to Design Robust Products, International Design Conference - Design 2012, pp.341-350, (2012).

[17] T. Freund et al., Modelling Use Phase Variability and its Application in Robust Design and Eco Design, International Design Conference - Design 2014, pp.1063-1072, (2014).

[18] T. Freund, Robust Design of Active Systems - An Approach to Considering Disturbances within the Selection of Sensors, Proceedings of the international symposium on robust design – ISORD'14, pp. 137-147, (2014).

CHAPTER 4:

Improved Product Quality by Online Monitoring and Closed-Loop Control of Manufacturing Processes

Applied Mechanics and Materials Vol. 807 (2015) pp 121-129
© (2015) Trans Tech Publications, Switzerland
doi:10.4028/www.scientific.net/AMM.807.121

Submitted: 2015-04-15
Accepted: 2015-08-04

Orbital forming of flange parts under uncertainty

Stefan Calmano[1,a *], Daniel Hesse[1,b], Florian Hoppe[1,c], Philipp Traidl[1,d], Julian Sinz[1,e] and Peter Groche[1,f]

[1]Technische Universität Darmstadt, Institute for Production Engineering and Forming Machines, 64287 Darmstadt, Germany

[a]calmano@ptu.tu-darmstadt.de, [b]hesse@ptu.tu-darmstadt.de, [c]hoppe@ptu.tu-darmstadt.de, [d]traidl@ptu.tu-darmstadt.de, [e]sinz@ptu.tu-darmstadt.de, [f]groche@ptu.tu-darmstadt.de

Keywords: orbital forming, FE modeling, mushroom effect, estimation model

Abstract

Uncertainty in the properties of semi-finished parts can cause fluctuations in the product properties, especially if they have a strong effect on the process and cannot be compensated by process adjustments. Incremental forming processes have the potential to react to changing conditions by adapting the tool movement during the process. This paper analyzes the feasibility of controlling material flow in an orbital forming process in order to selectively fill those geometric elements which were specified with narrow tolerances by the designer. The effect of different process parameters on the mushroom effect and the degree of mold filling are analyzed by FEM simulations and experiments. In order to realize online monitoring and control, an estimation model is introduced, which maps signals from sensors and the process control to the geometric target values.

Introduction

Metal forming operations are characterized by high productivity and a good utilization of material. In the production of net-shape parts, the challenge is a constant process behavior and product quality, even if the properties of the semi-finished parts are affected by uncertainty. Uncertainty can occur in upstream processes which are necessary to produce the semi-finished parts.

The initial situation in this paper is the challenge of achieving tolerances in the geometric specifications of gear components if the volume of the semi-finished parts is fluctuating. When using single stroke bulk metal forming processes like closed-die forging, the geometric properties of the product are strongly depending on the volume of the blank.

If the blank volume is too small, a poor mold filling can be expected. If the blank volume is too high, forming the part without flash is hardly possible [1]. In case the required overall precision of a product is coarse and only selected dimensions are of stricter tolerance, coarse specifications for the geometric dimensions of the blank can be specified in order to reduce material cost. Accordingly, the material flow in the process must be controlled to ensure mold filling at the location where higher precision is required. This can be realized by the incremental forming process of orbital forming.

A schematic representation of an orbital forming process is shown on the right side of Figure 1. The significant influence factors for the orbital forming process are the inclination angle, which describes the angle between the upper die and vertical machine axis, and the infeed per revolution of the relative movement of the dies. In contrast to the upsetting process (Figure 1, left) there is only a partial contact surface between the upper die of the tool and the workpiece. Therefore the yield stress (k_{f}) is only exceeded in a small area which causes a decrease of the maximum stress (G_{zmax}). Decreased contact area and stresses result in reduced forming forces [2]. Further advantages of the orbital forming process are an increasing utilization of material and the possibility of achieving an increased plastic deformation [2]. The longer duration for this incremental forming process can be seen as a disadvantage, compared to an upsetting process.

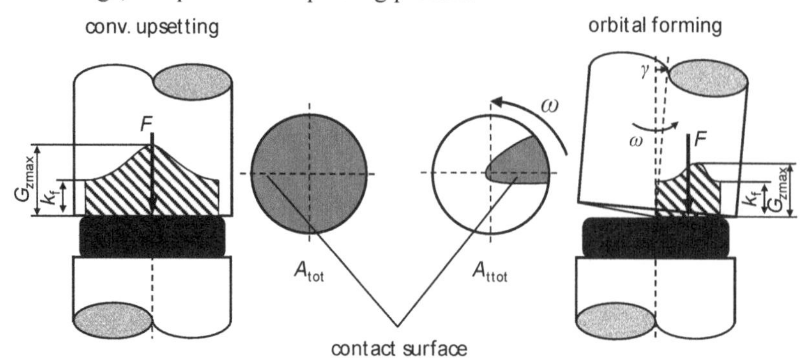

Figure 1: Comparison of conventional upsetting (left) and orbital forming (right) according to [2]

An analysis of the most important influence factors, like inclination angle and infeed per revolution, is shown in [3, 4]. To quantify the effect of the mentioned parameters on the geometry of a finished part, the mushroom effect is used. This is a dimensionless value which describes the maximum difference of diameters of the workpiece in the final state relative to the initial diameter [5].

Apart from the influence factors described by [3, 4] the aim is to examine if the movement pattern has an influence as well. In this paper, the mushroom effect is utilized to produce different kinds of mold filling. The examined geometry is shown in Figure 2. Starting from a cylindrical semi-finished part, a shaft with two gears is produced. As only for this examination the diameters of the two gears matter (Figure 2, right), they are substituted by two flanges to simplify the parametric study.

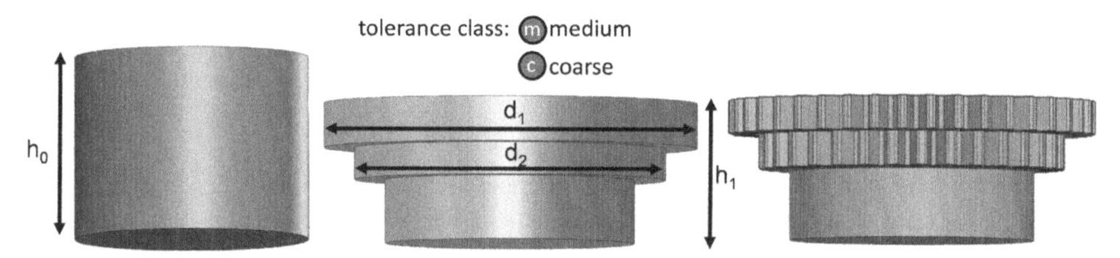

Figure 2: left: Semi-finished part with initial height h_0; **centric:** simplified part with two flanges including the diameters d_1, d_2 and the final height h_1; **right:** part with target geometry

Achieving a decreased tolerance of the diameters which are influenced by the fluctuations of the semi-finished parts it will be required to design a control strategy using the three process parameters. The following describes the approach how to develop the flanges according to the tolerance classes depending on the initial state of the semi-finished part.

Strategy and process sequence

Assuming that the initial states of the semi-finished parts differ in their heights, the question is if every required tolerance class of the geometric specifications can be achieved. Obviously, only an ideal initial height $h_{0,\text{set}}$ can results in an ideal finished part. Otherwise the specification of at least one of the two flanges or the final height cannot be fulfilled. In this approach two tolerance classes, medium (m) and coarse (c), are considered for the product. Those are selected by the designer because they seem sufficient referring to the application case. First, h_0 is measured and compared with $h_{0,\text{set}}$. Figure 3 shows the situations $h_0 < h_{0,\text{set}}$ and $h_0 > h_{0,\text{set}}$. Second, the process has to react to the actual state of the semi-finished part by choosing the right strategy for the actual set of specifications. If all variations are examined, it can be seen that in case of $h_0 > h_{0,set}$ a medium tolerance cannot be achieved for the final height h_1 (Figure 3, bottom center). Otherwise, if $h_0 < h_{0,set}$, all considers combinations of tolerance classes can be achieved.

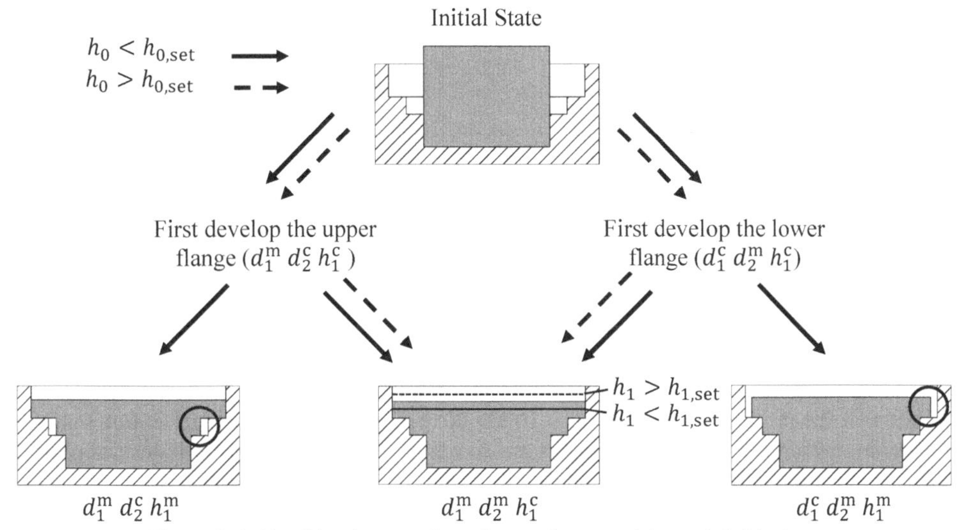

Figure 3: Achievable tolerances depending on the state of the semi-finished part

In this case, the control of material flow is used to force mold filling of the flange whose diameter is specified with medium tolerance. Afterwards, the forming process is continued until the desired height is achieved. The result is, that the flange which is specified as coarse tolerance achieves a more or less exact shape, depending on the available material (Figure 3, circles on bottom left and right). The next section analyses in which way the influencing factors of the orbital forming process help to achieve those tolerance classes for the different parameters.

Examined process parameters

Three influencing factors are considered in order to examine the effects of the forming process on the parameters h_1, d_1 and d_2, which were defined as objective for the different tolerance combinations mentioned above. Thereby, the initial height of the semi-finished part is supposed as constant. The influencing factors are the infeed per revolution, the inclination angle and the movement pattern. The movement pattern describes the pathway of the contact area between the upper die and the workpiece in the orbital forming process. The different patterns are generated by different angular functions of the tilting angle of the x-axis and y-axis of the press ram and can be seen in Figure 4.

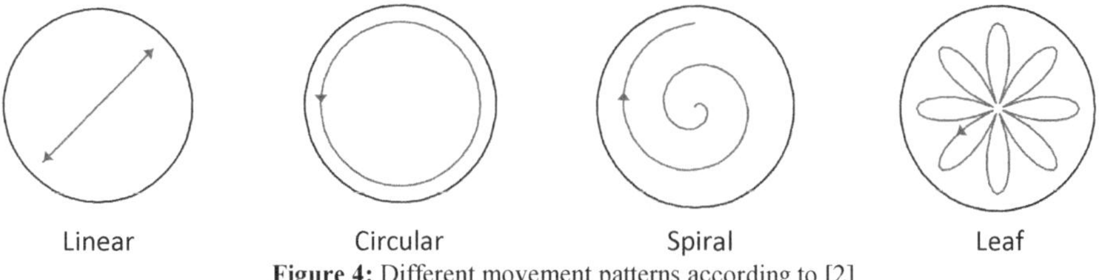

Figure 4: Different movement patterns according to [2]

In the examinations infeed values of 0.02 mm/rev, 0.1 mm/rev and 1.0 mm/rev as well as inclination angles of 1°, 2° and 3° are considered. The combination of circular movement pattern, an infeed of 0.1 mm/rev and an inclination angle of 2° is defined as the initial state to start the examinations. From this basis, two of the three initial state parameters are fixed while one is varied respectively. The examination plan can be seen in Table 1. The modified parameters of each examination series are marked in gray.

Table 1: Examination plan

	Infeed per revolution in mm/rev	Movement pattern	Inclination angle in °
Examination series 1	0.02; 0.1; 1.0	Circular	2
Examination series 2	0.1	Linear, Spiral, Leaf	2
Examination series 3	0.1	Circular	1; 3

The experiments will be performed on the prototype of the 3D Servo Press [6], which enables a freely programmable 3-DoF ram motion which is necessary to execute the different movement patterns. Before that, the effects of the influencing factor are examined by FEM simulations. In order to meet the requirements of the 3D Servo Press, the upper limit for the infeed was defined as 1.0 mm/rev and accordingly the limit was set to 3° for the inclination angle for the simulations. Due to the fact that the prototype only provides 10 kN of nominal pressing force, it is not possible to form materials which are commonly used in massive forming. For that reason industrial plasticine will be used in the experiments and simulations accordingly.

Simulation model

A simulation model was developed to analyze the effect of the process parameters on mold filling. For the simulations the program MSC Marc Mentat was used. The setting and the specifications of the model are defined as following.

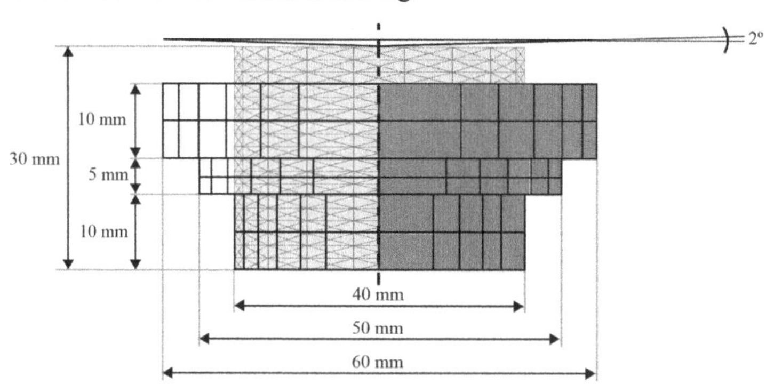

Figure 5: Dimensions of the simulation model

Figure 5 shows the dimensions of the simulation model. The semi-finished part is illustrated in light gray. The diameter of the part is 40 mm and its height is 30 mm. The mesh was created using three-dimensional four-node tetrahedrons. Due to the large plastic strain that occurs while forming, a global remeshing using the PATRAN TETRA method was applied to the model. Thereby, a strain change of 0.4 was the criterion for the creation of a new mesh. In the elastic/plastic/isotropic material model a Young's modulus of 1.166 N/mm² and a Poisson's ratio of 0.35 was used. For plastic deformation, the flow curve of the industrial plasticine was taken from [7].

The mold is composed of three sections. The semi-finished part is placed in the lower section which has a diameter of 40 mm and a height of 10 mm. The middle and upper section are limiting the first and second flange. The diameter and height of the middle section are 50 mm and 5 mm and accordingly 60 mm and 10 mm for the upper section. The upper die is cone shaped with a slope angle of 2°. The mold is fixed while the part is standing loose in it. For the contact a friction coefficient of 0.1 was chosen. The control of the infeed per revolution, the inclination angle and the movement pattern is executed by the upper die.

Results and control strategy

Simulations have been performed to analyze the influence of process parameters on mold filling, based on the set of parameters given in Table 1. The diameters d_1 and d_2 have been taken from the simulations at 0.5 mm above the corresponding flange edge, which is the tolerance in flange height (see Figure 9).

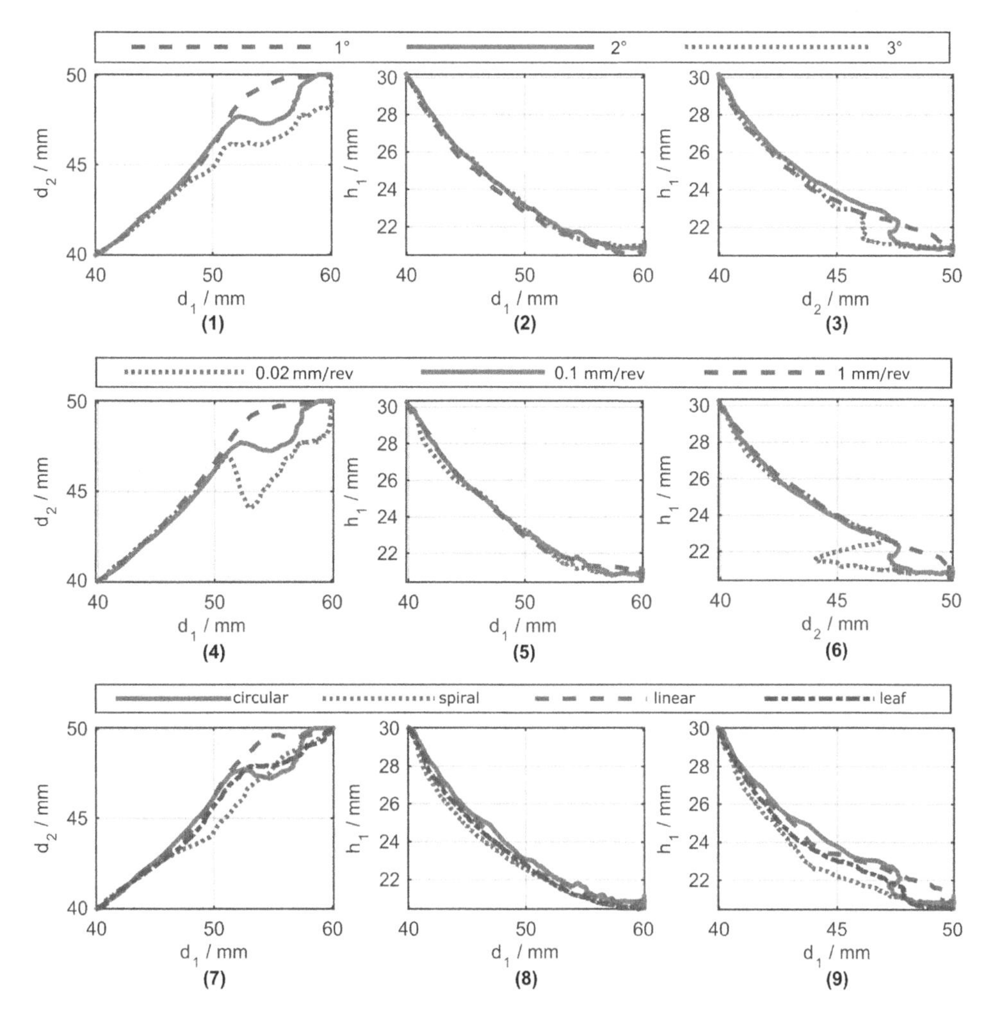

Figure 6: Influence of process parameters on mold filling

High inclination angle and low infeed rate are known to increase the mushroom effect [2] and therefore speed up mold filling of d_1. On the contrary, low inclination angle and high infeed rate should result in faster mold filling of d_2. The aim is to fill d_1 and d_2 selectively, depending on the desired tolerances. In terms of control theory, adjusting just one system variable without affecting the other is called decoupling. Since no mathematical relation between d_1 and d_2 can be deployed beforehand, a decoupling strategy is being searched empirically.

Figure 6 shows the simulation results for variations in process parameters and their influence on the control variables d_1, d_2, h_1. The first row shows the variation in inclination angle. It can be seen in (1) that an inclination angle of 1° causes d_2 to be filled first because the workpiece tends to bulge. An increasing inclination angle enhances the mold filling capability of the upper flange d_1. The influence of the inclination angle on the part height h_1 is shown in (2) and (3). While h_1 and d_1 cannot be decoupled by a variation of the inclination angle, the mushroom effect gives the possibility to approach the tolerance of h_1 before d_2 is filled completely. This gives the opportunity to handle parts with a lower initial height $h_0 < h_{0,set}$. Nevertheless, the possible tolerance combinations of $(d_1\ h_1)$ are limited to the initial height h_0 and cannot be influenced by the inclination angle.

The results for a variation in infeed per revolution are illustrated in the second row of Figure 6. Just as the inclination angle, the infeed per revolution governs the mushroom effect and hence gives the possibility to control mold filling of d_1 and d_2. It can be clearly seen in (4) that a lower infeed rate causes d_1 to be filled faster than d_2. Furthermore, it can be seen that the upper flange has a major impact on the evolution of d_2. Once the part touches the upper flange edge during a low infeed per revolution, the material tends to be pulled up such that d_2 is being reduced and d_1 filled faster. This phenomenon enhances the possibility to fill d_1 and d_2 separately.

The influence on the progress of h_1 is shown in (5) and (6). Unlike d_2 and h_1, $d_1$and h_1 cannot be decoupled and thus their tolerance combinations are as limited as for a varying inclination angle.

A minor role is attributed to the movement pattern. Its effect on d_1, d_2, h_1 is pictured in the third row of Figure 6. The largest difference can be seen from a linear to a circular motion pattern but not as clear as for varying inclination angle and infeed rate.

From the results shown above, two strategies for filling d_1 and d_2 can be framed, one for filling d_1 first with a motion that expands the upper part of the workpiece and one for filling d_2 first with a motion that is close to upset forming:

- Expanding strategy → filling d_1 first: spiral motion, 0.02 mm/rev, 3° inclination angle
- Upsetting strategy → filling d_2 first: linear motion, 1 mm/rev, 1° inclination angle

As presented in Figure 7, the superposition of process parameters enhances the decoupling of d_1, d_2 and h_1 in the expanding and upsetting strategy. These two strategies can be used to control mold filling of d_1 and d_2 separately.

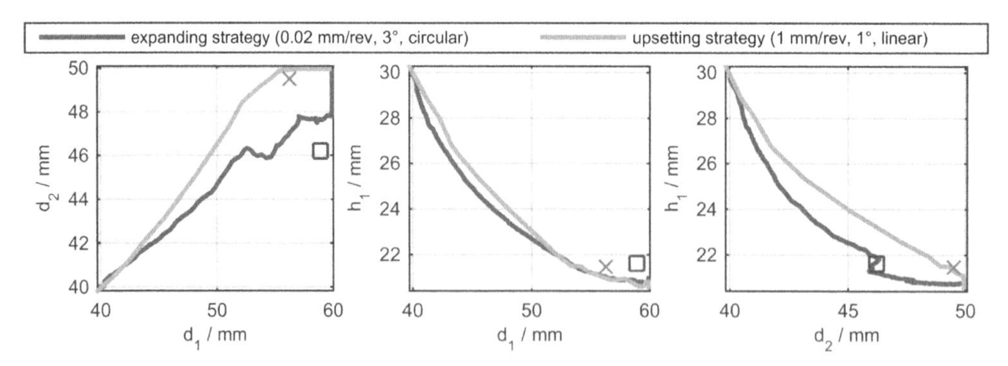

Figure 7: Comparison of expanding vs. upsetting strategy (**line graph**: simulation; **cross and box:** experiment)

First experimental investigations using plasticine as model material confirm the results gained by the simulations. It can be shown, that the material flow can be influenced by the movement strategy of the tool. The geometric evaluation of the parts is given in Figure 7, marked as cross and box.

However, there are some shortcomings in the experimental setup, which lead to cracking of the samples. Further development concerning material choice, die geometry and movement pattern is necessary in order to realize a reproducible production. Figure 8 shows a sample resulting from the experiments.

Figure 8: Photo of experimentally produced sample

Besides the modifications mentioned above, the extended experimental setup contains sensors and an online model in order to estimate mold-filling during the process. The diameters d_1 and d_2 cannot be measured directly. However, diameters at positions s_1, s_2 as well as the forming force F (see Figure 9) are being measured. Due to the installation height of the sensors, s_1 and s_2 are located above d_1 and d_2 and have to be used to estimate d_1, d_2. In order to estimate d_1 and d_2, a feedforward estimation model can be used which maps all process information such as s_1, s_2, tool position, inclination angle on the diameters d_1, d_2.

An estimation model requires the input variables

- measurements: diameters s_1, s_2, workpiece height h_1 and forming force F
- machine and tool parameters: rotary angle φ, infeed rate per revolution $\dot{z}$, infeed z
- process parameters: friction coefficient μ, motion pattern m_p, inclination angle α

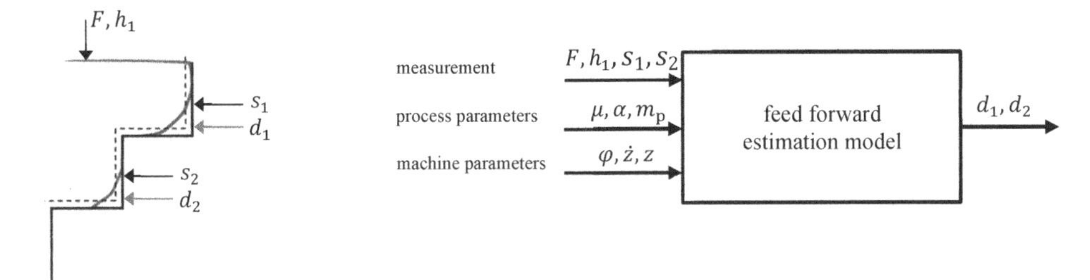

Figure 9: Measured and estimated variables

A black-box model which has been trained with simulation data can be suitable to map all dependencies of input to output variables d_1, d_2. These estimated variables have to be fed back into a process feedback controller which adjusts the desired tolerances using the expanding and upsetting strategies. The process controller uses the desired tolerances and current diameters to calculate the desired process parameters for the orbital forming process, i.e. inclination angle, infeed per revolution and motion pattern. As orbital forming machine, the prototype of the 3D Servo Press is used. Therefore the desired process parameters are fed into a machine controller which translates the desired process parameters into motor positions. The whole process model is shown in **Figure 10**. Initial dimension of the semi-finished part are its diameter d_0 and height h_0. Desired dimensions are the final diameters d_1, d_2 and the final height h_1.

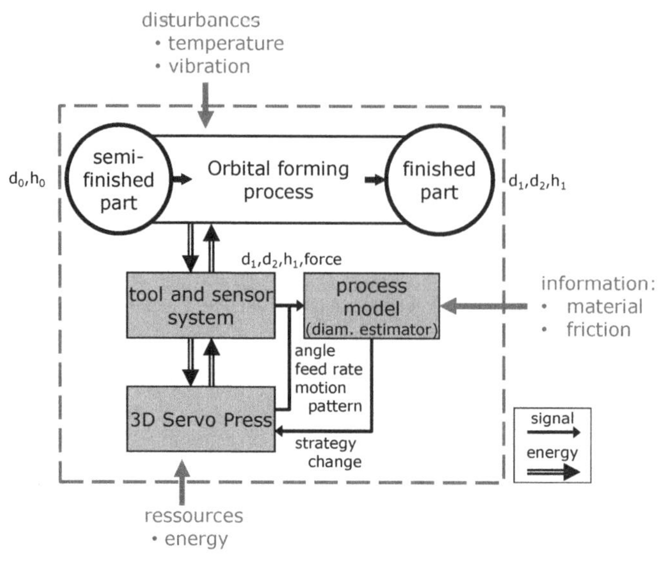

Figure 10: Process model according to [8]

Conclusion

Forming technology often has to deal with uncertainty in the properties of semi-finished parts, leading to fluctuations in process behavior and product quality. In order to cope with these uncertainties, processes which have multiple degrees of freedom become advantageous to influence process behavior and part properties during the process. In this case, an orbital forming process is analyzed concerning its possibility to influence the material flow in order to react to a lack or excess of material volume. Simulations show that the tool motion parameters of the orbital forming process have a significant influence on the mushroom effect, which is used to promote mold-filling of the upper section of the work piece, compared to a bulging form, which leads to mold-filling in the lower section of the work piece. In order to monitor the mold-filling online, an estimation model is proposed, which maps values gained by tactile position sensors on d_1 and d_2. These are the actual diameters of the part in the concave edges of the mold and can be seen as target values for geometric quality. The results are confirmed in experiments using plasticine material and the 10 kN prototype of the 3D Servo Press.

Acknowledgements

The results of this paper are achieved in the Collaborative Research Centre SFB 805 "Control of uncertainty in load-carrying mechanical systems" in subproject "B2: Forming – Production families at equal quality" and in transfer project "T3: The 3D Servo Press – from a research version to an industrial standard machine" funded by Deutsche Forschungsgemeinschaft (DFG).

References

[1] Lange, K.; Pöhlandt, K., Handbook of metal forming, McGraw-Hill, 1985.
[2] Schondelmaier, J., Grundlagenuntersuchung über das Taumelpressen (in German, Basic Investigantions on Orbital Forming), Springer-Verlag, Berlin Heidelberg New York, 1992.
[3] Munshi, M.; Shah, K.; Cho, H.; Altan, T., Finite element analysis of orbital forming used in spindle/inner ring assembly, in: Proc. ICTP, 2005.

[4] Han, X.; Hua, L., Effect of size of the cylindrical workpiece on the cold rotary-forging process, Materials & Design, Vol. 30 (2009), pp. 2802-2812 (doi:10.1016/j.matdes.2009.01.021).
[5] Liu, G.; Yuan, S. J.; Wang, Z. R.; Zhou, D. C., Explanation of the mushroom effect in the rotary forging of a cylinder, Journal of Materials Processing Technology, Vol. 151 (2004), pp. 178-182 (doi:10.1016/j.jmatprotec.2004.04.035).
[6] Groche, P.; Scheitza, M.; Kraft, M.; Schmitt, S., Increased total flexibility by 3D Servo Presses, CIRP Annals - Manufacturing Technology, Vol. 59 (2010), pp. 267-270 (doi:10.1016/j.cirp.2010.03.013).
[7] Brill, K., Modelwerkstoffe für die Massivumformung von Metallen (in German, Model Materials for Bulk Metal Forming), Technische Hochschule Hannover, Hannover, 1963.
[8] Bretz, A.; Calmano, S.; Gally, T.; Götz, B.; Platz, R.; Würtenberger, J., Darstellung passiver, semi-aktiver und aktiver Systeme auf Basis eines Prozessmodells (Representation of Passive, Semi-Active and Active Systems Based on a Prozess Model), unreleased paper by the SFB 805 on http://www.sfb805.tu-darmstadt.de/media/sfb805/f_downloads/150310_AKIII_Definitionen_aktiv-passiv.pdf, 2015.

Applied Mechanics and Materials Vol. 807 (2015) pp 130-139
© (2015) Trans Tech Publications, Switzerland
doi:10.4028/www.scientific.net/AMM.807.130

Submitted: 2015-04-15
Accepted: 2015-08-04

Data-based support in the development of press systems using the example of sheet metal forming

Julian Sinz[1,a*], Daniel Hesse[1,b], Sebastian Öchsner[c] and Peter Groche[1,d]

[1]Institute for Production Engineering and Forming Machines,
Technische Universität Darmstadt,
Otto-Bernd-Strasse 2, Darmstadt, Germany

[a]sinz@ptu.tu-darmstadt.de, [b]hesse@ptu.tu-darmstadt.de, [c]sebastianoechsner@googlemail.com,
[d]groche@ptu.tu-darmstadt.de

Keywords: press system, development, design

Abstract. The market pressure and the products of competitors are decreasing the duration of product life cycles. Besides that, special customer requirements and rising product complexity result in an increase of the effort and the costs of the development process. In order to minimize the effort of the development process as well as to handle the uncertainties which come along with the planning and selection of manufacturing systems a data-based configuration tool was developed for the use in the field of sheet metal forming. Based on the product data of existing press systems, the tool is able to predict the fundamental configuration of a newly developed press system. The utilization and functionality of the tool is illustrated on two exemplary press systems.

Introduction

The demand for the reduction of the effort in press system development is based on many reasons. Forced by the market and the products of competitors, the duration of a product life cycle is decreasing [1]. A study by Berger [2] showed that the duration of a product life cycle has shortened by nearly 25% in the years from 1997 to 2012. In the meantime, the product variety across all industries has more than doubled in that period. Besides the shortened duration of the product life cycle and the growing product variety, manufacturers have to deal with further challenges such as an increasing product complexity and a decreasing batch size [3]. The higher product complexity and special customer requirements also result in a more complex planning of products and higher process costs caused by that [2]. Technological differentiation such as creating customer specific products with high quality, multifunctional features and high flexibility, is essential to obtain an international economic competitiveness [4]. The increasing complexity and the shortened development times should be handled. Thus, the support of data processing and the utilization of virtual prototyping tools will be more and more important [5].

Different pathways can be pursued in order to manage the requirements for high variability and complexity as well as to reduce the effort in the design of products. In the automotive industry predominantly standardization and modularization strategies are used to decrease the internal production effort and to increase the cost effectiveness [2]. Thereby platform construction is an essential approach to develop products with high variety within short cycle times [6]. Platform construction has to be distinguished from modular construction. While a basic variant-neutral product platform with product-specific additions is used in platform construction [7], predefined modules are used to build up product variants in modular construction [6]. The aim of modularization is the optimization of the product architecture concerning product requirements [8] as well as the rationalization of the production process [6]. The complexity of a product development process can be reduced by modularization also [9]. Thereby the duration and the effort of the development process can be reduced through the parallelization caused by the decoupled arrangement as well as the reuse of existing modules [9].

Besides modularization, good planning has high relevance to reduce the effort of product development. According to Groche et al. [10], the planning and selection of manufacturing systems is afflicted with uncertainties. This is because future events, which could have an impact on the development and selection of a manufacturing system, can only be assumed [10]. The uncertainties manufacturers are faced with can be divided into four main uncertainties [11]. These include the market acceptance of kinds of products, the length of the product life cycles, the specific product characteristics and the aggregate product demand [11]. Not only manufactures but also developers have to deal with these uncertainties in the development process of a press system. Besides these four main uncertainties, more aspects have to be considered. Some of these aspects are exemplary illustrated in Fig. 1 on the basis of the SFB 805 process model [12].

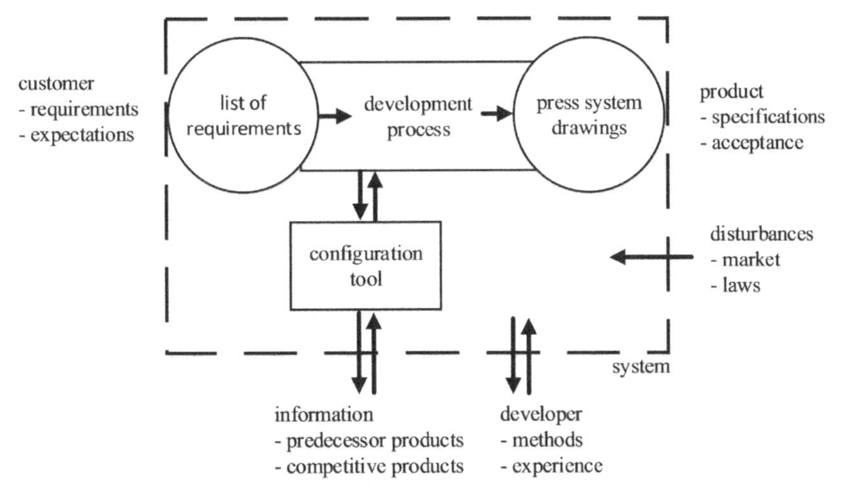

Fig. 1: Exemplary uncertainties in a press system development process on the basis of [12]

Depending on the development task, the development process of a press system is more or less afflicted with uncertainty. Hereby the individual qualification of the developer, such as the knowledge about methods or the experience, is more or less suitable to handle the problem. Furthermore, the information the developer can use such as predecessor or competitive products depend on the requested task. The developer can only resort to information if an existing product has to be modified. There is no or just very little information if the task leads to the development of a new product. To determine whether usable information can be extracted from existing products or not, a good communication between customers and developers is important. In this way, the development task can be exactly defined, in order to transfer the customer requirements into a technical solution. According to Röder et al. [13], many errors that occur while requirement acquisition more often can be referred to communication problems between the customer and the developer than to technically wrong decisions or to the lack of expertise. On the one hand, the customer is not able to define every single requirement the developer would need to build up a press system. On the other hand, the customer sometimes expects certain properties of the product, which are not defined specifically as a requirement. Apart from good communication, precise planning is important to speed up the duration of the development process as well as to reduce the development costs. Precise planning requires, among other things, the knowledge of the market and technological trends, the differentiation to competitive products, the consideration of possible niches as well as the knowledge about restricting laws.

A supporting tool was developed in order to handle the challenges mentioned above. With that, the surrounding conditions of a technical solution can automatically be generated. In doing so, different aspects have to be considered. Firstly, the tool has to be helpful to reduce the effort and the duration of the development as well as to give information about the position of the product within the market. Secondly, information about existing products should be analyzed and be provided to the developer to reduce the uncertainty in planning which comes along with the development of new products. The developed configuration tool uses the example of a press system for sheet metal forming. The functionality and its utilization are described below.

Configuration tool

The procedure and the placement of the press configuration tool are shown in Fig. 2. The configuration tool and thus the configuration process are placed before the actual development process. Usually the requested development task of the customer is written down in the list of requirements. This list represents the start of the configuration process.

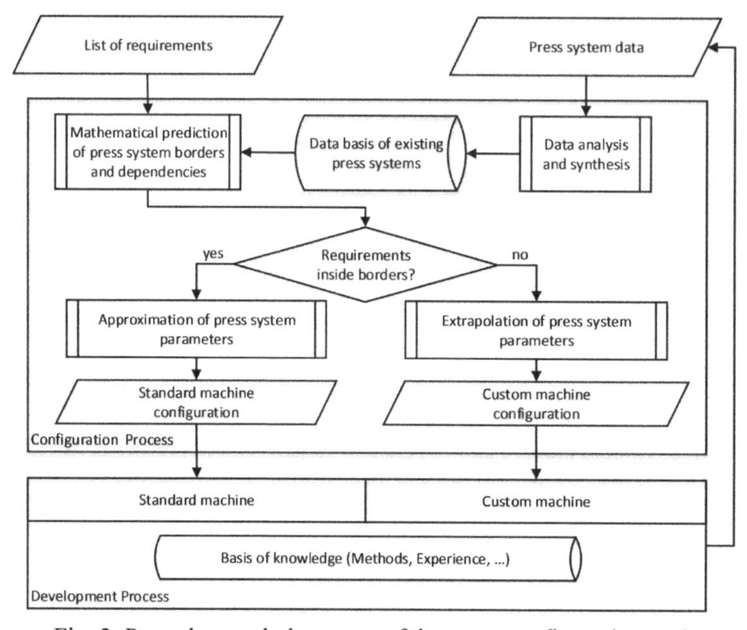

Fig. 2: Procedure and placement of the press configuration tool

In the first sub-process of the configuration process, the borders of the press system data basis as well as the dependencies between the data are calculated. Hereby a mathematical prediction model is used, as discussed later in this paper. The data basis is deduced by an analysis and synthesis of the existing press system data. For the calculation of the dependencies between the data different approximation approaches are used in the prediction model. In the next step, the requirements are compared to the calculated borders. If the requirements arrange inside the borders the requested task leads to the development of a standard machine. Thereby the configuration tool is able to approximate the standard machine configuration in a second sub-process using the data dependencies. In contrast, if the requirements are arranged outside the borders, the requested task leads to the development of a custom machine. In that case, the configuration tool is only able to extrapolate the calculated dependencies to predict the machine parameters. As a result of the configuration process an approximated standard machine configuration or an extrapolated custom machine configuration is given. This represents the start of the actual development process. In that, the developer provides a certain basis of knowledge which is necessary to fulfil the development process such as experience or knowledge about methods. As soon as the development of a standard or a custom machine is finished the data of the developed system is returned to the data basis of existing press systems by the analysis and synthesis process. As a result the data basis of existing press systems enlarge.

Mathematical prediction model. The prediction model uses data of existing press systems as well as the dependencies between the data to predict the press system configuration for the required development task. Requirements are defined by the customer. Depending on that, different parameters for the development have to be considered. For instance, the customer focuses on the recycling phase of the product life cycle. Then requirements like an easy disassembly lead to the consideration of parameters that have a lower relevance for the developer to build up a technical basis of a press system. The first version of the configuration tool only considers fundamental parameters, such as nominal pressing force, maximum press stroke rate or tool installation space. For a more detailed prediction of a press system configuration the fundamental parameters partially mentioned above are not sufficient. Hence, further parameters have to be considered in the prediction model. However, this is not purpose of the paper.

The product data was obtained from the industrial partner of the project and from a research carried out at considerable manufacturers of presses for sheet metal forming. After collecting the data, every parameter is opposed to every other parameter. Moreover, the dependency between the parameters is determined by using different approximation approaches. The fundamental parameters as well as the different approximation approaches used to determine the dependency between product data are shown in Table 1.

Table 1: Fundamental parameters and approximation approaches

		NPF	RT	APB	MPSR	MH	MM	TIS	DP	MSL	SA	AP	LPB
nominal pressing force	NPF												
ram traverse	RT	lin											
area of press bed	APB	lin	lin										
maximum press stroke rate	MPSR	pow	pow	pow									
machine height	MH	lin	lin	lin	pow								
machine mass	MM	lin	lin	lin	pow	lin							
tool installation space	TIS	lin	lin	lin	pow	lin	lin						
driving power	DP	lin	lin	lin	pow	lin	lin	lin					
maximum stroke length	MSL	lin	lin	lin	pow	lin	lin	lin	lin				
stroke adjustment	SA	exp	exp	lin	pow	exp	pow	exp	exp	exp			
average price	AP	lin	lin	lin	pow	lin	lin	lin	pow	lin	pow		
length of press bed	LPB	lin	lin	pow	pow	pow	lin	pow	qua	qua	log	qua	

lin: linear approximation function qua: quadratic approximation function exp: exponential approximation function

pow: power approximation function log: logarithmic approximation function

Fig. 3 exemplary shows two different approximation approaches. Every cross in the charts represents the combination of the opposed parameters of one single existing press system. The cross is designated as machine data point. On the left side of Fig. 3 a linear function is used to calculate the approximation curve. In contrast, an exponential function is used on the right side. The approximation is pictured as a continuous black curve. Thereby the best matching approximation approach is selected respectively. The defined parameter is always applied on the x-axis. By a vertically parallel translation of the approximation curve to the highest and lowest available machine data point the upper and lower borders can be deduced. The left and right borders are determined by the translation of a line parallel to the y-axis of the charts to the outermost available machine data points. The areas inside the borders are designated as the areas of standard machines, which are marked in gray in Fig. 3.

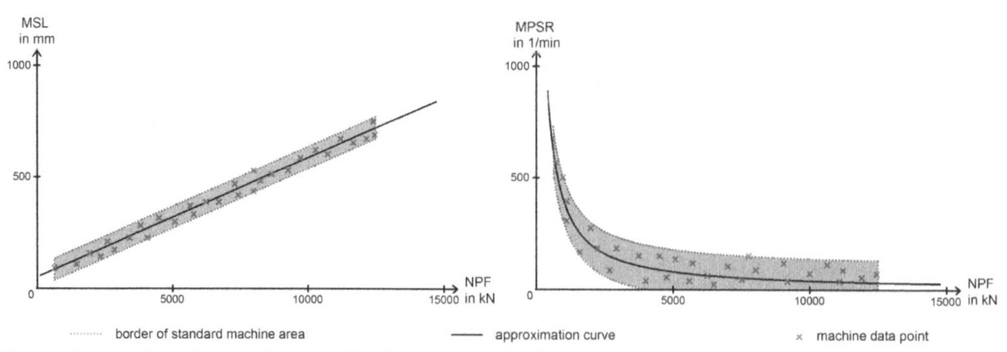

Fig. 3: Approximation models and borders of standard machine areas exemplified in the comparison of Nominal Pressing Force (NPF) with Maximum Stroke Length (MSL) and Maximum Press Stroke Rate (MPSR)

The values of the parameters defined by the developer are subsequently designated as input requirements. To decide whether the required task leads to the development of a standard machine or a custom machine, the configuration tool compares the input requirements with the borders calculated by the prediction model. If the input requirements arrange inside the borders, the task for the development of a standard machine is given. If one input requirement is located outside the borders, the task leads to the development of a custom machine.

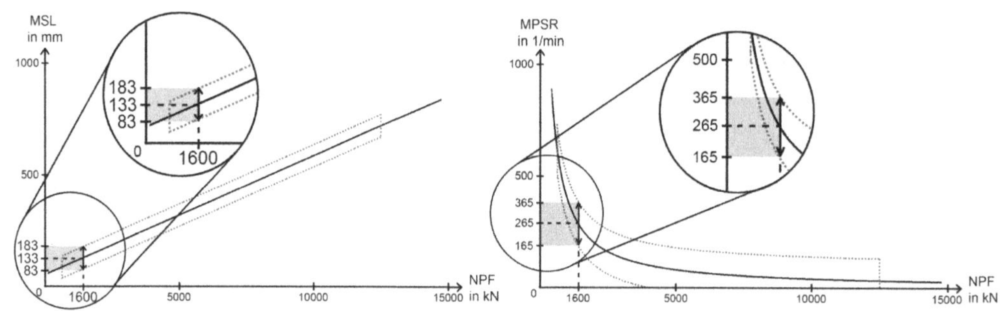

Fig. 4: Procedure of predicting the parameters Maximum Stroke Length (MSL) and Maximum Press Stroke Rate (MPSR) as well as their limits by only defining an input requirement for Nominal Pressing Force (NPF)

Fig. 4 exemplary shows the procedure of predicting parameters if only one input requirement is given. On both sides, the parameter NPF (nominal pressing force) is applied as the input requirement. With that, the lower and upper limits as well as the predictions for the parameters MSL (maximum stroke length) and MPSR (maximum press stroke rate) can be deduced. Thus, the predictions are calculated by projecting the intersection points of the applied input requirement with the respective approximation curve onto the y-axis. The possible range in which the parameters MSL and MPSR can arrange in order to stay in the field of standard machines are calculated by projecting the intersection points of the input requirement with the upper and lower borders onto the y-axis. On the left side of Fig. 4 the exemplary value of 1600 kN for NPF respectively results in the prediction of 133 mm for MSL as well as the upper and lower limits of 183 mm and 83 mm. Accordingly the prediction for MSPR is 265 1/min and the limits are 165 1/min and 365 1/min. This calculation is done with every combination of the parameter NPF and the remaining parameters.

In case that not only one input requirement is given, an adjustment of the possible areas of standard machines as well as an adjustment of the lower and upper limits has to be done. This is shown in Fig. 5 using two exemplary input requirements for the parameter MSL. In contrast to Fig. 4 MSL and MPSR are compared in both charts. The lower and upper limits for these two parameters, which were calculated when only NPF was defined (see Fig. 4), are plotted on the x-axis and y-axis of Fig. 5. The adjusted area of standard machines can be determined by calculating the intersection between the area spanned by these limits (light gray) with the area inside the borders of the comparison between MSL and MPSR. This area is marked in dark gray in Fig. 5.

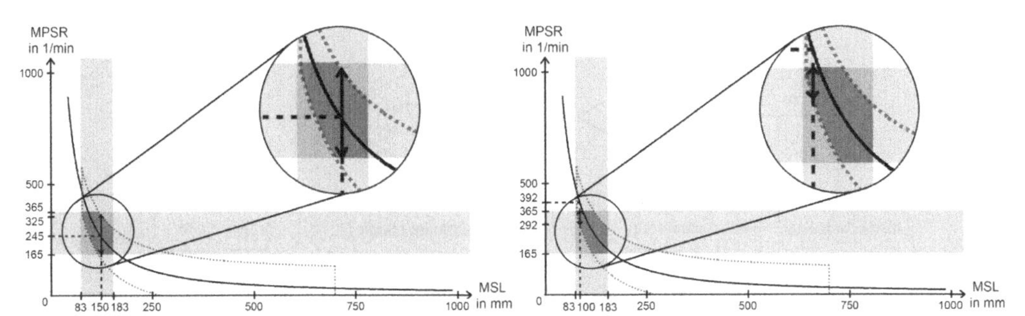

Fig. 5: Procedure of predicting Maximum Press Stroke Rate (MPSR) and its limits by defining input requirements for Nominal Pressing Force (NPF) and Maximum Stroke Length (MSL)

After calculating the adjusted area of standard machines, an input requirement for MSL can be applied. On the left side of Fig. 5 the value of 150 mm is chosen for MSL. As a result, the adjusted lower and upper limits as well as the prediction for MPSR are deduced in the same way as shown in Fig. 4. It can be seen that the upper limit is reduced from 365 1/min to 325 1/min while the lower limit stays at 165 1/min. The value of the prediction is 245 1/min and arranges inside the limits. On the right side, the value of 100 mm is chosen for MSL. The calculated lower limit for MPSR increases from 165 1/min to 292 1/min, the upper limit stay at 365 1/min. Compared to the left side of Fig. 5 the predicted value for MPSR (392 1/min) arranges outside the calculated limits. It is important that the predicted value of a remaining parameter is not always located inside the area of standard machines, as it is the case in the right side of Fig. 5. If the developer defines that prediction as an input requirement for MPSR, the task will lead to the development of a custom machine. If the input requirement is chosen inside the limits for MPSR, the task will still lead to the development of a standard machine.

Utilization of the configuration tool. In order to use the configuration tool, a software program was developed by using Matlab; the definition of the input requirements for the prediction model is done with a graphical user interface (GUI). The possible standard ranges as well as the predicted values for the remaining parameters are also displayed in the GUI. Fig. 6 illustrates the process chart of the configuration tool.

After the start of the program, the area of application has to be chosen. In the context of this paper, the area of sheet metal forming is selected. In the second step, the first input requirement is entered. Thereby the nominal pressing force has always to be entered as the first input requirement caused by the functionality of the program. If the defined input requirement is located inside the area of standard machines, the prediction of the remaining parameters and possible ranges is executed. If the value of the input requirement is arranged outside the area of standard machines, a warning is displayed to inform the developer that the required task will lead to the development of a custom machine. After that, the remaining parameters and ranges are also predicted and a graphical illustration of the press system configuration is given in the form of a spider net chart (see Fig. 7). Subsequently, the next input requirements can be entered and the calculation process repeats. If no further requirements are given and the predicted values for the remaining parameters are satisfactory, the program ends, otherwise the program restarts.

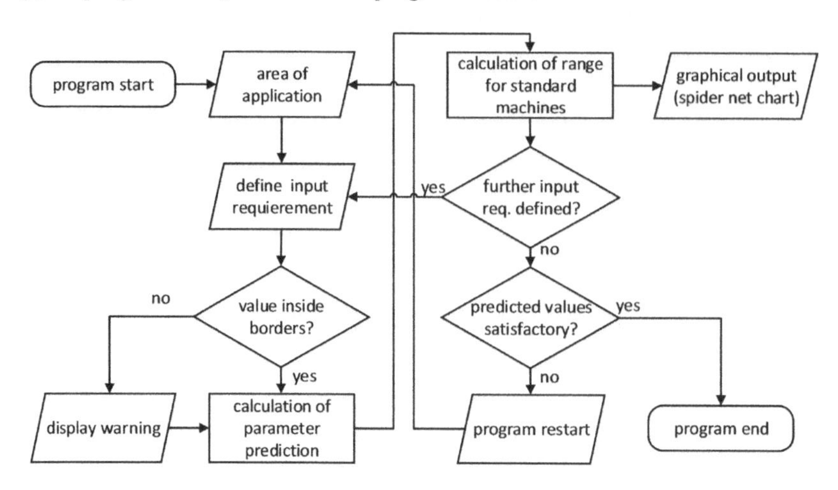

Fig. 6: Process chart of the configuration tool

Examples of use. Two examples are given below, to illustrate how the configuration tool can be used in practice. On the one hand, the configuration tool is used on an existing press system of the industrial partner. On the other hand, it is used on the 3D Servo Press which was developed by the PtU in order to handle the uncertainties, which come along the planning and selection of manufacturing systems [10]. The 3D Servo Press provides 1600 kN of nominal pressing force. To start with the same initial situation, a press system of the industrial partner with the same nominal pressing force was considered. After feeding the configuration tool with that value the graphical output is given, as shown in Fig. 7. This spider net chart is valid for both examples mentioned above. The outside edge of the chart represents the upper border of the remaining parameters to stay inside the area of standard machines. The lower border is represented by the outside edge of the dark gray area in the middle of the chart. The black curve illustrates the predicted value of the remaining parameters. Due to the fact that the parameter NPF is defined as an input requirement, the upper and lower borders match in this point. The calculated area of standard machines for the remaining parameters is illustrated in light gray.

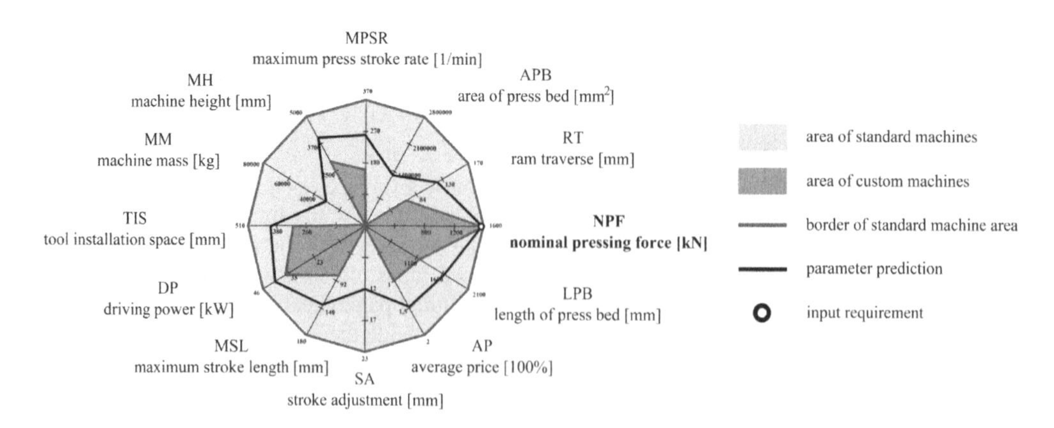

Fig. 7: Spider net chart with Nominal Pressing Force (NPF) as input requirement

By defining more input requirements, the area of standard machines is successively reduced. On the left side of Fig. 8 the combination of the press system of the industrial partner is illustrated when six parameters are fixed as input requirements. As before, the upper and lower borders match at these parameters. It becomes apparent that the area of standard machines is reduced while the area of custom machines enlarges. The black crosses represent the actual press system specifications of the chosen industrial example. As expected, the values of the remaining parameters are all located inside the area of standard machines. Moreover, the predicted values for the remaining parameters nearly match the actual specifications except the value for the stroke adjustment (SA).

On the right side of Fig. 8 the second parameter combination, which represents the 3D Servo Press, is illustrated. The same six parameters as those as in the industrial example are fixed. It is obvious that the four parameters maximum stroke length, tool installation space, machine height and ram traverse are located outside the area of standard machines, which was predicted after the nominal pressing force had been defined (see Fig. 7). Because of that, the configuration tool displays a warning that the requested combination leads to the development of a custom machine.

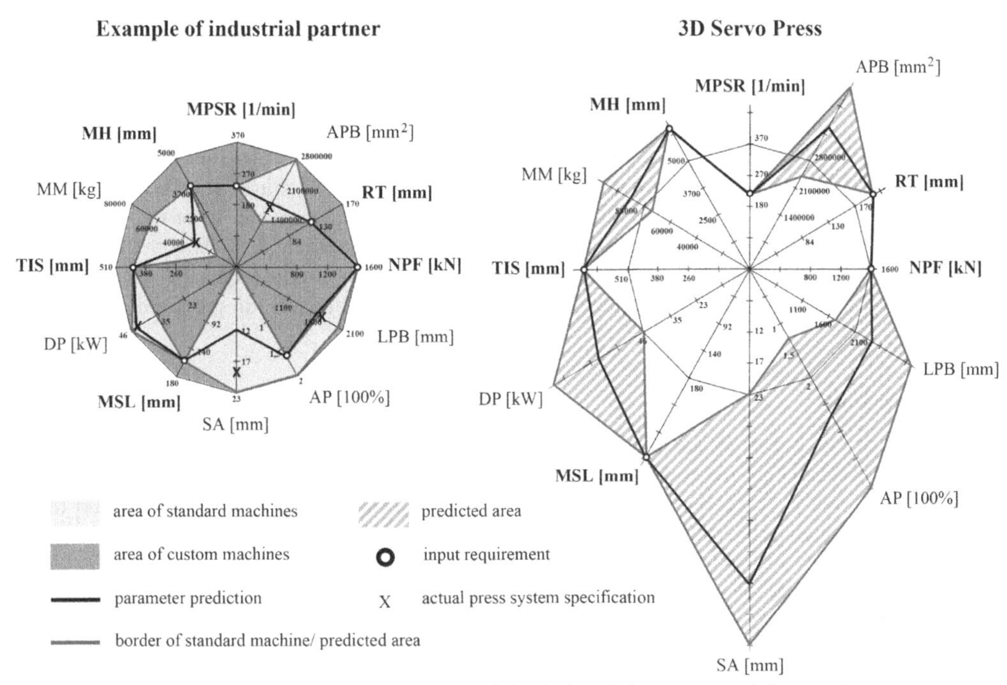

Fig. 8: Spider net chart for the example of the industrial partner and the 3D Servo Press

Consequently, neither the area of standard machines nor the area of custom machines is illustrated in the right spider net chart. Even though the requested parameter combination leads to the development of a custom machine, the configuration tool still calculates a prediction for the remaining parameters and ranges. This is done by an extrapolation of the existing product data and its dependencies. The predicted area is illustrated in a light gray hatching.

Conclusion and further research

The benefit developers can obtain by the utilization of the configuration tool includes different aspects. First, the communication between the customer and the developer can be improved. With the configuration tool, the developer is able to retrieve the required information of the customer more easily. Second, the discussion about the requirements is facilitated because the developer can directly estimate the basic conditions of the requested task, and then can give the customer a feedback whether the requested task leads to the development of a standard or custom machine. This helps to define the development task more precisely and builds up the basis for a good planning and execution of the development process. Additionally, the effort and duration of the development process can be reduced. While feeding the configuration tool with input requirements, the developer directly receives a feedback after every new entered parameter. In doing so, the developer always knows the possible range in which a further input requirement has to be arranged to keep the development task inside the area of standard machines. By staying inside the area of standard machines, the effort of the development is smaller than in the area of custom machines because of the availability of existing solutions. Furthermore, the effort inside the area of custom machines can be reduced in some cases also because the configuration tool calculates a prediction of possible parameter combinations outside the area of standard machines.

In order to give a more detailed prediction of areas as well as to enable the prediction in further application areas, the data basis of the configuration tool has to be enlarged with more product data. With an advanced analysis of the product data, the additional benefit of the configuration tool can be raised further. On the one hand, the prediction of the future technological trends of the different parameter combinations could be enabled by combining the product data with those of the developments. On the other hand, an assignment of existing technological solutions to different parameter combinations is contemplated. With that, the developer could not only be provided with a combination of values but also with a possible technological solution. Last, the utilization of the configuration tool for the support of the development of custom machines should be improved further. Therefore, the configuration tool should be able to recommend a certain design or development method, which should be used in the particular development task. This method should also be deduced by the analysis of the dependency between the product data of existing press systems in combination with size range or modularization methods from the field of product development.

Acknowledgements

The results of this paper are achieved in the Collaborative Research Centre SFB 805 "Control of uncertainty in load-carrying mechanical systems" in subproject "B2: Forming – Production families at equal quality" and in transfer project "T3: The 3D Servo Press – from a research version to an industrial standard machine" funded by Deutsche Forschungsgemeinschaft (DFG)

The authors would like to gratefully thank Andritz Kaiser GmbH for the collaboration within the framework of the transfer project "T3: The 3D Servo Press – from a research version to an industrial standard machine".

References

[1] B. Möslein-Tröppner, Produktionswirtschaftliche Flexibilität in Supply Chains mit hohen Absatzrisiken; University of Bamberg Press, Bamberg, 2010.

[2] R. Berger, Mastering Product Complexity, Roland Berger Strategy Consultants, Düsseldorf, 2012.

[3] D. Schmoeckel. Developments in Automation, Flexibilization and Control of Forming Machinery. CIRP Annals – Manufacturing Technology. 40/2 (1991) 615-622

[4] G. Schuh, Zukunftsperspektive des deutschen Maschinenbaus, first ed., Apprimus, Aachen, 2012

[5] R. Neugebauer, B. Denkena, K. Wegener, Mechatronic Systems for Machine Tools, CIRP Annals – Manufacturing Technology. 56/2 (2007) 657-686

[6] G. Pahl, W. Beitz, J. Feldhusen, K.H. Grote, Engineering Design, third ed., Springer, London, 2007

[7] H. Haf, Plattformbildung als Strategie zur Kostensenkung. VDI Berichte 1645 (2001) 121-137

[8] I. Baumgart, Modularisierung von Produkten im Anlagenbau; Rheinisch-Westfälische Technische Hochschule Aachen, Aachen, 2004

[9] G. Pahl, W. Beitz, J. Feldhusen, K.H. Grote, Konstruktionslehre, 8th ed., Springer, Berlin, 2013

[10] P. Groche, M. Scheitza, M. Kraft, S. Schmitt, Increased total flexibility by 3D Servo Presses, CIRP Annals – Manufacturing Technology. 59/1 (2010) 267-270

[11] D. Gerwin, Manufacturing Flexibility: A Strategic Perspective, Managing Science. 39/4 (1993) 390-410

[12] A. Bretz, S. Calmano, T. Gally, B. Götz, R. Platz, J. Würtenberger, Darstellung passive, semi-aktiver und aktiver Systeme auf Basis eines Prozessmodells (Representation of Passive, Semi-Active and Active Systems Based on a Prozess Model), unreleased paper by the SFB 805 on http://www.sfb805.tu-darmstadt.de/media/sfb805/f_downloads/150310_AKIII_Definitionen_aktiv-passiv.pdf

[13] B. Röder, C. Dietrich, H. Birkhofer, A. Bohn, Multidimensinal Systems of Concepts – An Approache for a better Communication in the Process of Requierement Aquisition, International Design Conference - Design 2012. (2012) 1091-1100

Applied Mechanics and Materials Vol. 807 (2015) pp 140-149
© (2015) Trans Tech Publications, Switzerland
doi:10.4028/www.scientific.net/AMM.807.140

Submitted: 2015-06-19
Accepted: 2015-08-04

Proved Quality by Online Monitoring and Closed-loop Control of Pin Insertion

Maximiliane Erhardt[1, a *], Carsten Kaschube[1,b] and Markus Menacher[2,c]

[1]Robert Bosch GmbH, 70442 Stuttgart, Germany
[2]Robert Bosch GmbH, 71301 Waiblingen, Germany

[a]maximiliane.erhardt@de.bosch.com, [b]carsten.kaschube@de.bosch.com, [c]markus.menacher@de.bosch.com

Keywords: Press in, metallic pin, polymer housing, automation, assembly

Abstract. Production and assembly processes need to be controlled in order to provide high quality. This paper analyzes the effect and necessity of online monitoring and closed-loop control of pin insertion processes. Pin insertion is an assembly process wherein metallic pins are pressed into polymeric structures. The investigations described in this paper are made at the Robert Bosch GmbH to meet current requirements towards production standards, quality and traceability.

Introduction

Companies are forced to guarantee a cost efficient production with high quality in order to produce in line with national and international market requirements [1]. This leads to highly automated and monitored processes to ensure the required precision and accuracy [2]. Monitoring requires measurements and inspections as well as identification of the parts [3]. The data gained via sensors or the processes and their machines can be fed back into the automation system for a closed-loop control.

The production facilities of the Robert Bosch GmbH are often highly automated and controlled to meet the customer expectations concerning quality and certainty. The gained data as well as the process data itself are collected and stored with the aim to control and improve the components or systems in terms of cost, performance and quality. This applies not only to assembly and production processes but also to the full life-time-cycle of components and systems.

Pin insertion is an assembly process where metallic pins are pressed into polymeric structures. This joint of conductive and isolating material can be applied for example to connectors. Nowadays, most of the connectors are produced with pins which are inserted into the injection molding machine. In the molding process the molten polymer flows around these pins. This insert molding of the pins is a difficult and challenging process. In comparison to insert molding the pin insertion process seems to be less complex and therefore economically advantageous. This paper discusses how the pin insertion process can be included in automated assembly systems via online monitoring and closed-loop control. The aim is to control the process and its uncertainty throughout the assembly to obtain high quality products.

State of the art

The insertion of metallic pins in different materials is investigated by several researchers. These investigations were motivated by the disadvantages of soldering, which is mostly used to create electrical connections [4]. The quantity of heat inserted during the soldering process is the main problem concerning the polymers used in insulation. Besides soldered connections, also press-fit connections can be used [4, 5]. Press-fit connections are solderless joints of pins inserted in plated through holes in circuit boards [4, 5]. To obtain good electrical and mechanical characteristics high radial forces between the press-fit portion and the plated through hole are important [4]. The pins can be rigid or compliant. In a press-fit pin connection both elements (press-fit pin and plated through hole) are exposed to elastic-plastic deformation [4]. In the interference of press-fit pins and plated through holes regions of high stresses appear [5]. Two analyses were performed to understand the source of these stresses: thermal stress was applied on the mounted connection and the process of insertion was analyzed [5]. The plated through hole is surrounded by a polymeric reinforced structure with long glass fiber. When the plated through hole deforms the softer polymer is also stressed. This leads to the assumption that the results of these researches can be transferred to insertion of metallic pin in short glass fiber reinforced polymers. The assembly process with a linear motion for pressing in the pin is similar to the one applied in this paper.

The press-fit connections led to corresponding inventions concerning insertion and assembly of pins. For example the U.S. patent describes the connection of two circuit boards arranged in 90° via bent pins with press-fit zones at each end [6]. The bent pins are fixed in an insulative body, which can carry more than one pin. This connection device does not seem to withstand high mechanical load. With the design of the later on described pin the mechanical linkage could be strengthened.

To assemble a connection system for printed circuit boards one or more contact pins are pressed into a separate flexible contact carrier [7]. These contact carriers can be applied between two circuit boards to connect them via press-fit contact pins. Therefore the press-fit contact pins are designed with two press-fit areas. How the pins are fixed in the contact carrier is not mentioned in detail.

One type of circuit carrier, the so called MID (Molded Injection Devices) is a three-dimensional molded circuit carrier of thermoplastic materials and solderless connections [8, 9]. Contact pins can realize the required electrical connection between electronic devices [8, 9]. These pins have an electrical contact area and a mechanical area without any soldering [8, 9]. They are designed to meet the specifications of MIDs [8]. For the press-fit pins very narrow production tolerances are necessary. The spring-like compressible designs of the pins avoids this requirement. The pin design applied in the investigations of this paper is designed to fit the corresponding polymeric holes and carry mechanical loads without the need of a complex pin design like spring profiles.

The insertion of metallic pins in polymeric structures is described in [10, 11, 12]. In the application of polymer matrix with continuous fiber reinforced composites interlaminar delamination is a major concern [10, 11]. To reinforce the thickness pinning in is an alternative [10]. Before the matrix is cured these so called "z-pins" are inserted orthogonally in the prepreg to pin together individual layers. The holes are not prefabricated. The z-pins need a foam to be supported while mounting and can simply be inserted. The process can also be supported by the use of an ultrasonic insertion gun. A fully automated and controlled joining is possible.

Z-pins are built out of stiff, high strength material such as titanium alloy, steel or fibrous carbon composite with a diameter of 0.2–1.0 mm [11]. They connect the layers like fine nails by a combination of friction and adhesion. The pins investigated in this paper have a rectangular shape with a thickness of 0.6 mm and a width of 1.4 mm. The same effects of friction and adhesion hold them in place but additionally form closure.

The international patent [12] refers to metallic pins which are pressed into polymeric holes. The core statement is an additional process to seal the connection between pin and hole. The sealing of the connection is done by a pin being at least partly coated.

There are only few publications describing details of the design as well as the assembly process of pressing the pin in short fiber reinforced polymers. For example, similar designs of the pins applied in this research can be found in patents [12, 8]. The focus of this research is on the utilization of different materials than in the examples mentioned before. The pin is inserted into short fiber polymers. Thereby the polymer is plastified and form closure is created.

The high quality requirements regarding position and mechanical loads require reliable processes. The purpose of this paper is to discuss how uncertainty, monitoring, occurrence of failure as well as closed-loop control in the pin insertion process can be handled.

Pin insertion

In this paper the term "pin insertion" is used for an assembly method wherein metallic parts are pressed into polymeric structures to create a permanent connection (Fig. 1). In this analysis the metallic parts are pins made out of copper (CuSn6). The housings are made of polymers which are reinforced with up to 40% of short glass fiber. The base polymers are PBT HR (polybutene terephthalate hydrolysis-resistant), PA66 (polyamide 66) or PC (polycarbonate). Possible applications are based on the material properties and the strength of the joint. The pins are conductive to electricity and the housings are electrically isolating. Therefore pin insertion can be applied in electrical connectors for example in electronic control modules.

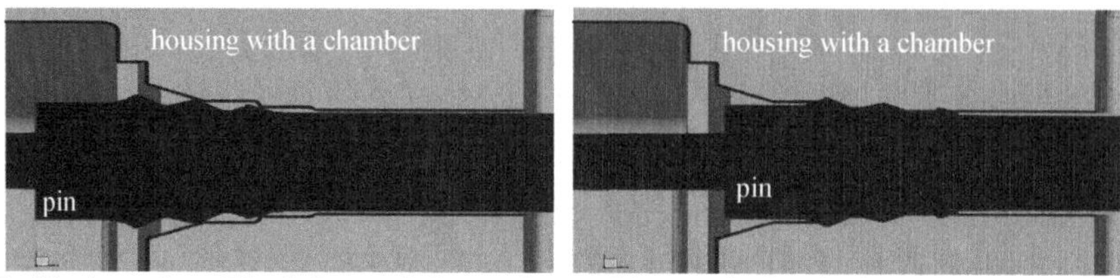

Figure 1: Pin insertion: The left panel shows the position at the beginning of the pin insertion and after the pin is inserted into the chamber positioned by its sidewalls; the right panel shows the position at the end of the pin insertion when the pin has been pressed into the plastified polymer.

Design. The pins (Fig. 2) are designed with two shoulders so the insertion force of the tool can be applied. The pin shaft is divided into two zones, a press-in zone and a fitting zone. The press-in zone is structured at its sides to create the required plastification of the polymeric chamber. The fitting zone is designed to position the pin straight in the camber. Both ends of the pin enable electrical connection. The housing is built with holes, so called "chambers", corresponding to the pin geometry.

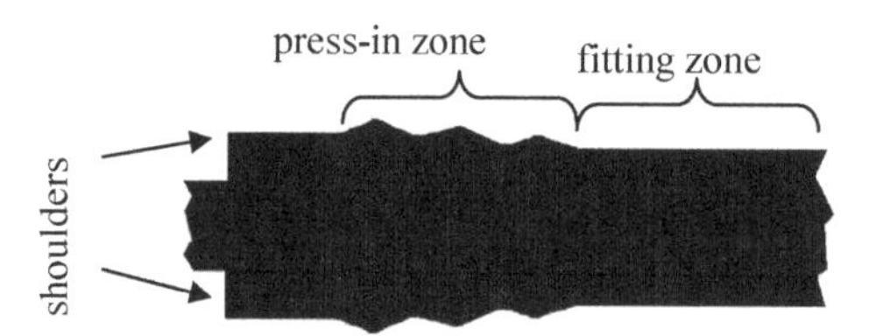

Figure 2: Sketch of the pin design

The pin needs to move linearly during the assembly process. In Fig. 1 the pin insertion starts at the left side. In the first step the pin is aligned to meet the chamber of the housing and to secure that the right position of the pin is guaranteed. The chamber fits the pin. When the pin is pressed further into the chamber the polymer is plastified. Thereby a characteristic pressing force along the press-in path can be observed. It is also possible to press in several pins at once.

Pin insertion as an automatic assembly process is connected to its supply (Fig. 3). The pins can be provided through coils and need to be handled. They need to be cut and fed into the tool for pressing in the pins. The housing can be put into its tool for example via robot. After positioning both tools cause the pin and the housing to meet with the required speed and force. It is not determined which tool is moved as long as the relative velocity of the parts is obtained. It is important that the nominal position is reached by a translational movement. The process can be varied in speed but it is controlled by position. It is possible to measure the force and the distance within the process. Thereby characteristic force-displacement-curves can be measured.

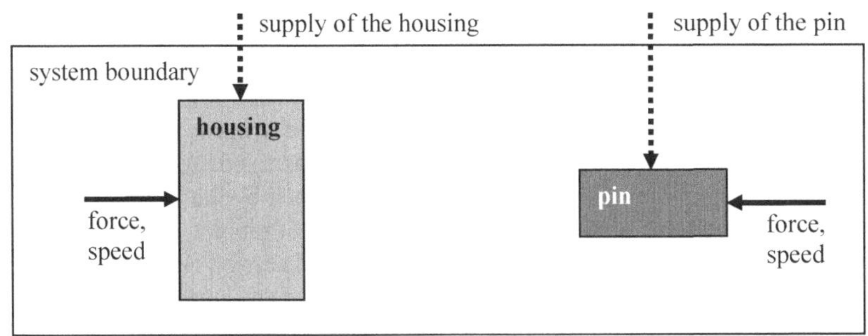

Figure 3: Schematically illustration of the pin insertion as an assembly process

Proved quality by online monitoring and closed-loop control of pin insertion

In this section the general possibilities of monitoring the pin insertion process are outlined as well as the setting of the process, online monitoring, possible defaults and exemplary application of process data directly after one mounting process. A set of experiments and tests was carried out leading to the subsequent preliminary conclusions.

Monitoring. The required data can be gained in two different ways during pin insertion. One way is to get information out of the machine itself. This means there are programs installed which can transform data of machines and sensors into information. A linear motor was used as a testing machine for the pin insertion. Information and diagrams about velocity and path of the axis have been obtained via the software "IndraWorks Ds 11V10.0152".

The other way is the application of additional sensors, which can be installed to gain additional data. For the testing with the linear motor a load cell and a potentiometric position sensor were applied ("DigiForce 9306", Burster Präzisionstechnik GmbH & Co. KG). Fig. 4 shows an example of the recorded process data with said linear motor and monitoring.

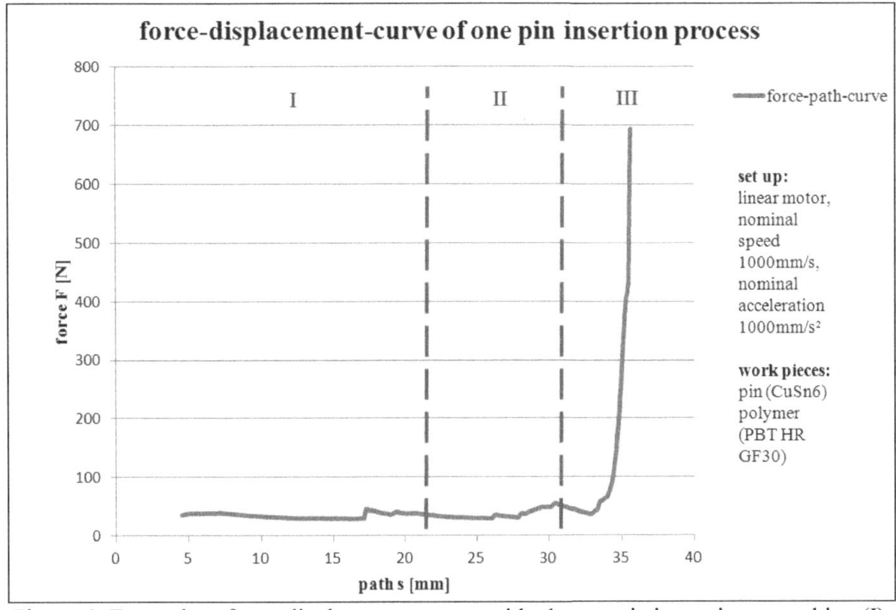

Figure 4: Exemplary force-displacement-curve with characteristic sections: reaching (I), fitting (II), plastification (III)

The data was measured while inserting a pin made of CuSn6 into a housing made of PBT HR reinforced with 30% of short glass fiber. The force-displacement-curve of an insertion process with nominal parts can be divided in three sections (Fig. 4). The first section (I) shows the friction of the tool and the pin as well as the inertia when the tool with the pin is moved towards the housing. In this test the tool with the pin moved while the tool holding the housing was fixed. When the second section (II) starts the pin is at the beginning of the chamber being positioned by the fitting zone. At the transition of the second (II) to the third section (III) the plastification of the polymeric chamber starts. This is the same state as shown in Fig. 1 at the left panel. The third section (III) is where the plastification provides a secure position for the pin in the chamber. At the end of the third section (III) the assembly of the pin is finished, which can also be seen at the right panel of Fig. 1.

Especially in production with high volumes to control the pin insertion process inspections, identification of the parts, sensors for force and position as well as machine data should be provided.

Set-up of the pin insertion process. Each machine and process has its own characteristics. For their setting it is possible to use the online monitoring data. The linear motor used for the investigations had process software installed to collect data concerning velocity and path. With this information it is possible to measure the real velocity within the process of pin insertion. The nominal velocity is reached for approaching the work piece but varied within the pin insertion process caused by deceleration. For example with a nominal acceleration of 1000 mm/s² and a nominal velocity of 1000 mm/s an actual velocity of 93.35 mm/s at the beginning of the pin insertion was measured.

The measurement of the path gave information about the overshoot (Fig. 5). To set the process correctly the overshoot must be considered. The adaption of the actual to the nominal value is possible. Some drivetrains like cam discs do not have a distinctive overshoot. Because pin insertion

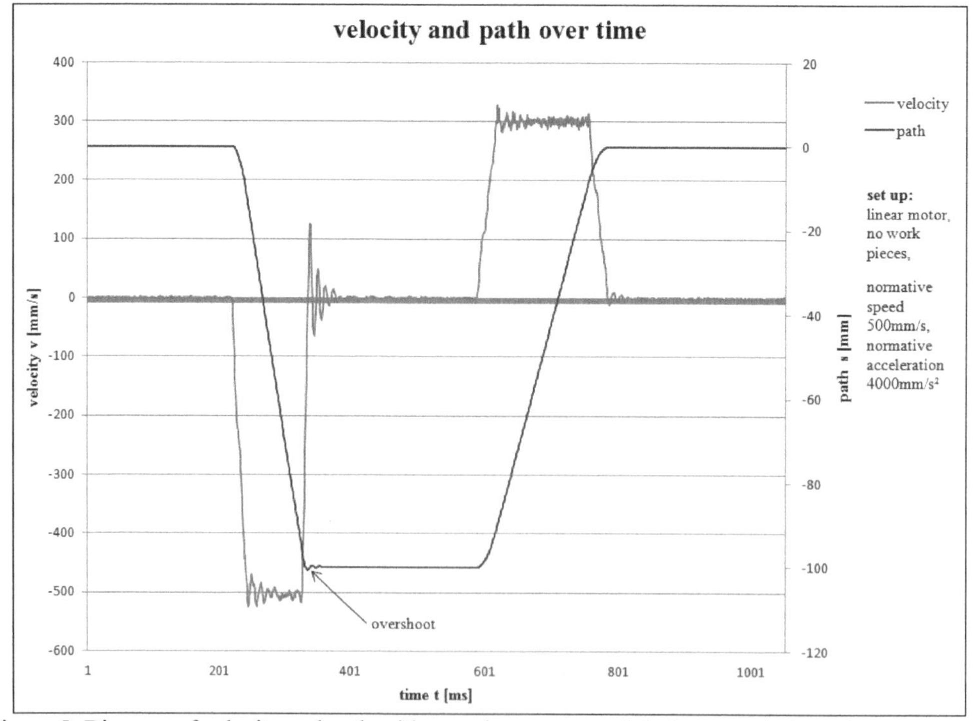

Figure 5: Diagram of velocity and path with overshoot measured during one pin insertion process

as a process is not bound to a specific drivetrain, each design has to be examined separately.

For the setting of the machine it is possible to perform the process without parts. The tests at the linear motor showed that there is nearly no difference in overshooting or speed when the same machine parameters are driven with or without parts. The overshoot without pins was 0.67 mm generated with a speed of 500 mm/s and an acceleration of 1000 mm/s². With the same machine parameters but with pins the overshoot was 0.63 mm. This leads to the assumption that the pin insertion itself only has a small influence on the overshoot.

Force and path of the pin insertion process have to be monitored (Fig. 4). If the depth for pressing in is not fully known there is the possibility to assess the full force-displacement-curve by pressing in the pin a bit too deep into the chamber and stop the process controlled by force. This results in a force-displacement-curve with high force values. Knowing the parameter of the pin insertion, also at the end of the pin insertion process, the depth of pressing in must be reduced. The only disadvantage of this setting method is that the work piece will be destroyed, caused by the pin being pressed in too deep, and therefore needs to be removed out of the process line afterwards.

The advantage is that if the process window is defined in the program, the closed-loop control could be a self adapting process also during production. Rising forces not being part of the process window could be a sign that the chamber in the polymer is not hit satisfactorily and therefore the position of the pin needs to be adopted. This could also be implemented by automatic processes via linear motors but is less reasonable in comparison to manual setting of this position. If the force or path varies from the nominal process window it is possible that the pins or housings are defect or out of required tolerances.

Online monitoring. Values can be measured online in the production [13]. Online measurements can be feed back to the automatic control through interfaces. This makes online adaption, error checking and corrections in ongoing processes possible [13].

The online monitoring of force and path provides several possibilities within the process to gain information about the state of the production (Fig. 6). If the process windows are defined via force and path, the online monitored data can be fed back into the system to adjust the following process of pin insertion. This makes continuous learning possible. Also within one pin insertion process itself closed-loop control makes a reaction possible if there are variations to the nominal force-displacement-curve. For example the depth can be adjusted or an adaption to small variations in the pins or housings can be compensated.

The press fit joining is controllable through force and path monitoring. The exemplary force-displacement-curve of scenario 1 in Fig. 6 shows the ideal force-displacement-curve. This curve is displayed as a watermark in the following scenarios to show the differences in comparison to the ideal one. The curves shown in Fig. 6 are examples to illustrate possible characteristic scenarios.

Scenario 1 is the reference, the plot of an ideal force-displacement-curve.

Scenario2 displays no force over the path at all. If this occurs, the supply did not provide a pin and/or a housing and therefore the assembly could not take place. Because section I is also influenced by the friction of the tool and the inertia, the pin being not present would not have a big influence on that part of the curve. The significant deviation will be seen in the curve in section II and III. If it is detected that no part is handled to be joined, an error message should occur in the system. This information is necessary for the operator of the machine to check for example if the supply machine failed or ran out of parts. This makes a surveillance of previous processes, in this case the supply, possible. If only one part, pin or housing, is in the process it can be reused and assembled as soon as the corresponding part is conveyed.

Scenario 3 is the case when one pin is already mounted and a second one is pressed into the same chamber. It was observed that the pin slides under the one already mounted and is inserted parallel to it. The displayed force-displacement-curve is exemplary. The deviance of the curve starts when meeting the end of the already assembled pin. Therefore the deviance depends on the length of the pin and no generalized characteristic section can be named.

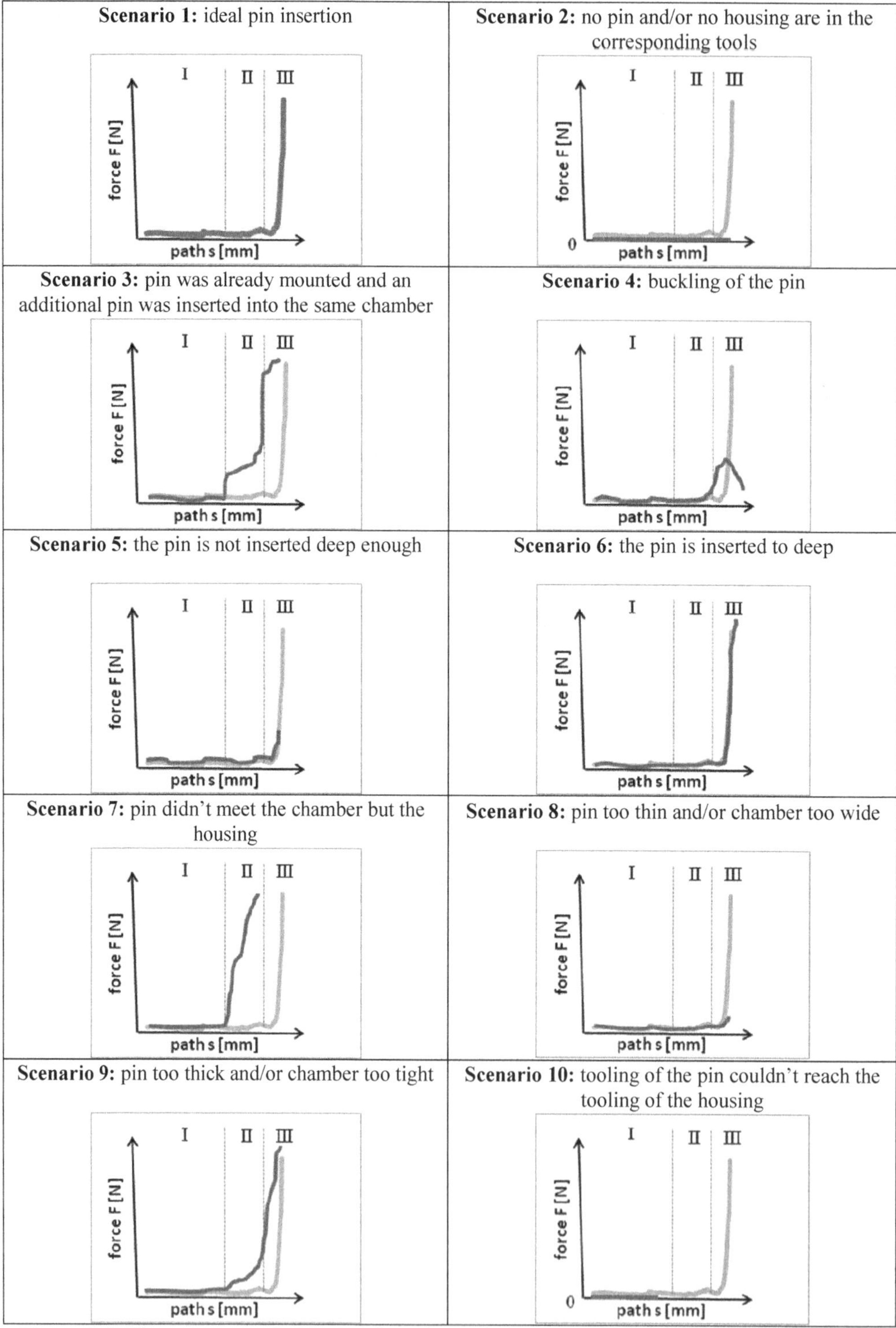

Figure 6: Different characteristic scenario and correlating exemplary force-displacement-curves while one pin insertion process

Scenario 4 shows the buckling of the pin. There are different reasons for buckling. It is quite likely that the pin did not hit the chamber in its correct position. Another explanation is that the chamber is not fabricated correctly, the pin tilted and then buckled. It can be assumed that the pin was not straight and in its pre-bent shape tended to buckle. The pin could also have other defects. The buckling creates a characteristic shape of the curve. A rising force occurs while bending the pin and then slopes down, caused by the folding of the pin. This can happen in section II or III.

Scenario 5 occurs if the pin is not inserted deep enough. This is visible in the section III. The assembly process does not need to stop because there is the opportunity to compensate depth by closed-loop control. But the pin not being pressed in deep enough can also have other reasons. One would be that the housing or the pin does not meet their specifications, for example pins are too short. If this occurs, even if the pin is inserted through feed-back control with the ideal force, the position tolerance might cause defaults. This could be controlled after the pin insertion process by monitoring the assembled parts.

Scenario 6 shows a pin being pressed in too deep, seen in section III. If this occurs then the insertion zone might get damaged. This force-displacement-curve can be applied for setting the insertion process as mentioned before.

Scenario 7 is characteristic for its rising force in section II. If this occurs, the pin cannot be inserted in the hole and therefore rams the housing.

Scenario 8 displays an ideal curve in section I and II. It is possible that the force in section II is already lower than the ideal on. Section III shows a curve with lower forces but the whole displacement. It is a sign that the fitting is not matching. This can be caused by a pin being too thin and/or chamber too wide.

Scenario 9 is opposite. The pin is too wide and/or chamber too tight. The forces in section II and III are significant higher than the ideal ones. It can also result in a buckling of the pin. In this case the force-displacement-curve looks more like scenario 4.

Scenario 10 is the braking and stopping of the tool on its way to insertion, which is shown by no force in section I. This might happen for example because one pin fell underneath the tooling, something got broke or the machine needs maintenance.

If parts, which are about to be mounted, are damaged in forehand or do not meet their required tolerances, it might result in many different force-displacement-curves. Via windows and envelops the force-path-curves can be interpreted for the assembly process to define parts as OK or not OK [1]. If the deviation is too big or the nominal values are not reached than the program has to tag these parts as not OK. Nearly all mentioned scenarios (Fig. 6), besides scenario 5, result in defective parts. Although the possible failures mentioned before can not all be prevented through monitoring, they can at least be identified. With the gained information corresponding actions can be initiated.

The monitoring of the path gives information about the press-in depth. This is directly transferable to the position of the pin tip concerning set points. This information is a proof for a part being in specification concerning the requirement of this position.

After pin insertion. After the assembly the mounted part can be passed to the following process station. The pin depth is measured in the insertion process and can be used for example to select a matching contact element for the pin. Therefore the data acquired during pin insertion is applied. Chances of process and equipment parameters can be monitored and fed into the control of the production system which must exhibit adaptability to fluctuations and therefore generates a self-learning process [14].

Testing the joint after the assembly process through tactile or visual sensors is also possible. A vision system can detect whether the pin is mounted or not. It can also detect if the pin is in the correct position and matches the demanded tolerances. Tactile sensors can check if the pin is present. If the tactile sensor would create too high forces on the pin while testing, a needle can be connected to the pin to measure the presence via electric resistance. The results from this testing can be fed back into the system to adjust the pin insertion of the following parts.

Summary

Pin insertion is an assembly process where control has to take place. Closed-loop control would help even more to guarantee high quality, high production standards and surveillance with the aim of optimization.

To generate the required traceability of each part online-monitoring and closed-loop control can be applied. The applications presented in this article show a rising effort of programming the feed-back control. On the one hand it increases process stability. On the other hand clear limits of requirements should be known and defined in the program. The accuracy and precision of processes is highly improved by using closed-loop control. A control system without feedback control is usually a cheaper alternative but has to be fed with algorithm concerning the mechanical structure, its deformation behavior and forces within the process [2]. To gain this information is also effortful, but needed for a standard control system. Therefore it should be balanced beforehand whether a control needs a feed-back or not.

Acknowledgements

The authors would like to acknowledge G. Finnah from the Center of Competences Plastics of the Robert Bosch GmbH for starting to develop the process of pin insertion. Also, the authors are grateful to B. Beiermeister and A. Ziegler from the Center of Competence Manufacturing Technology Plastics of the Robert Bosch GmbH. The corresponding project team of the Robert Bosch GmbH with special thanks to B. Riethmüller, M. Reinhard, A. Lux, V. Barnstorff, J. Vollert and S. Althaus for their support. The authors also would like to acknowledge T. Schmidt-Sandte, P. Traub and R. Maier of the Robert Bosch GmbH and P. Groche of the Technische Universität Darmstadt for their support. H. Schaffert provided expertise for automation in production and processes.

References

[1] B. Lotter, H.P. Wiendahl, Montage in der industriellen Produktion, Springer-Verlag, Berlin, Heidelberg, 2012, pp. 222-227

[2] F. Klocke, T. Kohmäscher, Autonomes Frässystem, in: T. Pfeifer, R. Schmitt (Eds.), Autonome Produktionszellen - Komplexe Produktionsprozesse flexibel automatisieren, Springer-Verlag, Berlin, Heidelberg, 2006, pp. 89-177

[3] S. Hesse, Automatisieren mit Know-how, Hoppenstedt Bonnier Zeitschriften, Darmstadt, 2002, pp. 21-316

[4] T. Kanai, Y. Ando, S. Inagaki, Design of a Compliant Press-Fit Pin Connection, IEEE Transactions on Components, Hybrids, and Manufacturing Technology, Vol. 8, 1985, pp. 40-45

[5] R. Goel, E. Guancial, Stress Distributions Around an Interference-Fit Pin Connection in a Plated Through Hole, IEEE Transactions on Components, Hybrids, and Manufacturing Technology, Vol. 3, 1980, pp. 392-402

[6] W. Woody, U.S. Patent 6,592,382 B2 (2003)

[7] E. Kotowicz, International Patent WO 08095816 A1 (2008)

[8] M. Reichenberger, M. Eisenbart, R. Meier, Alternative interconnection technologies for MID, MID, Molded Interconnect Devices, International Congress 4, Bamberg, 2000, pp. 285-297

[9] R. Sander, U.S. Patent 7,946,861 B2 (2011)

[10] A.P. Mouritz, Review of z-pinned composite laminates, Composites Part A: Applied Science and Manufacturing, Vol. 38 (12), 2007, pp. 2383-2397

[11] I.K. Partridge, D.D.R. Cartié, Delamination resistant laminates by Z-Fiber® pinning: Part I manufacture and fracture performance, Composites Part A: Applied Science and Manufacturing, Vol. 36 (1), 2005, pp. 55-64

[12] R. Sander, International Patent WO 077819 A1 (2009)

[13] E. Hering, K. Bressler, J. Gutekunst, Elektronik für Ingenieure und Naturwissenschaftler, fifth ed., Springer-Verlag, Berlin, Heidelberg, 2005, pp.298-319

[14] D. Stokic, S. Scholze, J. Barata, Self-learning embedded services for integration of complex, flexible production systems, IECON 37th Annual Conference of IEEE Industrial Electronics, Melbourne, 2011, pp. 415-420

CHAPTER 5:

Uncertainty in High Precision Manufacturing Processes

Applied Mechanics and Materials Vol. 807 (2015) pp 153-161
© (2015) Trans Tech Publications, Switzerland
doi:10.4028/www.scientific.net/AMM.807.153

Submitted: 2015-04-15
Accepted: 2015-08-04

Control of uncertainty in high precision cutting processes: Reaming of valve guides in a cylinder head of a combustion engine

Christian Bölling[1,a *], Sebastian Güth[1,b] and Eberhard Abele[1,c]

[1]Institute of Production Management, Technology and Machine Tools (PTW),

Technische Universität Darmstadt, Otto-Berndt-Str. 2, 64287 Darmstadt, Germany

[a]Boelling@ptw.tu-darmstadt.de, [b]Gueth@ptw.tu-darmstadt.de ,[c]Abele@ptw.tu-darmstadt.de

Keywords: uncertainty, automotive industry, hole finishing, reaming

Abstract: In production processes uncertainty has a great impact on the product quality as well as production costs. In automotive industry the reaming of valve guides in a cylinder head of a combustion engine is a quality determining process. Due to the force fitting of the valve guides into the cylinder head the final reaming process has to deal with increased uncertainty. On the other hand, the finished hole is closely tolerated. To ensure the process reliability the admissible tolerance must be strictly met even in case of uncertainty. This paper presents a possibility to achieve process reliability by a modified process chain with an additional pilot reaming tool. Thereby, the effect of different cutting edge preparation is also analyzed. Further, the influence of the pilot reamer geometry on the final hole quality is investigated.

Introduction

Uncertainty in machining processes significantly influences the product characteristics during its lifecycle. In automotive industry hole making and finishing are frequently used machining processes. Those operations consume 89% of overall production time of a cylinder head whereby half of them are classified as fine machining operations [1]. Especially in fine machining the control of uncertainty is an important key to ensure and even enlarge the product lifetime. These machining processes are often carried out in the final stages of the overall production process. Any variances or failed tolerances due to uncertainty lead to waste of product respectively – if this has not been recognized by the quality assurance – to a failure during the life cycle. This always leads to a loss of money.

The Collaborative Research Center (CRC) 805 which is sponsored by the German Research Foundation (DFG) deals with the analysis and control of uncertainty in load-carrying structures in mechanical engineering. At the Institute of Production Management, Technology and Machine Tools (PTW) the research focuses on uncertainty within the process chain hole making and hole finishing. Thereby, the attention is turned to the reaming process. It is the quality determining process and due to the foregoing pre-manufactured hole and its process errors the final reaming process has to deal with increased uncertainty.

This article addresses uncertainty in valve guide reaming in a cylinder head of a combustion engine. In a first step, experimental tests are performed to verify the influence of process faults on the quality of the reamed valve guide. Based on those results, the process chain and the machining strategy are modified to increase the process reliability. Furthermore, different reaming tool geometries are investigated. The result is an increased control of uncertainty in the machining process of valve guides in a cylinder head.

State of research

In general, uncertainty during the production process can be classified into different categories which can be derived from the cause and effect diagram according to Ishikawa [2]. The main categories, as shown in Fig. 1, are 'milieu', 'man', 'method', 'machine' and 'material' and are often named as the 5Ms.

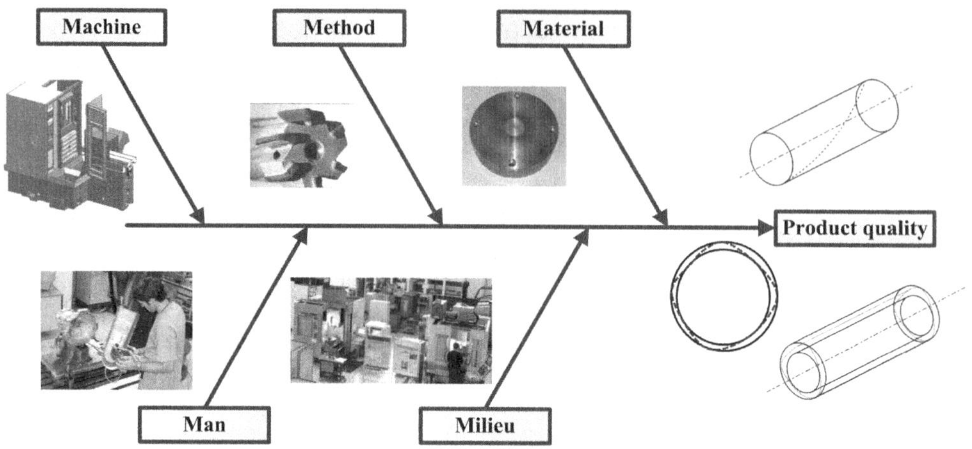

Figure 1: Main uncertainty factors during production process (5Ms)

Thereby, the category 'milieu' includes the uncertainty from the environment such as temperature gradients in plants, excitations by other machines or air humidity effects. On the other hand, all effects caused by human actions like programming errors or workpiece clamping errors are summarized by the category "man". Meanwhile, the research at PTW focuses on the three categories 'method', 'machine' and 'material'. Within the category 'method' uncertainty correlated to the production operation, e.g. reaming, drilling, the variation of process parameters like cutting speed and feed and also the use of different tool geometries is included. Another source of uncertainty is the machine tool itself. Here, the axes positioning and repeatability accuracy, internal heat and vibration sources have to be taken into account. Finally, the workpiece material properties have to be treated as variables due to material inhomogeneity and strength gradients.

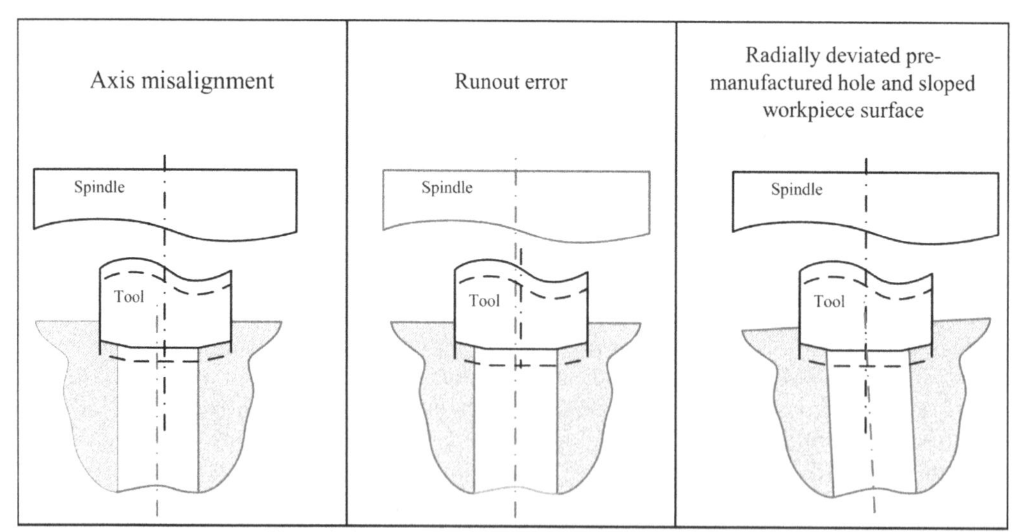

Figure 2: Uncertainty factors during the reaming process

The uncertainty within the process chain hole making and hole finishing can be mainly described by three occurring process errors (Fig. 2). Firstly, an axis misalignment between the axis of the pre-manufactured hole and the spindle-tool-system can appear. This leads to a revolving chip cross-section at the cutting edge during the machining process and finally to varying cutting forces. Furthermore, a runout error, which is classified as an alignment between the spindle-workpiece-axis and the tool axis, affects a constantly changed chip cross-section during the cutting process. Today, this process error can be eliminated by using steerable tool holders in industrial applications. These two uncertainty factors have been considered and investigated in various research projects [1,3,4,5]. The third main uncertainty is a radially deviated pre-manufactured hole. This leads to a varying, increasing chip cross-section across the reaming depth. According to standard literature, a reaming process cannot influence the position or form errors caused by previous machining steps, e.g. drilling [6,7]. Investigations executed within the CRC805 show that this assumption is not applicable for all cases [8,9]. At the moment, the occurrence of a radially deviated pre-hole in combination with a sloped workpiece surface is also under investigation within CRC805.

Influence of an axis misalignment on the hole quality in case of an ordinary process chain

From further investigations, which have been carried out in CRC805, it is known that an occurring axis misalignment deviates the reaming tool significantly, especially during the entrance phase of the tool. Later on, this deviation spreads nearly linear along the machining depth. With this knowledge an optimized reaming tool geometry has been developed which reduces the deviation of the tool even in case of a present axis misalignment. The investigation has shown that a lead angle of κ = 90° provides the best results regardless of the number of blades [5]. This finding is transferred to the following test series. The ordinary process chain in case of reaming of valve guides in a cylinder head is shown in Fig. 3. The representation is based on the process model developed within the CRC805 [10,11,12].

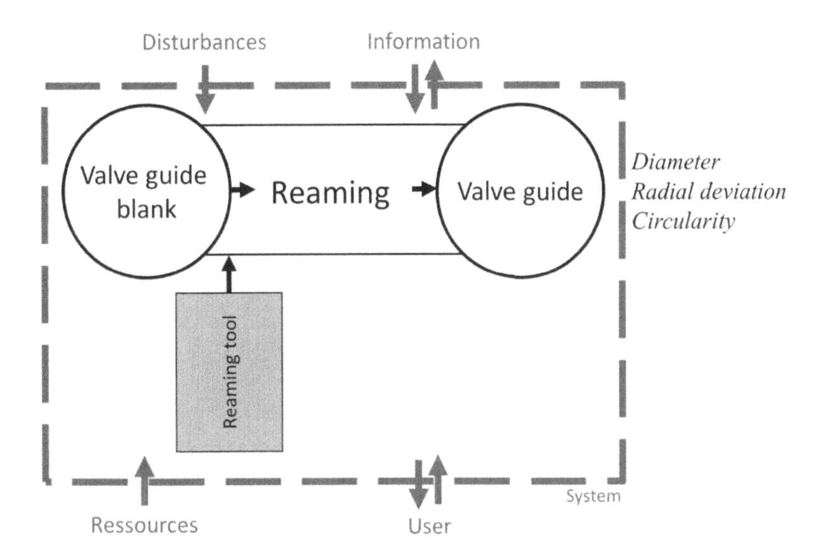

Figure 3: Machining strategy in case of an ordinary reaming process

All experimental investigations are carried out with reaming tools possessing six blades and a slightly uneven spacing according to [1]. The tools are made of cemented carbide and coated with a TiAlN-based coating. The cutting edges are prepared in two different ways: The cutting edges of the first tool are ground sharply and the cutting edges of the second one have an edge hone of about 10 µm. From investigations done with twist drills it is known that an edge hone can improve the stability of a tool and therefore results in an increased tool life [13]. In case of the reaming process the influence of an edge hone on the quality of the reamed hole is currently part of the investigation.

As a process fault an axis misalignment of 80 µm in x-direction between the spindle-tool axis and the workpiece axis is applied. The choice to select this value is made due to specifications in industrial applications. Usually, values up to 0.1 mm in any axial direction are allowed in machining of valve guides in the cylinder head. On the other hand, the radial deviation of the reamed hole has to be below 20 µm measured from top to bottom. Furthermore, the entrance strategy is varied by an adaption of the feed rate during the entrance period of the tool into the workpiece (Fig. 4). Until a depth of 5 mm the feed is reduced on the one hand while the second setup provides a constant feed. A reduced feed rate also decreases the mechanical load at the cutting edges during the unsteady phase at the tool entrance which could help to lower the radial deviation. The diameter of the reamed holes is expected to be 6 mm. All tests are carried out on material GJL-250 which has similar material properties compared to the common valve guide. Those are mainly made of sintered steel alloys with embedded graphite. In the forefront of the investigation benchmark tests with both materials have been carried out and the results show that the tool behavior in both materials is comparable. For the assessment of the tools and their process capability, the resulting hole quality is evaluated by geometric product specifications (diameter, roundness, radial deviation) according to DIN EN ISO 1101 [14].

The results show in case of an ideal process that the radial deviation of the reamed hole achieves values of about 18 µm (Fig. 4). The circularity of the reamed holes reaches values of about 4 µm. Although, the edge preparation does not have any influence on both the radial deviation and the circularity. The diameter on the other hand differs significantly dependent on whether the tools have a sharp or rounded cutting edge geometry. In case of sharp cutting edges the diameter of the reamed hole reaches values between 5.994 and 5.996 mm while for rounded cutting edges the diameter is only 5.986 mm. The diameters of both tools in new condition have been measured to 6.000 mm.

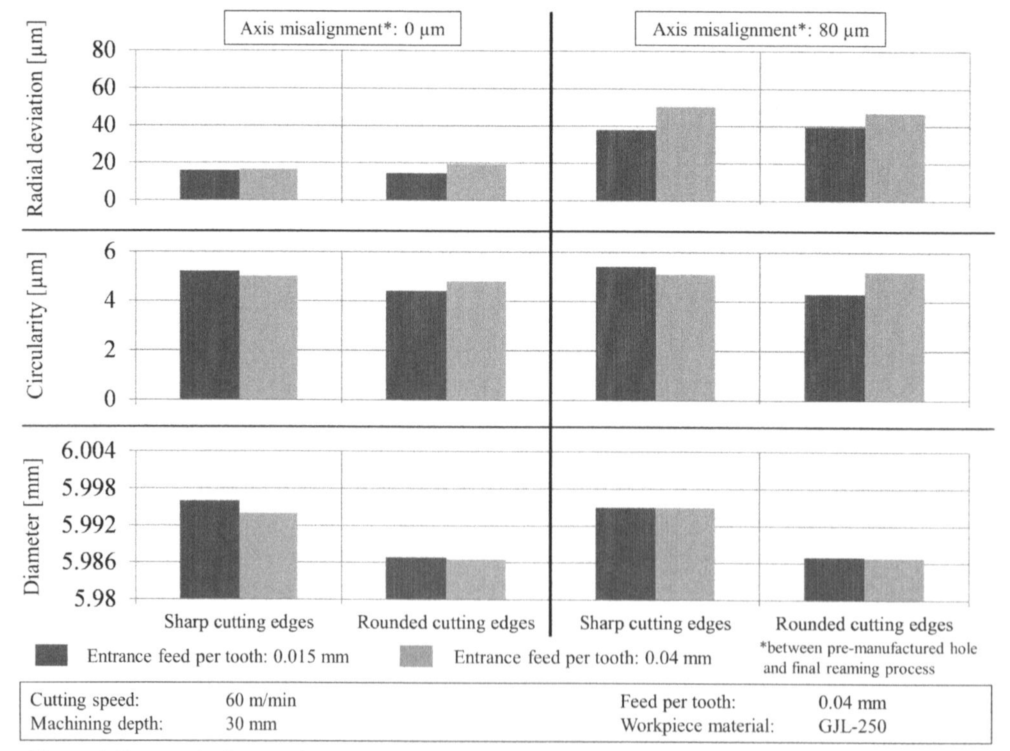

Figure 4: Test results for reaming process in case of an ideal process and with presence of axis misalignment

However, the measured tool diameter after machining tests differs considerably. The diameter of the tool with sharp cutting edges is surveyed to 5.996 mm and therefore matches the measured

diameters of the reamed hole. On the other hand the diameter of the tool with round edges decreases to 5.991 mm. The measured diameter of the reamed hole on the other hand is only 5.986 mm. This can be explained by taking the elasticity of the workpiece material into account. The workpiece material is partly formed instead of cut. This causes a resilience of the material after the machining process and therefore decreases the effective diameter of the reamed hole. It can also be stated that the entrance feed does not have any significant influence on the reachable quality. Taking the axis misalignment into account the result varies significantly in case of the reached radial deviation while the resulting diameter and the circularity are not being influenced. However, the radial deviation of the reamed hole increases considerably to values of about 40 to 50 μm. This is caused by the uneven stock at the cutting edges due the axis misalignment which leads to a resulting radial force. This force is bending the tool during the entrance of the tool into the workpiece. In consequence, the radial deviation rises with increasing machining depth. The cutting edge preparation has no influence on the deviation of the tool while an increased entrance feed generally leads to slightly worse results. This can be explained by considering the cutting forces during the entrance phase of the tool. An increased feed causes also higher cutting forces. In case of a misalignment the resulting radial force thus increases and leads to a grown radial deviation.
From the results of the experimental investigation can be concluded that the ordinary process chain with only a final reaming process is not resistant against process faults like an axis misalignment. Even the optimized reaming tool geometry according to [5] does not improve the results concerning radial deviation in a satisfactory manner.

Extension of process chain hole making – hole finishing

One possibility of answering the problem is the implementation of an additional pilot reaming process into the process chain (Fig. 5). This additional process is carried out before the final reaming process.

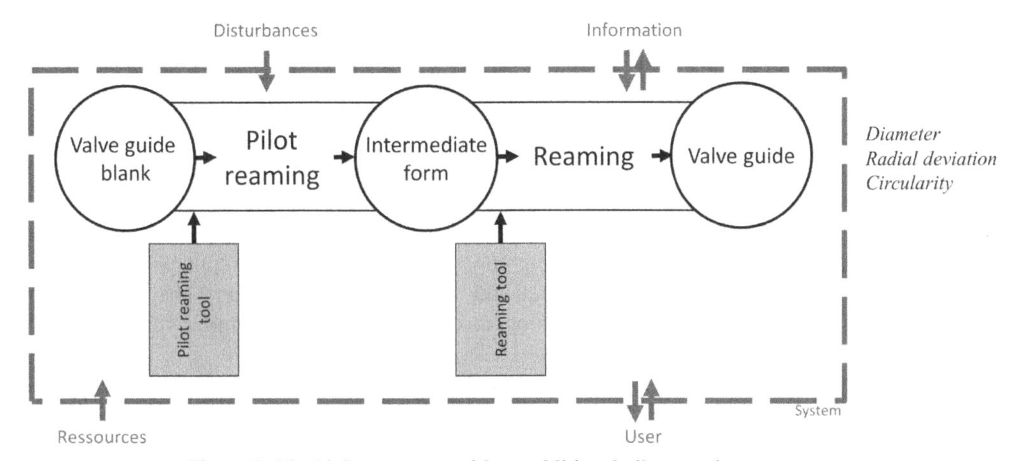

Figure 5: Machining strategy with an additional pilot reaming process

The characteristics of this process are the application of a brief reaming tool especially developed for this task. The unsupported length l of the tool is only 17 mm compared to 60 mm of the final reaming tool (Fig. 6). The deflection of the tool at the top due to a resulting radial force can be approximately calculated by using the Euler-Bernoulli beam theory:

$$w(l) = \frac{F_{res} \times l^3}{3 \times E \times I}$$

The resulting radial force F_{res} arises due to an uneven stock at the cutting edges of the tool due to the axis misalignment (Fig. 6). The length l of the tool influences the deflection with the power of three. Therefore, the deflection of the pilot reaming is in theory about 40 times smaller in comparison to the final reaming tool.

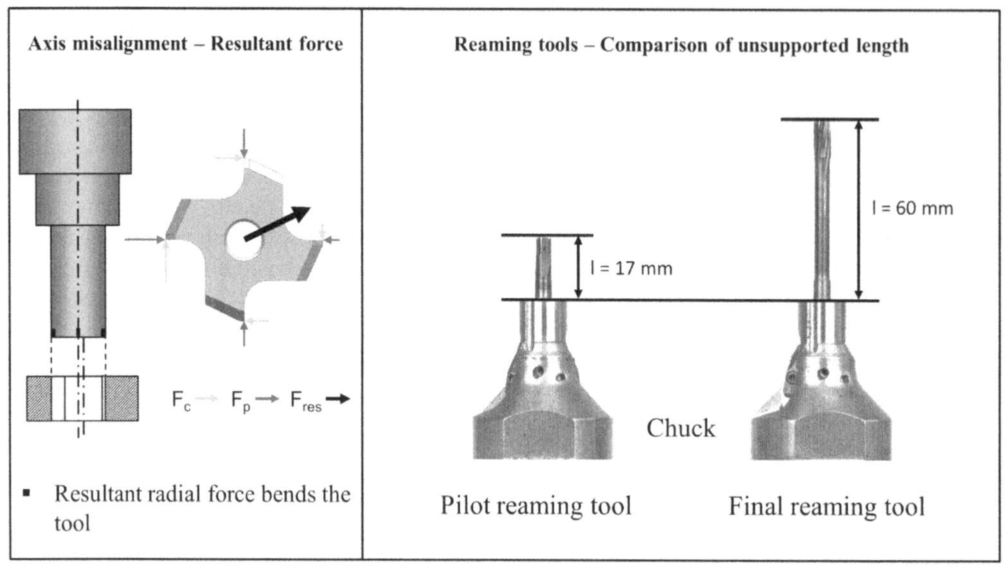

Figure 6: Resulting radial force and comparison of unsupported length of reaming tools

The depth of the pilot hole is set to 4 mm and furthermore the diameter of the pilot reaming tool is slightly smaller compared to the final reaming tool. The results from experimental investigations (Fig. 7) indicate that the application of a pilot reaming has a significant influence on the quality of the reamed hole. The radial deviation achieves values noticeably below 10 µm in case of an ideal process. This is just half of the value reached in an ideal process without using a pilot reaming tool. The most significant improvement is attained in the event of an axis misalignment of 80 µm. Here, the presence of an axis misalignment does not result in an obvious increase of the radial deviation of the reamed hole. The values still do not exceed the boundary of 12 µm even if it can be stated that rounded cutting edges show slightly worse results. The adjustment of the stock to be removed by the final reaming tool in combination with reduced process forces during the entrance phase decreases the tool bending significantly. The sharp cutting edges of the reaming tools apparently also lead to an improved circularity in both cases without and with axis misalignment. The final diameter however is not influenced by the pilot reaming tool and also reaches similar results as in the previous experimental run.
Taking those results into account it can be reasoned that the realized modification of the process chain hole making and hole finishing by adding an additional pilot reaming process increases the process reliability in an arbitrative way. The uncertainty concerning the process fault of an axis misalignment between the pre-manufactured hole and the reaming process is reduced greatly. A disadvantage of this extension on the other hand is the increased cycle time in production due to the additional machining process.

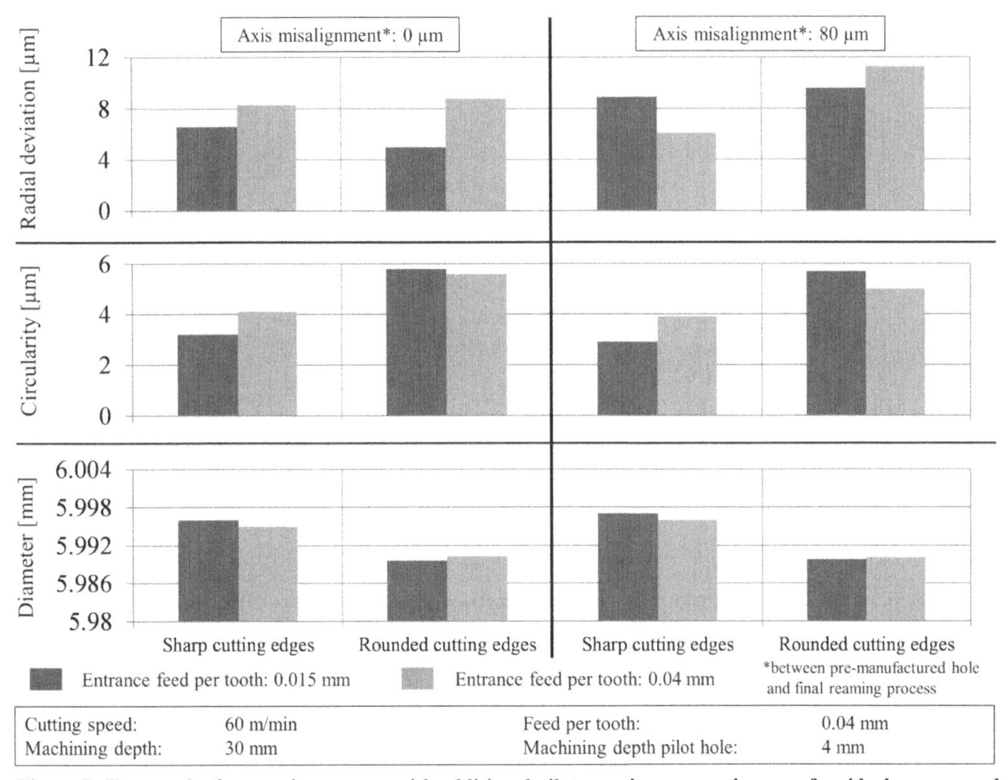

Figure 7: Test results for reaming process with additional pilot reaming process in case of an ideal process and with presence of axis misalignment

Influence of machining parameters and pilot reamer geometry on the quality of the reamed hole

In a further step the influence of machining parameters, e.g. the feed and the pilot reamer geometry on the quality of the reamed hole, have been studied. The feed per tooth of the final reaming tool is varied from 0.04 mm up to 0.16 mm and the pilot reaming tool geometry is changed from a straight to a tapered geometry with varied taper angles (Fig. 8). From the first to fourth geometry the taper angle is increased stepwise. All experimental investigations are performed with an axis misalignment of 80 µm between the pre-manufactured hole and the pilot reaming and final reaming process.

The results show that the pilot reaming tool geometry itself has only a minor influence on the radial deviation of the reamed valve guide (Fig. 8). The radial deviation of the reamed hole, also evaluated in a machining depth of 30 mm, is in every studied case below 15 µm. These results match with the observation made in the previous section. The cutting speed of the final reaming tool also has only little influence on the radial deviation. A slight increase in radial deviation from low cutting speed to the point of high cutting speed can be noticed. More substantial influence on the other hand has the feed per tooth of the final reaming tool on the radial deviation of the reamed hole. Here, a correlation between increasing feed per tooth and at once ascending radial deviation is clearly visible for all tested pilot reaming tools.

Summarized, the geometry of the pilot reaming tool and also the cutting speed of the final reaming tool have an minor significance compared to the chosen feed per tooth of the final reaming tool. The machining parameters therefore need to be selected carefully to reduce uncertainty within the valve guide reaming process in cylinder heads.

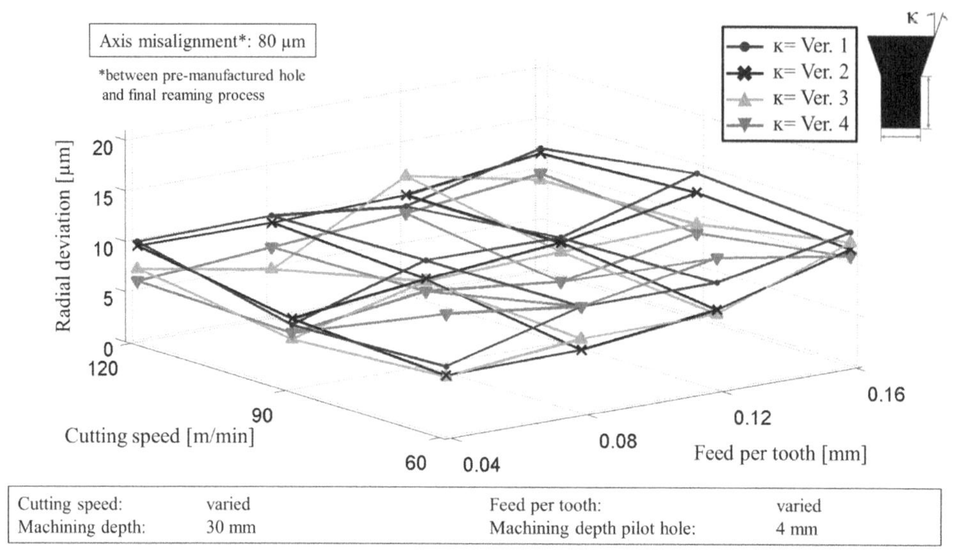

Cutting speed:	varied	Feed per tooth:	varied
Machining depth:	30 mm	Machining depth pilot hole:	4 mm

Figure 8: Influence of different pilot reaming tool geometries on radial deviation of the reamed valve guide

Conclusion and outlook

The presented method of extending the process chain hole making and hole finishing by an additional pilot reaming process increases the process reliability in valve guide reaming significantly also in case of a present process fault. The radial deviation, which is one of the key quality parameter, is unaffected by occurring process faults when using a pilot reaming tool. This enables the operator to reduce and finally control the uncertainty within the reaming process. An optimized reaming tool geometry on its own does not allow to control of uncertainty in case of present process faults. Furthermore, the additional quality parameters circularity and diameter of the reamed holes are not influenced by a process fault axis misalignment. The investigation also shows the dependence of the reamed hole quality on the shape of the cutting edges. Generally, tools with a sharp cutting edge geometry offer more advantages compared to tools with rounded edges. In this case, the diameter of the reamed holes and the tool accord to each other and the circularity is especially in case of an additional pilot reaming process considerably improved.
Based on the achieved results further research will be conducted. Within the CRC805 the reaming process based on realized research is applied to manufacturing processes of components for the common SFB demonstrator. Next to the valve guide reaming process the valve seat plunging is under investigation and an extension to a parallel machining process will be carried out. Thereby, also a dynamic simulation of the plunge cutting process in addition to the already existing valve guide reaming model will be developed.

Acknowledgement

The authors would like to thank the German Research Foundation (DFG) for funding the research activities at the Collaborative Research Centre (CRC) 805 – Control of Uncertainty in Load-Carrying Structures in Mechanical Engineering.

References

[1] F. Koppka: A contribution to the maximization of productivity and workpiece quality of the reaming process by analyzing its static and dynamic behavior, Shaker Verlag, Aachen (2009)

[2] K. Ishikawa: Guide to Quality Control, Asian Productivity Organization, Tokyo (1992)

[3] O. Bhattacharryya, S.G. Kapoor, R.E. DeVor: Mechanistic model for the reaming process with emphasis on process faults, International Journal of Machine Tools and Manufacture, 46 (2006), page 836-846

[4] O. Bhattacharryya, M.B. Jun, S.G. Kapoor, R.E. DeVor: The effects of process faults and misalignment on the cutting force system and hole quality in reaming, International Journal of Machine Tools and Manufacture, 46 (2006), page 1281-1290

[5] T. Hauer: Modellierung der Werkzeugabdrängung beim Reiben – Ableitung von Empfehlungen für die Gestaltung von Mehrschneidenreibahlen (engl. Modeling of tool deflection in reaming – Derivation of recommendations for the design of multi-blade reaming tools), Shaker Verlag, Aachen (2012)

[6] E. Pauksch, S. Holsten, M. Linß, R. Tikal: Zerspantechnik (engl. Cutting Technology), Vieweg + Teubner, Wiesbaden (2008)

[7] F. Klocke, W. König: Fertigungsverfahren I (engl. Manufacturing Processes 1), Springer-Verlag, Berlin Heidelberg (2011)

[8] E. Abele, T. Hauer, M. Haydn, C. Bölling: Reduzierte Unsicherheit bei der Bohrungsfeinbearbeitung - Neue Erkenntnisse zum Vorbohrungseinfluss auf den Reibprozess (engl. Reduced uncertainties during hole-finishing – New knowledge concerning the influence of pre-drilled holes on reaming process), wt Werkstattstechnik online, 101 (2011), page 81-87

[9] M. Haydn, T. Hauer, E. Abele: Methods for the control of uncertainty in multilevel process chains using the example of drilling/reaming, Applied Mechanics and Materials, 104 (2012), page 103-113

[10] R. Engelhardt, J. F. Koenen, G. C. Enss, A. Sichau, R. Platz, H. Kloberdanz, H. Birkhofer, H. Hanselka: A model to to categorise uncertainty in load-carrying systems, 1st MMEP International conference on modelling and management processes, Cambridge/UK, 19-20 July 2010, page 53-64

[11] H. Hanselka, R. Platz: Ansätze und Maßnahmen zur Beherrschung von Unsicherheit in lasttragenden Systemen des Maschinenbaus (Approaches and procedures to control uncertainty in load-carrying structures in mechanical engineering), Konstruktion, 11/12 (2010), page 55-62

[12] A. Bretz, S. Calmano, T. Gally, B. Götz, R. Platz und J. Würtenberger: Darstellung passiver, semi-aktiver und aktiver Maßnahmen auf Basis eines Prozessmodells (Representation of passive, semi-active and active systems based on a process model), unreleased paper by the SFB 805 on http://www.sfb805.tu-darmstadt.de/media/sfb805/f_downloads/150310_AKIII_Definitionen_aktiv-passiv.pdf

[13] K. Risse: Einflüsse von Werkzeugdurchmesser und Schneidkantenverrundung beim Bohren mit Wendelbohrern in Stahl (engl. Influence of tool diameter and cutting edge preparation when drilling steel using twist drills), Shaker Verlag, Aachen (2006)

[14] DIN EN ISO 1101: Geometrical Product Specifications (GPS) – Geometrical tolerancing – Tolerances of form, orientation, location and run-out, Beuth Verlag, Berlin (2013)

Applied Mechanics and Materials Vol. 807 (2015) pp 162-168
doi:10.4028/www.scientific.net/AMM.807.162
Submitted: 2015-04-15
Accepted: 2015-08-04

Control of uncertainty based on machining strategies during reaming

Sebastian Güth [1,a], Andreas Bretz [1,b *], Christian Bölling [1,c], Andreas Baron [1,d] and Eberhard Abele [1,e]

[1] Institute of Production Management, Technology and Machine Tools (PTW),

Technische Universität Darmstadt, Otto-Berndt-Straße 2, 64287 Darmstadt

[a] gueth@ptw.tu-darmstadt.de, [b] bretz@ptw.tu-darmstadt.de, [c] boelling@ptw.tu-darmstadt.de, [d] andreas.baron@stud.tu-darmstadt.de, [e] abele@ptw.tu-darmstadt.de

Keywords: Reaming, Uncertainty, Machining Strategy

Abstract. The bore quality is influenced by machine's accuracy, work piece and tool errors as well as handling errors. This uncertainty has a huge impact on the quality of the finished bore. During reaming, the majority of tool deflection arises during the unsteady process phase. The entry phase is regarded as the biggest influencing factor for the reamer's deflection with increasing bore depth. This paper examines the influence of various entry strategies on the bore quality during the reaming process.

Introduction

In order to meet the emission limits, the automotive industry focuses more and more on downsizing [1]. Improving engine's control, adding a boosting device, reducing number of cylinders and increasing the surface properties accompanies downsizing. To gain smoother surfaces, especially the reaming process for bore machining is a factor that has to be taken into account. Tolerance limits for highly stressed cylinder heads allow a maximum deviation for the valve guides circularity for a single measuring circuit of 5 µm and a maximal deviation of 8 µm regarding the cylindrical shape [2]. Therefore, in most cases industrial serial production uses single-edged reamers. Consisting of a cutting edge and several guide rails, they generate a smooth surface while machining takes place with high rotational speed but low feed rate [3]. Due to increasing sales figures focus is on the reduction of cycle times. This is the reason why nowadays multibladed reamers are tested and used. Multibladed reamers consist of more than one cutting edge, whereas the biggest difference to single-edged reamers is that there are no external guide rails. Instead, circular grinding chamfers on the edges are used to guide the reamer. Multibladed tools are characterised by main cutting edges and secondary cutting edges for material removal. These cutting edges are arranged on the circumference. The simultaneous involvement of all edges during the reaming process leads to a distribution of the total feed per rotation on all edges. A distribution of the functions (rough machining, guiding, finishing) on different edges leads to improved results [4]. Additionally the circularity defects can be reduced with an unequal division of the cutting edges (see Fig. 1) [5].

Occurring uncertainty during the reaming process

Within the Collaborative Research Centre "SFB 805 – Control of Uncertainty in Load-Carrying Structures in Mechanical Engineering", the subproject B3 addresses the two-stage process chain boring and reaming. This process chain is shown accordingly to the SFB-model in Fig. 2. In the first step a pilot bore is drilled which is then reamed to the final diameter. Due to uncertainty during the boring and reaming process the quality of the work piece can vary. This can be caused by interferences such as insufficient position repeatability of the machine, defects of the tools and chucks as well as faulty positioning of the work piece caused by the operator.

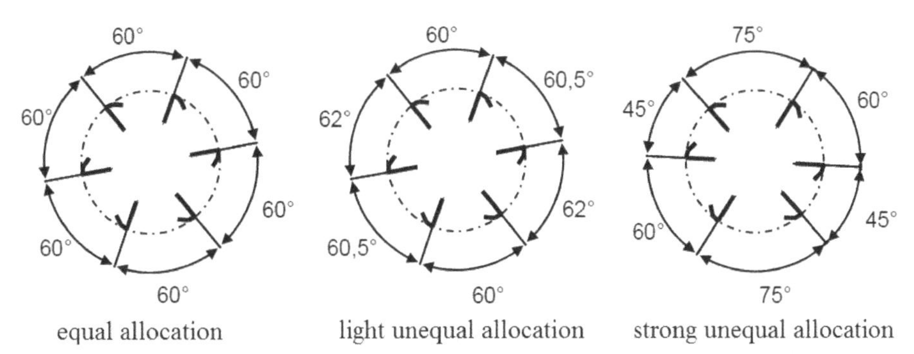

Fig. 1: Various types of cutting edge spacing in multibladed reamers [6]

This paper deals with examining how this uncertainty can be controlled through an adjustment of the machining strategy during the reaming process. The main focus is on the deflection of the reamer with increasing bore depth. The basic types of the occurring uncertainty of the two-staged process chain are explained in the following:

Axis offset. A parallel displacement of the tool axis along the boring axis is called an axis offset. During the tool's entry not all edges enter the workpiece simultaneously and the resulting radial forces deflect the reamers from the ideal course. This error is caused by the lack of positioning repeatability concerning the positioning of tools and/or workpieces.

Eccentricity error. Eccentricity errors on the other hand are caused by an offset of the tool's rotation axes in relation to the motor spindle. Therefore, the edges do not rotate on the same axis as the spindle and cause varying undeformed chip widths. This leads to higher stress and increased signs of wear. Most commonly eccentricity errors are caused by outdated bearings or impurities. Nowadays, hydraulic compensation chucks are used, which can reduce errors to values in the micrometre range.

Sloped pre-drilled bores. Since each machining process is affected by undesired disturbances, deviations concerning the target geometry in the area of the pre-drilled bores can occur [8]. Sloped or uneven surfaces, material inclusions, different hardness gradients and other material defects can cause a sloped pre-drilled hole. The sloped axis of the bore causes changes in chip thickness which generates varying forces at the cutting edges. These process errors can cause a deflection of the reamers, varying levels of stress and increased wear.

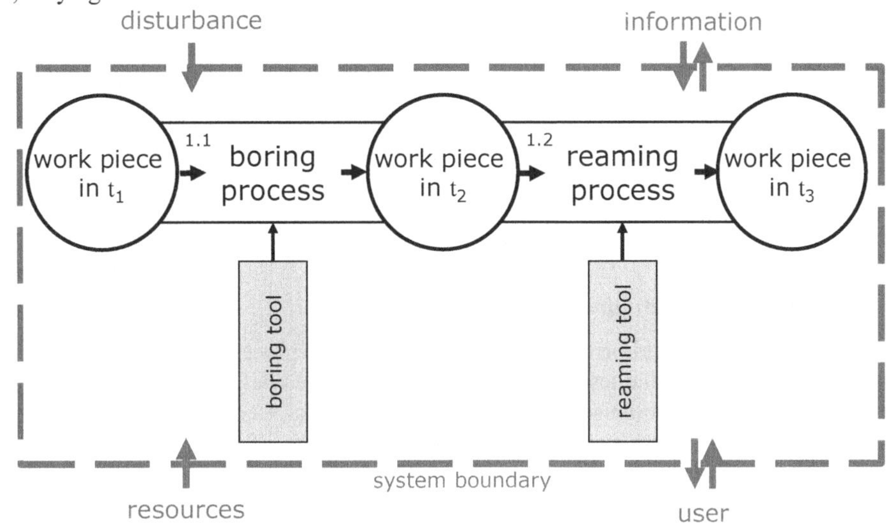

Fig. 2: Process chain boring and reaming according to the SFB 805 process model [7]

Relevance of the unsteady process phase during reaming. Hauer's research indicates that the deviation during the reaming process can be described as a linear function [2]. Furthermore, according to Hauer, the majority of tool deflection arises during the unsteady process phase. The entry phase is regarded as the biggest influencing factor for the reamer's deflection over the bore depth. The unsteady process phase begins with the entry of the rough cutting edges into the workpiece and is finished when the initial cutting area of the rough cutting edges enters the material. Thus, the maximum chipping thickness or the maximum feed per tooth for each individual cutting edge is achieved. Therefore, in this paper will be presented how the process uncertainty can be decreased by adjusting the machining strategy during the entry of the reamer.

General test conditions

In order to examine the influence of the entry strategy during the reaming with multibladed reamers the material 42CrMo4V (see Table 1) according to DIN EN 10083 was used. This quenched and tempered steel is characterised by its high stability with a high tenacity at the same time.

Table 1: Material properties of used 42CrMo4V

$R_{p0,2\%}$	R_m	A	Z	hardness
672 N/mm^2	900 N/mm^2	19%	57%	280 HV

For these experiments tests with two different bore diameters were conducted. To bore the pre-drilled holes, drillers with the diameters $d_{1,1}$ = 13.8 mm and $d_{2,1}$ = 19.8 mm were used. The reamers have a diameter of $d_{1,1}$ = 14 mm and $d_{2,1}$ = 20 mm with a l/d ratio of 9.1 respectively 10. According to the manufacturer's information, tolerances between IT 6 and IT 7 can be achieved. Further processing parameters are shown in Table 2. To achieve a robust evaluation, each experimental set-up was repeated ten times.

Table 2: Used tools and processing parameters

	feedrate	lubricant pressure	cutting speed
reamer 1: $d_{1,1}$ = 14 mm	0.1 – 1.2 mm/R	60 bar	80 m/min
reamer 2: $d_{2,1}$ = 20 mm			60 m/min

The insertions of the bore as well as the reaming tests were carried out on a 5-axis machining centre "Hermle C32 U". Bores were drilled in the test-workpiece with twisted carbide drills using internal cooling. Test parameters are shown in Table 2. The finished reamed bores were measured with a coordinate measuring machine type Leitz PMM 864. Measured data sets were processed and analysed for each individual bore. Each bore was divided up in eight measuring circles which are allocated in an equal distance along the depth of the bore. The pre-drilled bores were also measured in three circles. Deflection, diameter and circular form of each measuring circle were gathered. The reaming process was not carried out along the entire bore depth. The pre-drilled bores were measured with three circles for reference at the bottom of the unprocessed pilot bore. The individual measuring circles allow statements about the deflection, diameter and circular form.

Machining tests without additional process errors

In the first test series, experiments with the smallest possible process errors were carried out. For this purpose the surfaces of the test pieces were precisely milled before the bores were machined. The eccentricity error was minimised by using adjustable chucks. Occurring axis offsets are caused by the position repeatability error of the machine and could not be influenced.

Constant feedrate. The first tests were carried out with a constant feedrate (0.3 mm/R or 1.2 mm/R) like it is normally used in production processes. Bores with a depth of 55 mm were drilled. Afterwards these bores were reamed to a depth of 50 mm. Comparative values for the deflection of the driller along the bore could be measured since the pre-drilled bore was deeper. The results of the deflection of the reamers and the deviation of circularity are shown in Table 3. The maximum value for the deflection as well as the mean of all ten measurements of deflection and the deviation of circularity are indicated. As expected the deflection of reamer 2 with a diameter of $d_{1,2} = 20$ mm is smaller than the one of reamer 1. The value of the process forces is identical because they share the same a_p. Therefore the bigger moment of inertia causes a smaller bending. Both reamers have in common that the deflection is smaller with a lower federate of $f_1 = 0.3$ mm/R.

Table 3: Results of the constant feedrate tests

	feedrate	maximum deflection	medium deflection	medium deviation of circularity
reamer 1	0.3 mm/R	16 µm	8 µm	3 µm
reamer 1	1.2 mm/R	25 µm	15 µm	6 µm
reamer 2	0.3 mm/R	12 µm	4 µm	10 µm
reamer 2	1.2 mm/R	17 µm	8 µm	6 µm

Figure 3 presents the deflection of reamer 1 along the drilling depth for both feedrates. The results of each single measurement (black) as well as the average deflection (grey) were recorded. The last three measuring points at a drilling depth ranging from 50 to 55 mm belong to the measurements of the unreamed pre-drilled bores and are used as reference points. The change in radial deflection for the feedrate $f_1 = 0.3$ mm/R clearly shows that deflections in the reamed areas are significantly lower in comparison to the primary pre-bores.

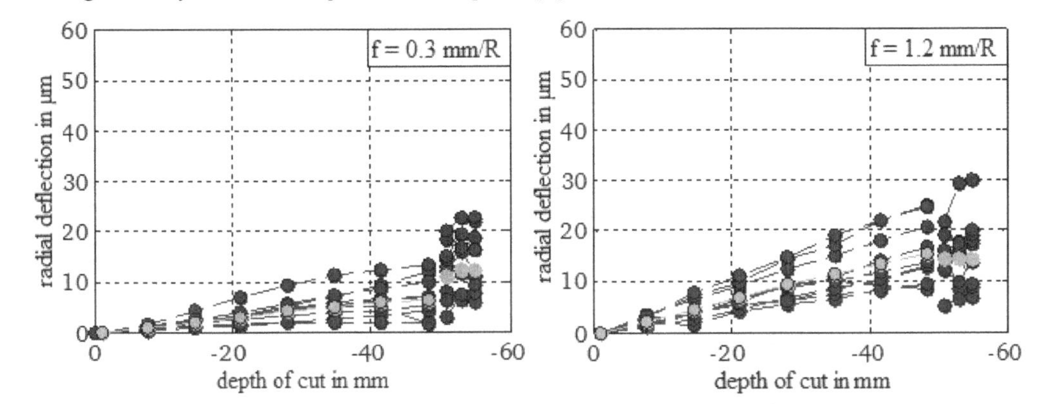

Fig. 3: Deflection of reamer 1 along the depth of cut

Linear feed increase. In case of the linear feedrate increase, the feedrates during the gating process starting at $f_{1,1} = 0.1$ mm/R and $f_{2,1} = 0.4$ mm/R are increased linear to $f_{1,2} = 0.3$ mm/R respectively $f_{2,2} = 1.2$ mm/R. The final feed was reached at a cutting depth of 4 mm. The results of these machining tests are shown in Table 4. In comparison with the constant feedrate tests, the maximum deflection for a lower feedrate is less, whereas the medium deflection is identical. No improvement could be achieved for the higher feedrate.

Table 4: Test results with linear feed increase

	feedrate	maximum deflection	medium deflection	medium deviation of circularity
reamer 1	0.1 - 0.3 mm/R	12 µm	8 µm	4 µm
reamer 1	0.4 - 1.2 mm/R	25 µm	14 µm	4 µm
reamer 2	0.1 - 0.3 mm/R	10 µm	4 µm	10 µm
reamer 2	0.4 - 1.2 mm/R	18 µm	10 µm	6 µm

The average deviations from the circular form for reamer 1 with a diameter of 14 mm are shown in Fig. 4. $R_{1,2}$ are the results from the reference test whereas $L_{1,2}$ are the results of the linear entry strategy. The deviations of the constant test for the lower feedrate (left) are less than the deviations of the linear strategy. The opposite occurs for a higher feedrate. A higher feed during the gating process leads to a better quality circular form.

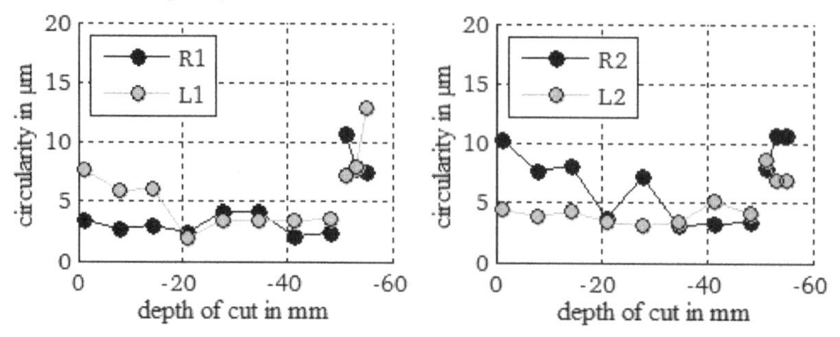

Fig. 4: Medium deflection from the circular form for f = 0.3 mm/R (left) and f = 1.2 mm/R (right) for reamer 1

Cubic feed increase. For the cubic feedrate increase, the values were increased analogously to the linear series of experiments from the initial value to the final value with a cutting depth of 4 mm. These changes were of a cubic nature, so that it results in a tangential gradient in connection with the preceding and following feedrate values [9]. The results of these experiments are shown in Table 5. As with the linear strategy, no improvements concerning the medium deflection could be achieved. The medium circularity is also almost identical. The deviations for reamer 1 are just 1 µm more (f = 0.3 mm/R) or 1 µm less (f = 1.2 mm/R) and are therefore to be considered as steady.

Table 5: Results of tests with cubic feed increase

	feedrate	**maximum deflection**	**medium deflection**	**medium deviation of circularity**
reamer 1	0.1 - 0.3 mm/R	15 µm	10 µm	4 µm
reamer 1	0.4 - 1.2 mm/R	25 µm	14 µm	5 µm
reamer 2	0.1 - 0.3 mm/R	11 µm	6 µm	10 µm
reamer 2	0.4 - 1.2 mm/R	20 µm	11 µm	7 µm

Abrupt feed increase. This strategy starts with a low steady feedrate ($f_{1,1}$ = 0.1 mm/R or. $f_{1,2}$ = 0.4 mm/R) until the reamer's gating area has completely entered the bore (cutting depth z = 0.75 mm). The feedrate will then be increased abruptly to the default value of 0.3 mm/R respectively 1.2 mm/R. In most cases the results for the maximum as well as the medium deflection are higher than the results of the reference tests. The circular form could only be improved in one case (reamer 2, f = 0.3 mm/R).

Table 6: Results of tests with sudden feed increase

	feedrate	**maximum deflection**	**medium deflection**	**medium deviation of circularity**
reamer 1	0.1 - 0.3 mm/R	16 µm	7 µm	4 µm
reamer 1	0.4 - 1.2 mm/R	33 µm	15 µm	5 µm
reamer 2	0.1 - 0.3 mm/R	18 µm	12 µm	6 µm
reamer 2	0.4 - 1.2 mm/R	18 µm	10 µm	8 µm

Summary of the experiments without incorporated process errors. None of the described special entry strategies (non-constant feedrate) resulted in significant reductions of the medium deflection. The medium deviation of the circular form could also not be improved. Nevertheless these experiments lead to the knowledge that a higher feedrate results in a higher deflection of the reamer and therefore result in lower quality of the reaming process.

Machining tests with incorporated process errors. In a second series of experiments, process errors were specifically incorporated to examine the performance of the reamers. Sloped pre-drilled bores were achieved by turning the machine table. The planned deviation of 50 µm in a drilling depth of 50 mm results in an angle of rotation of 0.057°. The rotation simultaneously results in an axis off set of ca. 39.9 µm in the y-direction. Thus, two of the three described types of uncertainty were examined during the process. The results of the test with constant feedrate are shown in Table 7. Due to the axis offset and the sloped pre-drilled bores, the deflections are higher than in the reference test without process errors (see Table 3).

Table 7: Results of the constant feedrate tests with sloped pre-drilled holes

	feedrate	**maximum deflection**	**medium deflection**	**medium deviation of circularity**
reamer 1	0.3 mm/R	39 µm	23 µm	10 µm
reamer 1	1.2 mm/R	41 µm	26 µm	8 µm
reamer 2	0.3 mm/R	20 µm	12 µm	8 µm
reamer 2	1.2 mm/R	43 µm	28 µm	8 µm

Like the experiments without incorporated process error, the examined entry strategies did not result in an improvement of the bore quality. Therefore, the results of any further experiments are summarised. As shown in Table 8, the tool deflection could only be improved when adopting a cubic strategy and applying a low feed for the reamer with a 20 mm diameter. In all other cases the tests with a steady feed lead to better results. The circular form improves in almost all experiments with a specific strategy. However, this happens at the expenses of the deflection.

The great differences in deflection of the reamers with a 20 mm diameter are striking. While the deflection with the smaller reamer for both feedrates is nearly identical, the values for the bigger reamer differ significantly. In this case a lower feed rate leads to noticeably better machining results.

Table 8: Results of the different machining strategies with sloped pre-drilled hole

	feedrate	**maximum deflection**	**medium deflection**	**medium deviation of circularity**	**strategy**
reamer 1	0.1 - 0.3 mm/R	51 µm	30 µm	7 µm	linear
reamer 1	0.4 - 1.2 mm/R	59 µm	42 µm	11 µm	linear
reamer 2	0.1 - 0.3 mm/R	21 µm	12 µm	9 µm	linear
reamer 2	0.4 - 1.2 mm/R	35 µm	27 µm	7 µm	linear
reamer 1	0.1 - 0.3 mm/R	57 µm	36 µm	9 µm	cubic
reamer 1	0.4 - 1.2 mm/R	61 µm	50 µm	9 µm	cubic
reamer 2	0.1 - 0.3 mm/R	19 µm	9 µm	7 µm	cubic
reamer 2	0.4 - 1.2 mm/R	40 µm	28 µm	7 µm	cubic
reamer 1	0.1 - 0.3 mm/R	43 µm	32 µm	7 µm	abrupt
reamer 1	0.4 - 1.2 mm/R	69 µm	55 µm	5 µm	abrupt
reamer 2	0.1 - 0.3 mm/R	28 µm	14 µm	6 µm	abrupt
reamer 2	0.4 - 1.2 mm/R	48 µm	33 µm	8 µm	abrupt

Summary

In this paper different entry strategies for reaming were examined. The feed rates for reamers with 14 mm as well as 20 mm diameter were changed during the gating process. As expected the deflections for the reamers with a larger diameter were smaller. Furthermore the process uncertainty in the form of an axis offset and a sloped pre-drilled hole have a significant influence on the reached machining results. These process errors could only be partially balanced out with the reaming.
As a result it can be noted that a constant feedrate during the entire process achieves the best results regarding the deflection of the reamer over the bore depth. Simultaneously it seems that lower feedrates lead to a higher bore quality. For future experiments the area of lower feed rates should be examined further, to determine the ideal process parameters.

Acknowledgement

The authors would like to thank the German Research Foundation (DFG) for funding the research activities at the Collaborative Research Centre "SFB 805 – Control of Uncertainty in Load-Carrying Structures in Mechanical Engineering".

References

[1] M. Krail, Reduktionspotenziale bei PKW, in: Alternative Antriebskonzepte bei sich wandelnden Mobilitätsstilen, KIT Scientific Publishing (2013) pp. 147-164.

[2] T. Hauer, Modellierung der Werkzeugabdrängung beim Reiben - Ableitung von Empfehlungen für die Gestaltung von Mehrschneidenreibahlen. Dissertation TU Darmstadt (2012).

[3] M. Hagedorn, K. Weinert, Schnittkraftmessung beim Reiben mit Mehrschneidenreibahlen, in: Werkstattstechnik online, Vol. 95 (2005) pp. 485-489.

[4] K. Weinert, and D. Meister, Machining and tool wear, in: Production Engineering, Vol. II (1995) pp. 1-6.

[5] K. Schmalz, Reibahle für hohe Kreisformgenauigkeit, in: Werkstatt und Betrieb (1970) pp. 313-318.

[6] K. Kuenanz and R. Hilke, Hochpräzisionszerspanen mit geometrisch bestimmter Schneide, Band III Fräsen, Ausbohren und Reiben, Forschungsgemeinschaft Qualitätssicherung e. V., Frankfurt am Main, (1994) pp. 58 – 68.

[7] T. Eifler, G.C. Enss, M. Haydn, L. Mosch, R. Platz, H. Hanselka, Approach for a consistent description of uncertainty in process chains of load carrying mechanical systems, in: Applied Mechanics and Materials, Vol. 104 (2011) pp. 133-144.

[8] Z. Katz and A. Poustie, On the hole quality and drill wandering relationship, in: International Journal of Advanced Manufactoring Technology, Vol. 17 (2001) pp. 233-237.

[9] N.N. (2013). Siemens Sinumerik 840D - Programmierhandbuch Arbeitsvorbereitung. CNC Version 4.5 SP2

Applied Mechanics and Materials Vol. 807 (2015) pp 169-180
© (2015) Trans Tech Publications, Switzerland
doi:10.4028/www.scientific.net/AMM.807.169

Submitted: 2015-04-29
Accepted: 2015-08-04

Uncertainty of Additive Manufactured Ti-6Al-4V: Chemical Composition, Microstructure and Mechanical Properties

Daniel Greitemeier [1, a *], Claudio Dalle Donne [1, b], Achim Schoberth [1, c], Michael Jürgens [1, d], Jens Eufinger [2, e] and Tobias Melz [2, f]

[1] Airbus Group Innovations, 81663 Munich, Germany

[2] Fraunhofer Institute for Structural Durability and System Reliability, Bartningstr. 47, 64283 Darmstadt, Germany

[a] daniel.greitemeier@airbus.com, [b] claudio.dalledonne@airbus.com, [c] achim.schoberth@airbus.com, [d] michael.m.juergens@airbus.com, [e] jens.eufinger@lbf.fraunhofer.de, [f] tobias.melz@lbf.fraunhofer.de

Keywords: Additive manufacturing, Ti-6Al-4V, DMLS, DMD, EBM, static properties, fatigue properties, microstructure, defects.

Abstract: Ti-6Al-4V is the most common used titanium alloy in aerospace. Parts are typically machined from wrought material with often high buy-to-fly ratios. Additive manufacturing, however, allows to build parts rapidly and directly from computer-aided design information, offering better material utilisation and lead time reduction. Despite the high potential for aerospace applications, the reliability of the mechanical properties is still at an early stage. This work gives an overview by determining and comparing the mechanical properties of DMLS, EBM and DMD processed Ti-6Al-4V material. Each process is compared, based on standardised post treatments, specimen geometries and test methods.
It can be seen, that the chemical composition, the microstructure, and the defect formation differs between the processes, which leads to a scatter in the experimentally determined static tensile, axial fatigue, and crack growth properties.

Introduction

Additive Manufacturing (AM) is a process of making parts rapidly and directly from computer-aided design information without the need for dies, form tools, or molds. The final parts are grown up, layer by layer, due to melting of powder or wire.

In the aerospace industry, better material utilisation and lead time reduction are the main drivers for AM, compared to metallic forged or cast components. Furthermore, the potential design freedom of powder bed based AM is an enabler for weight savings through topology optimisation.

Despite the high potential for aerospace applications, the reliability of the mechanical properties is still at an early stage.

Within the literature, static properties of milled AM specimens are reported. By comparison of the published work, scatter can be seen. For powder bed processes, a scatter in yield strength (YS) of ~300 MPa and elongation values from 10 to 17 % are reported [1-9]. The same trend can be seen for wire feed processes, where the YS vary by ~60 MPa with elongation values from 6 to 10 % [3, 10, 11].

For the fatigue performance of milled specimens, high scatter can be seen for both, powder bed [3, 6, 8, 12] and wire feed processes [3, 11], compared to wrought material.

Despite the current uncertainty of the mechanical properties, it has been shown, that AM Ti-6Al-4V can be comparable to wrought material. However, the mechanical properties can be influenced by microstructure, chemical composition, as well as process inherent defects. All these influences have not been investigated thoroughly yet. It will therefore be one key factor for future structural applications, to characterise these factors and to understand the current uncertainty in mechanical properties.

This work gives an overview by determining and comparing different AM technologies. Each process will be compared by there chemical composition, microstructure and mechanical properties, based on standardised post treatments (heat treatment, surface quality), specimen geometries, and test methods.

Experimental Details

Manufacturing Processes. Four different manufacturing processes are investigated. Two powder bed processes, where a laser or an electron beam selectively melts the powder, and two wire feed processes, where a wire is continuously melted by a laser beam or a plasma arc source.

Due to the size limitations of the build platforms, the builds take place in several jobs. Some important parameters of the build process can be find hereafter.

- Direct metal laser sintering (DMLS): The parts are produced in an EOSINT M 280 (EOS) machine with 370 W, a scan speed of max. 7000 mm/s and a layer thickness of 60 µm. The scan strategy is based on a shell and core concept, rotating with an angle of 30° for each layer. Builds take place under protective argon atmosphere with a process chamber temperature of ~35 °C.
- Electron beam melting (EBM): An ARCAM A2 (ARCAM AB) machine is used to build the parts with 50-3500 W, a scan speed of max. $8*10^6$ mm/s and a layer thickness of 50 µm. The used scan strategy is based on a shell-core concept, which alternates each layer (0°, 90°). The process is carried out under vacuum (~$5x10^{-3}$ mbar) with a process chamber temperature of ~620 °C.
- Direct metal deposition by laser (DMD-L): A laboratory setup was developed at Airbus Group Innovations to build up parts. The setup consists of a robot (KUKA KR 100HA), a wire feeder (Weldaix), a diode-pumped disc Yb:YAG laser (TrueDisc 8002) with a wavelength of 1030 nm and a box, flooded with argon. A ø1.2 mm Ti-6Al-4V ELI (extra-low interstitials) wire is laterally feeded into the melt pool. The process chamber temperature is ~23 °C.
- Direct metal deposition by plasma arc (DMD-P): A plasma arc process from Norsk Titanium is used to build up test blocks. Therefore, a Ti-6Al-4V wire with a diameter of 1.6 mm is melted under argon atmosphere.

Heat Treatment and Final Machining. All parts are heat treated under vacuum at 710 °C for 2 h, followed by furnace cooling under argon atmosphere. This heat treatment reduces residual stress. Final machining takes place after heat treatment to avoid any contamination.

Microstructural and Chemical Characterisation. The samples are embedded in Epofix resin and grinded/polished (80 – 4000 grit) on a RotoPol 31 (Struers). The microstructure is investigated by an optical light microscope type Polyvar (Reichert-Jung) in the polished and etched (Kroll reagent) condition.

Aluminum (Al), Vanadium (V) and Iron (Fe) are analysed by X-ray fluorescence spectroscopy (XRF). The gases Oxygen (O), Nitrogen (N) and Hydrogen (H) are determined by fusion analysis and Carbon (C) by means of combustion analysis.

Mechanical Characterisation. All specimens are tested in the machined condition. The load is applied perpendicular to the layer direction. All tests are performed at ambient temperature.

Static tensile tests are performed with flat specimens on a servo hydraulic test machine Z 250 (Zwick) in accordance to EN 2002 [13]. A strain rate of 0.5 %/min. is applied for yield stress at 0.2 percent ($R_{p0.2}$) and 2 %/min. for tensile strength (R_m). An extensometer gauge length of 20 mm is used to calculate the elongation (A). The average value is based on 3-4 test samples.

Axial fatigue tests are performed in accordance to EN 6072 [14] on a resonance test machine type Microtron (Rumul). Flat specimens (K_t=1) are tested with a constant load ratio of 0.1.

Fatigue crack growth tests are performed on a servo hydraulic test device PC 160N (Schenck) in conformity to ASTM E 647 [15]. C (T) 40 specimens and SENB specimens (single-edge-notched bend) are used under a constant load ratio of 0.1.

Drawings of all specimens are shown in Fig. 1.

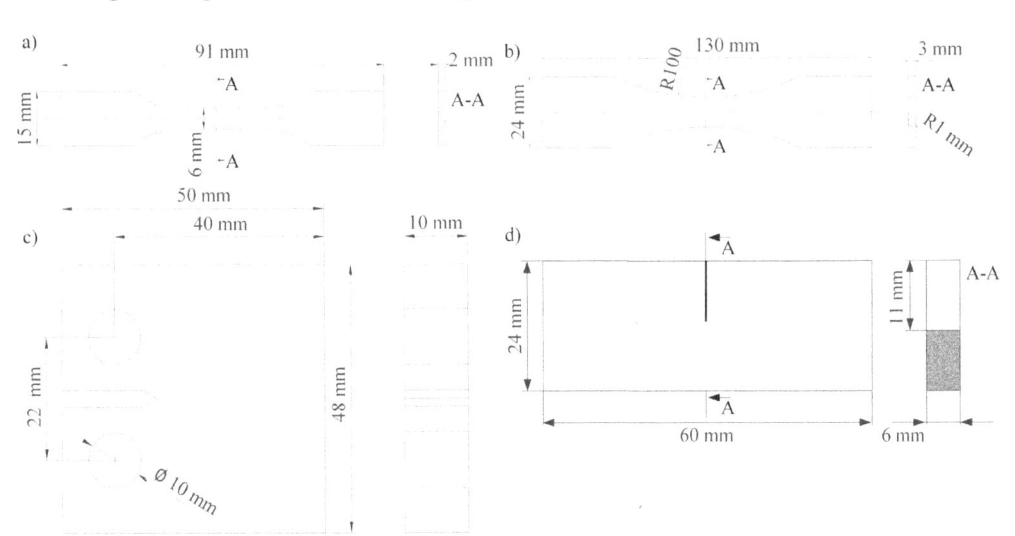

Figure 1: Specimen geometries: a) static tensile test specimen, b) axial fatigue test specimen, c) fatigue crack growth test specimen, C(T) 40, d) fatigue crack growth test specimen SENB

Results

Chemical Characterisation. All processes fulfill the requirements of ASTM F 2924 [16]. However, there are differences between the processes (Table 1). It is noticeable, that the Aluminum content differs between the processes, within the range of ASTM F2924, by approx. 12 %.

Due to the use of wire with extra low interstitial (ELI) for the DMD-L material, the Oxygen and Iron content is reduced, compared to the other processes. The chemical analysis is in agreement with the ASTM F3001 [17] for Ti-6Al-4V ELI.

Table 1: Chemical composition out of different AM processes in comparison to ASTM F2924 and ASTM F3001

Chemical analysis in weight-%	Process	Al	V	O	N	C	H	Fe
ASTM F2924 Ti-6Al-4V		5.5 – 6.75	3.5 – 4.5	<0.2	<0.05	<0.1	<0.015	<0.3
ASTM F3001 Ti-6Al-4V ELI		5.5 – 6.5	3.5 – 4.5	<0.13	<0.05	<0.08	<0.012	<0.25
Ti-6Al-4V	DMLS	6.47	3.99	0.165	0.022	0.007	0.0005	0.20
Ti-6Al-4V	EBM	5.92	4.13	0.126	0.015	0.010	0.0007	0.19
Ti-6Al-4V ELI	DMD-L	6.22	3.99	0.067	0.040	0.010	0.0014	0.07
Ti-6Al-4V	DMD-P	5.76	4.21	0.150	0.011	0.021	0.0037	0.16

Microstructural Characterisation. Process inherent defects can have a significant influence on the mechanical properties. Therefore, the material is investigated in the polished condition. It can be seen, that characteristics like pores or defects are mainly a phenomenon of the powder bed processes (Fig. 2a: DMLS; Figure 2b: EBM). Their distribution is homogeneous over the build height.

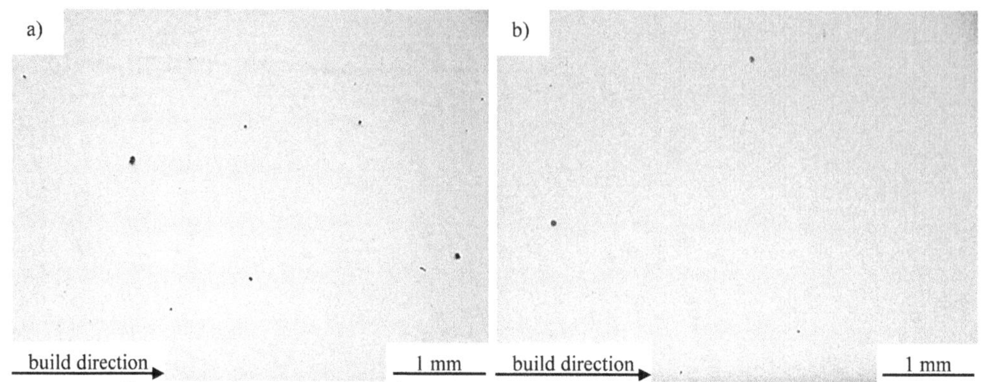

Figure 2: Representative defects (e.g. pores) formation within the specimens : a) DMLS; b) EBM

In contrast, only a few pores or defects could be detected for the wire-feed processes. It can be seen in Figure 3a, that the DMD-P material showed nearly no pores. In contrast, a few pores or defects could be detected in the DMD-L material (Figure 3b).

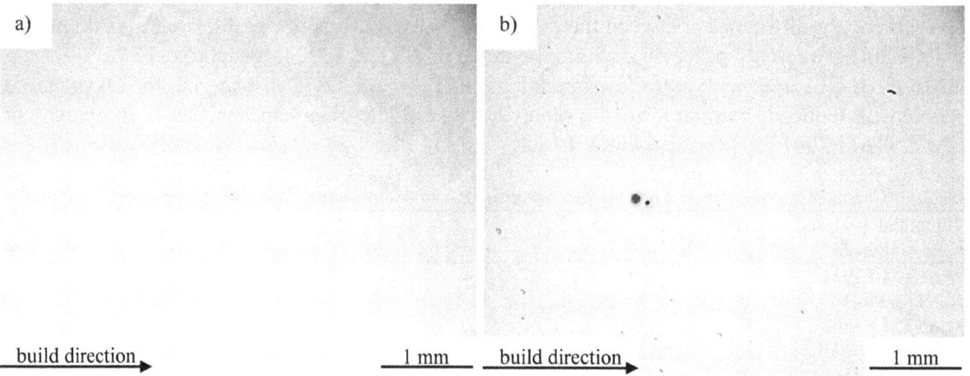

Figure 3: Representative defects (e.g. pores) formation within the specimens: a) DMD-P; b) DMD-L

In addition to the polished condition, specimens are etched to investigate their microstructure (Figure 4, Figure 5). All processes show columnar prior β grains, growing perpendicular to the layer direction. Differences can be seen between the processes in the α-lamella size, which coarsened from DMLS over EBM and DMD-L to DMD-P.

The DMLS material, displayed in Figure 4 a, show a mixture of α and β. The α is presented by a fine acicular morphology, which indicates, that it mainly consists of α or α′ (martensite). A clear distinction between both phases cannot be made, because of their similar lattice structure.

In contrast, the EBM material (Figure 4b) shows a slightly coarser microstructure due to the process temperature of ~620 °C. The α phase form is mainly acicular or plate-like with a Widmanstätten orientation. The α-platelets vary in orientation and size. Grain boundary α can be observed at the prior β grains.

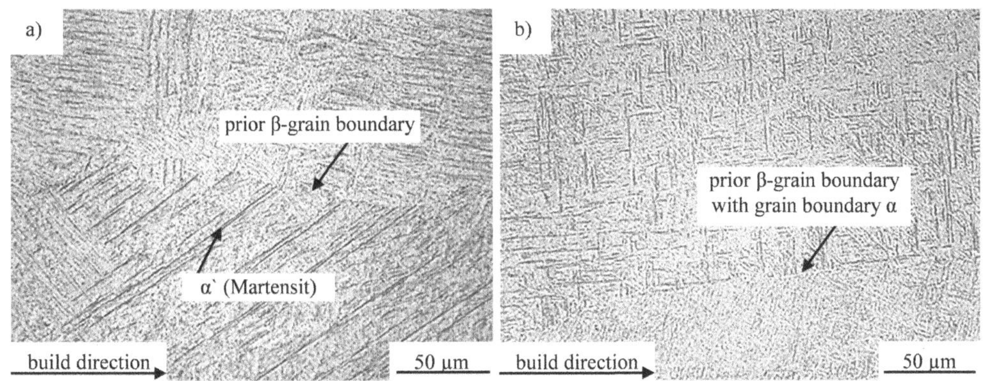

Figure 4: Typical microstructure: a) DMLS; b) EBM; heat treatment: 710 °C/2 h

Both wire feed processes (Figure 5) displays a coarse microstructure consisting of grain boundary α and a α plate like structure with a Widmanstätten orientation. It can be noted that especially the DMD-P process (Figure 5b) shows a scatter within the α lamellar width of ~1-5 µm.

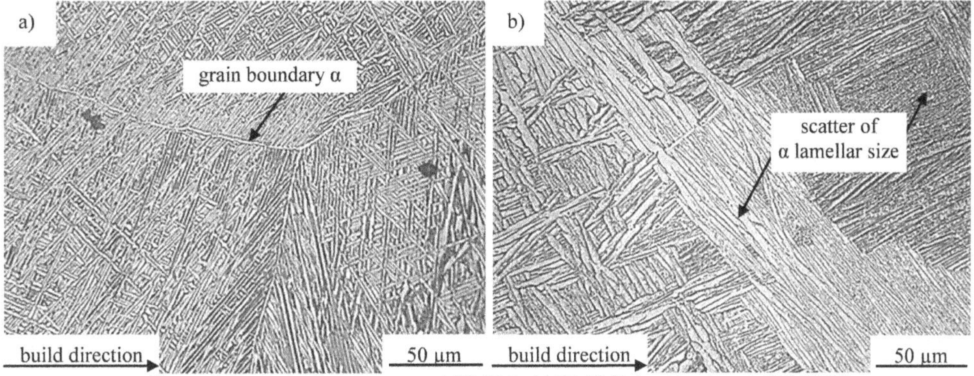

Figure 5: Typical microstructure: a) DMD-L; b) DMD-P; heat treatment: 710 °C/2 h

Tensile Properties. Figure 6 and Figure 7 summarizes the tensile properties including the standard deviation in comparison with the AM specification for Ti-6Al-4V (ASTM F2924) and Ti-6Al-4V ELI (ASTM F3001).

Both powder bed processes (DMLS, EBM) show values in accordance to ASTM F2924. However, the DMLS material displays higher ultimate tensile strength (+19 %) and yield strength (+25 %), with elongation values comparable with the EBM specimen.

The wire feed processes (DMD-L, DMD-P) show reduced ultimate tensile strength and yield strength, which do not meet the requirements of ASTM F2924 and ASTM F3001. It should be noted that the highest scatter in elongation (±22 %) of all processes can be displayed for the DMD-P process.

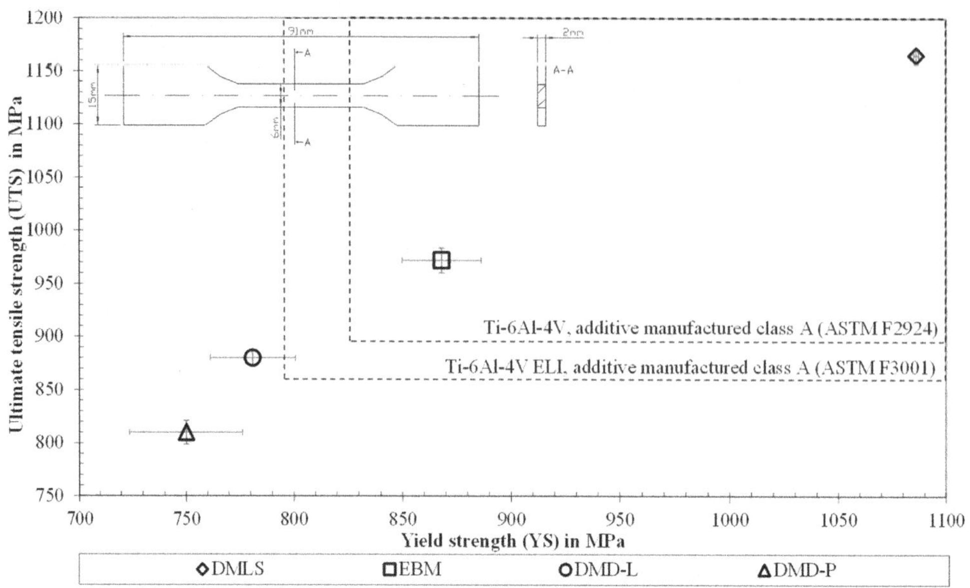

Figure 6: Ultimate tensile strength plotted against yield strength for Ti-6Al-4V material, manufactured by DMLS, EBM, DMD-L and DMD-P process, compared to the standards ASTM F2924 and ASTM F3001

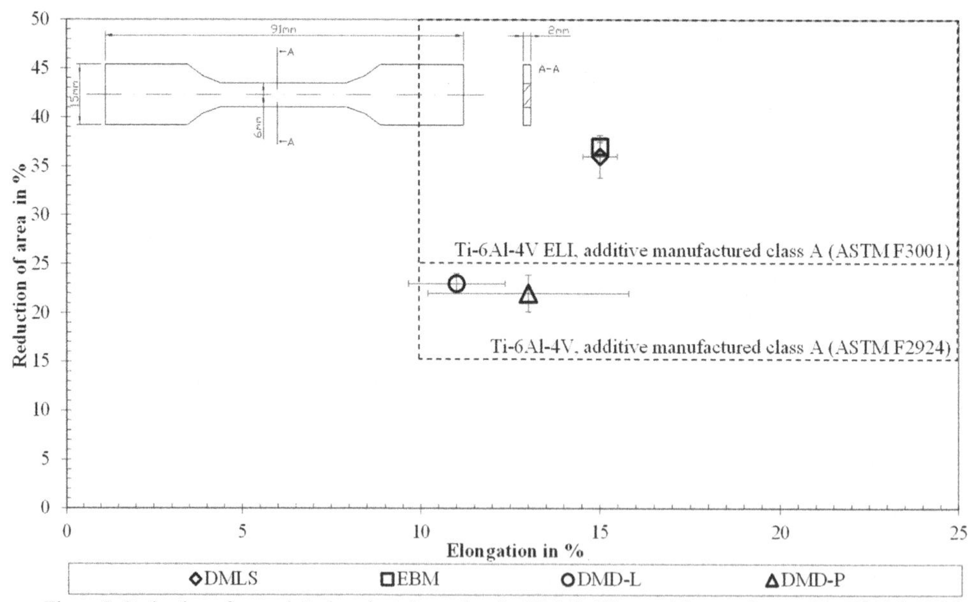

Figure 7: Reduction of area plotted against elongation for Ti-6Al-4V material, manufactured by DMLS, EBM, DMD-L and DMD-P process, compared to the standards ASTM F2924 and ASTM F3001

High Cycle Fatigue Properties. High cycle fatigue (HCF) behavior of all process routes are plotted in Figure 8 and Figure 9. To avoid plastic deformation during testing, a maximum upper stress level is determined with respect to the measured yield stress. A Weibull distribution is used to interpolate the S-N curves with a failure probability of 50 %. For comparison reasons, S-N curves for cast and wrought material according to [18] are added. Even though the load ratio differs slightly (cast: R=0.06; wrought: R=0.06 – 0.1), a rough classification of the values can be given [18].

Figure 8a) show the HCF-properties of DMLS specimens. The values are within the range of wrought material. Scattering appears over the entire measuring range. Fracture surface analysis reveals isolated defects (Ø < 100 μm).

The HCF-properties of EBM specimens are shown in Figure 8b. Compared to the DMLS-specimens, significantly lower cycles to failure can be seen, similar to cast material. A variety of defects (Ø >100μm) can be detected on the fracture surface.

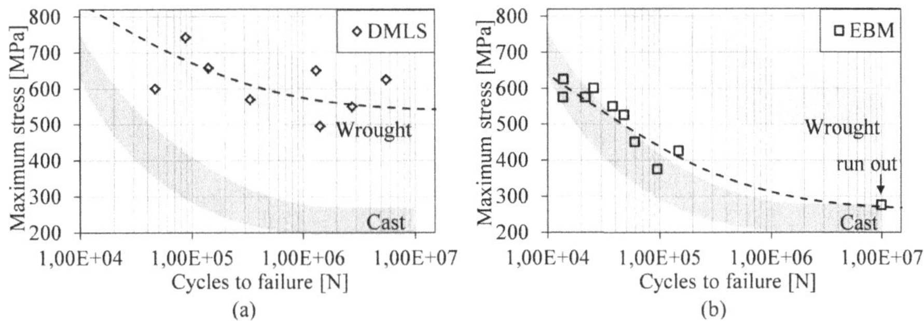

Figure 8: S-N curves with Weibull distribution (50 % probability of failure); R=0. 1; K_t=1: a) DMLS [19]; b) EBM

Figure 9a) displays the HCF-properties of DMD-L specimens. The values are comparable to wrought material, even though the range of applied stress is narrow. Isolated defects can be found on the fracture surface with scatter up to Ø 1 mm.

In comparison to DMD-L, scatter is reduced for DMD-P (Figure 9b). Low yield strength (750 MPa), however, limits the maximum stress to 650 MPa and therefore the cycles to failure to >60.000.

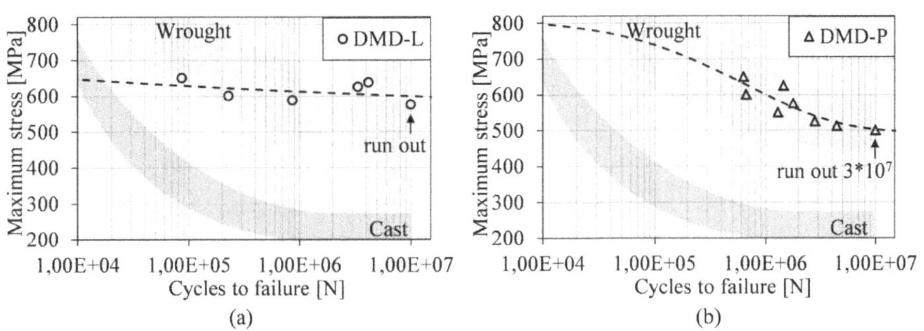

Figure 9: S-N curves with Weibull distribution (50 % probability of failure); R=0. 1; K_t=1; a) DMD-L; b) DMD-P

Fatigue Crack Growth Properties. To understand the differences between the S-N curves, crack growth rates of all four processes are measured and plotted logarithmically as a function of stress intensity range (Figure 10, Figure 11). Ti-6Al-4V plate material (heat treatment: ~730 °C for 2 h) is used for comparison purposes [20]. It can be seen, that the crack growth resistance is increased for the wire feed processes. However, all curves are comparable or better than the reference for plate material.

It is worth to mention, that the microstructure seems to have a strong impact on the crack growth near the threshold, as reviewed by Lütjering et al. [21]. A finer microstructure therefore tends to increased crack propagation, in comparison to a coarser one. This can also be seen for the experimentally determined crack growth resistance of AM Ti-6Al-4V.

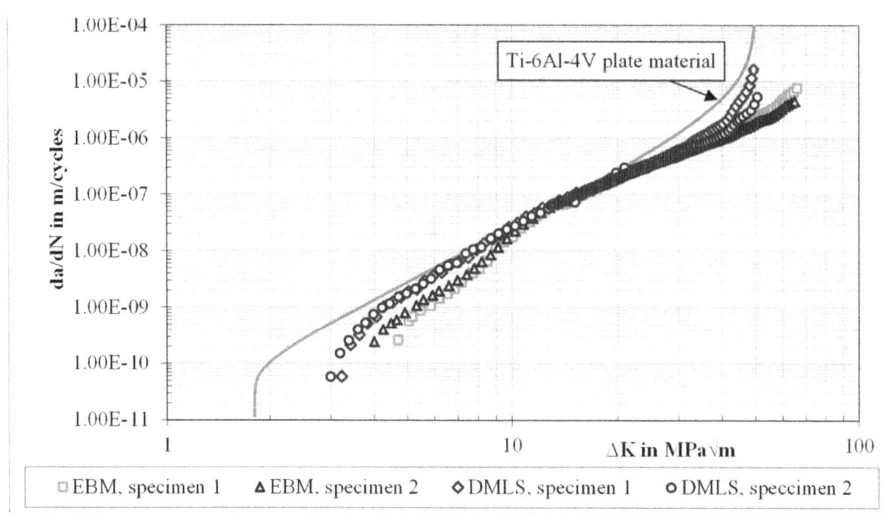

Figure 10: Fatigue crack growth properties for Ti-6Al-4V material manufactured by DMLS (C (T) 40) and EBM (C (T) 40) [22], compared to plate material [20], R=0.1, heat treatment: 710 °C/2 h

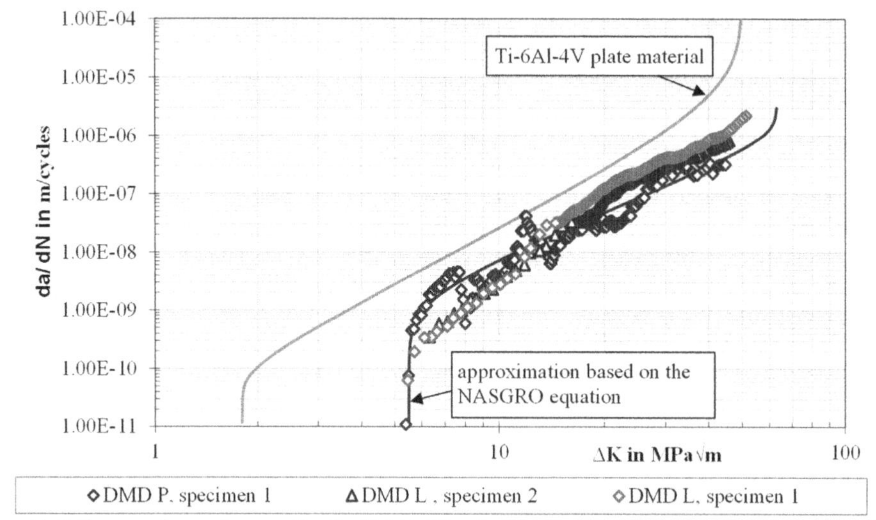

Figure 11: Fatigue crack growth properties for Ti-6Al-4V material manufactured by DMD L (C (T) 40) and DMD P (SENB) compared to plate material [20], R=0.1, heat treatment: 710 °C/2 h

Discussion

Static Tensile Properties. Static properties are influenced by a variety of factors, interacting with each other. For that reason, differences in the microstructure, chemical composition, and the influence of pores are discussed below.

Within the results, a scatter can be seen in the chemical composition between the four processes. An overall variation of the α stabilising elements Aluminum by ~12 % and Oxygen by ~46 % can be seen. The influence of these elements on the mechanical properties is known for e.g. plate material. Interstitial Oxygen stabilises the α phase and raises the hexagonal lattice parameter c/a [23]. Oxygen therefore tends to increase the strength [24].

Gysler and Lütjering [25] reported an increase in YS by ~170 MPa for lamellar Ti-6Al-4V due to a rise in the Oxygen proportion of 0.11 wt.-%. The same trend can be seen in this study. An increased Oxygen content of 0.11 wt-%. tends to a significant increase in YS by 336 MPa. It is assumed, that also the differences in Aluminum and Iron content have an impact on the YS. Aluminum stabilises the α phase and stimulates solid solution hardening. The ductility therefore decrease while strength increase [26]. Harwig et al. [27] reported an increase of strength in commercial pure Titanium with increased Iron content. The effect of Oxygen, Aluminum and Iron could be one explanation for the differences within the static properties of the four platforms (Figure 6, Figure 7). Detailed chemical analysis of the powder, wire and the final parts could help to reduce the current uncertainty. Particularly possible contamination and evaporation effects during the build, as well as the powder storage and handling could change the chemical composition and must therefore be monitored.

All alloying elements listed (Table 1) are influencing the microstructure by shifting the β transus temperature and therefore the β/α phase transformation [28]. The microstructure is therefore another factor affecting the tensile properties. Lütjering et al. [21] showed, that a coarser microstructure tends to decrease the yield strength. They also indicated lower elongation with increased lamella size. This effect on the static properties can also be seen by comparing the powder bed processes (DMLS, EBM) with the wire feed results (DMD-L, DMD-P). The influence of heat input during the build process and therefore the cooling rates through the β-phase field is reflected by the α-lamellar size, which increases with increased heat input. This can be described as a starting point for further lamellar growth due to post heat treatments (Figure 4, Figure 5).

A correlation between α lamella size and tensile values should however be interpreted carefully. The proportion of α and β seems to be important, too, due to the different crystal lattice (α: hex, β: bcc). Process related heat treatments could help to homogenies the microstructure and therefore reduce the uncertainty related to microstructural variations.

Another factor are defects. Material defects (pores, blowholes, lack of fusion) reduces the cross section area, which leads to local stress concentrations. It therefore can reduce the strength and ductility. The tensile strength is basically dependent on the number, size, and distribution of the defects. According to AMS 499 A, the maximum tolerable defect size must not achieve 33 % of the thinnest cross sections. Therefore, the influence of defects in this investigation seems to be negligible due to the amount and size of defects, shown by the microsections (Figure 2 and Figure 3).

High Cycle Fatigue Properties. S-N curves have shown a high scatter of fatigue strength within one process and even differences between the four manufacturing technologies (Figure 8, Figure 9). In terms of the investigated AM technologies, defects seem to be the dominant feature. From cast Titanium it is well known, that fatigue strength is affected by pores [24, 29]. However, considerable differences can be seen in the type of defects between cast and AM Ti-6Al-4V. Especially "lack of fusion" defects, caused by non-melted powder, can have an impact on the fatigue performance of AM Ti-6Al-4V [8]. Based on their size, geometry, distance to surface, and orientation to the load direction, defects lead to different stress concentrations. The consequence is an increased crack initiation compared to defect free material and a possible scatter in the fatigue performance. It can be seen at the fracture surface, that the defect size increases from DMLS to EBM, which results in lower fatigue performance. The results are in accordance to literature data from EBM specimens, where hot isostatic pressing reduces the defect size and tends to an increased fatigue limit of ~80 % [3]. The wire feed process (DMD-P, DMD-L) samples preferentially failed at microstructural inhomogeneities.

From a fracture mechanical point of view, a reduced uncertainty of the fatigue properties can be achieved by either reduced defect size, load optimised defect alignment, or improved material properties.

- Reduced defect sizes should be achieved by either optimised processes or post treatments. Especially hot isostatic pressing (HIP) can significantly reduce the defect size and therefore increase the fatigue properties [3]. However, there are some limitations. Surface defects cannot be reduced by HIP. Therefore a quality control has to be developed to account for this factor of uncertainty.

- Optimised defect orientation should be achieved by load oriented part alignment within the build. Examples are lack of fusion defects, mainly occurring parallel to the layer direction within the EBM process. The stress concentration could therefore be reduced by applying load parallel to the layer direction, rather than perpendicular.
- According to Lütjering et al. [30], the most important microstructural parameter concerning the mechanical properties is the α colony size. This can be achieved by appropriated heat treatments with respect to the thermal history of the part and impurity levels (O, N, Fe, H). Defect sensitivity could therefore be reduced, resulting in an extended crack initiation stage. In addition, a coarser lamella size could help to shift the threshold to higher values for long cracks (R=0.1), as shown in Figure 10 and Figure 11.

Fatigue Crack Growth. In the presence of large cracks, fatigue crack growth resistance can partly associate to microstructure. It is known from the literature for Ti-6Al-4V (R~0.1), that crack deflection and therefore roughness-induced crack closure could be responsible for increased fatigue-crack growth resistance [21, 24]. A coarser lamella size, therefore, tends to higher crack growth resistance. Similar conclusions can also be made out of the investigated AM material.

Conclusion

Four different AM platforms have been investigated for comparison reasons. An experimental test program was performed to determine their chemical, microstructural and mechanical properties. The following conclusions can be drawn:

- The uncertainty within the static tensile properties are mainly influenced by the chemical composition and differences within the microstructure, depending on the thermal history of the specimens.
- High cycle fatigue properties are dominated by defects resulting in high scatter and low fatigue properties based on the amount, size and orientation of defects.
- Fatigue crack growth rates are partly influenced by the microstructure for stress ratio of 0.1, which can be related to crack closure effects depending on the microstructure.

Acknowledgment

The author would like to thank Dr. Erhard Brandl, Vitus Holzinger, Christian Plander, Frank Palm and Birgit Vetter for their support.

Literature References

[1] H. K. Rafi, N. V. Karthik, H. Gong, T. Starr, B. Stucker, Microstructures and mechanical properties of Ti6Al4V parts fabricated by selective laser melting and electron beam melting, Journal of Materials Engineering and Performance, 22 (2013), p. 3872-3883.

[2] L. Facchini, E. Magalini, P. Robotti, S. Höges, K. Wissenbach, Ductility of a Ti-6Al-4V alloy produced by selective laser melting of prealloyed powders, Rapid Prototyping Journal, 16 (2010), p. 450 - 459.

[3] D. Greitemeier, K. Schmidtke, V. Holzinger, C. Dalle Donne, Additive layer manufacturing of Ti-6Al-4V and Scalmalloy^{RP}: fatigue and fracture, 27th Symposium of the International Committee on Aeronautical Fatigue and Structural Integrity (2013).

[4] ARCAM AB, Ti6Al4V Titanium alloy, 23.03.2015.

[5] S. S. Al-Bermani, M. L. Blackmore, W. Zhang, I. Todd, The Origin of microstructural diversity, texture, and mechanical properties in electron beam melted Ti-6Al-4V, Metallurgical and Materials Transactions A, 41 (2010), p. 3422-3434.

[6] A. A. Antonysamy, Microstructure, texture and mechanical property evolution during additive manufacturing of Ti6Al4V alloy for aerospace applications, Faculty of Engineering and Physical Sciences, 2012.
[7] S. Leuders, M. Thöne, A. Riemer, T. Niendorf, T. Tröster, H. A. Richard, H. J. Maier, On the mechanical behaviour of titanium alloy TiAl6V4 manufactured by selective laser melting: fatigue resistance and crack growth performance, International Journal of Fatigue, 48 (2013), p. 300-307.
[8] G. Kasperovich, J. Hausmann, Improvement of fatigue resistance and ductility of TiAl6V4 processed by selective laser melting, Journal of Materials Processing Technology, 220 (2015), p. 202-214.
[9] EOS GmbH - Electro Optical Systems, Material datasheet, EOS Titanium Ti64, EOS Titanium Ti64, AD, WEIL / 10.2011, 2011.
[10] L. Eriksen, Combined EBSD-investigations and in-situ tensile tests of a direct metal deposited Ti6Al4V-alloy, Materials Science and Engineering, 2013.
[11] E. Brandl, Microstructural and mechanical properties of additive manufactured titanium (Ti-6Al-4V) using wire, evaluation with respect to additive processes using powder and aerospace material specifications, Faculty of Mechanical, Electrical and Industrial Engineering, 2010.
[12] P. Edwards, M. Ramulu, Fatigue performance evaluation of selective laser melted Ti–6Al–4V, Materials Science and Engineering: A, 598 (2014), p. 327-337.
[13] EN 2002-001:2005, Aerospace series - metallic materials - test methods - part 1: tensile testing at ambient temperature, G. I. f. Standardisation, 2006.
[14] EN 6072:2010, Aerospace series - metallic materials - test methods - constant amplitude fatigue testing, Association of European Aircraft and Component Manufacturers, 2008.
[15] ASTM E647-00, Standard test method for measurement of fatigue crack growth rates, ASTM International, 2000.
[16] ASTM F2924-12, Standard Specification for Additive manufacturing titanium-6 aluminum-4 vanadium with powder bed fusion, ASTM International, 2012.
[17] ASTM F3001-14, Additive manufacturing titanium-6 aluminum-4 vanadium ELI (extra low interstitial) with powder bed fusion, ASTM International, 2014.
[18] U.S. Department of Defense, Military handbook: titanium and titanium alloys, MIL-H DBK-697A, Washington, D.C., 1974.
[19] D. Greitemeier, V. Holzinger, C. Dalle Donne, T. Melz, Fatigue prediction of additive manufactured Ti-6Al-4V for aerospace: Effect of defects, surface roughness, in press, 28th ICAF Symposium, (2015).
[20] ESTEC, ESA, ESACRACK user´s manual, ESA Publications Division, ESTEC Netherlands, 1995.
[21] G. Lütjering, J. C. Williams, Titanium, Springer, Berlin, Heidelberg, New York, 2003.
[22] D. Greitemeier, C. Dalle Donne, F. Syassen, J. Eufinger, T. Melz, Effect of surface roughness on fatigue performance of additive manufactured Ti-6Al-4V, in press, Materials Science and Technology, (2015).
[23] H. Kellerer, Übersicht über die Wärmebehandlung von TiAl6V4, Härterei-Technische Mitteilungen, 25 (1970), p. 242-253.
[24] R. Boyer, G. Welsch, E. W. Collings, Materials properties handbook, titanium alloys, ASM International, Ohio, 1998.
[25] A. Gysler, G. Lütjering, Influence of test temperature and microstructure on the tensile properties of titanium alloys, Metallurgical Transactions A, 13 (1982), p. 1435-1443.
[26] S. R. Lampman, ASM Handbook, fatigue and fracture, ASM International, Ohio, 1997.
[27] D. D. Harwig, W. Ittiwattana, H. Castner, Advances in oxygen equivalent equations for predicting the properties of titanium welds, Welding Journal, 80 (2001), p. 126-136.

[28] J. A. Hall, Fatigue crack initiation in alpha-beta titanium alloys, International Journal of Fatigue, 19 (1997), p. 23-37.

[29] J. Oh, N. J. Kim, S. Lee, E. Lee, High-cycle fatigue properties of investment cast Ti-6Al-4V alloy welds, Journal of materials science letters, 20 (2001), p. 2183-2187.

[30] G. Lütjering, Influence of processing on microstructure and mechanical properties of (α+β) titanium alloys, Materials Science and Engineering: A, 243 (1998), p. 32-45.

CHAPTER 6:

Modelling Uncertainty Information by Means of Semantics

Applied Mechanics and Materials Vol. 807 (2015) pp 183-192
© (2015) Trans Tech Publications, Switzerland
doi:10.4028/www.scientific.net/AMM.807.183
Submitted: 2015-04-16
Accepted: 2015-08-04

Representation of Human Behaviour for the Visualization in Assembly Design

Maximilian Zocholl[1, a], Felix Heimrich[1,b], Marius Oberle[2,c], Jan Würtenberger[3,d], Ralph Bruder[2,e], Reiner Anderl[1,f]

[1]TU Darmstadt, DiK, Germany

[2]TU Darmstadt, IAD, Germany

[3]TU Darmstadt, PMD, Germany

[a]zocholl@dik.tu-darmstadt.de, [b]heimrich@dik.tu-darmstadt.de, [c]m.oberle@iad.tu-darmstadt.de, [d]wuertenberger@pmd.tu-darmstadt.de, [e]bruder@iad.tu-darmstadt.de, [f]anderl@dik.tu-darmstadt.de

Keywords: OWL 2 Ontology, UMEA, CAD, human uncertainty, aviation, process-chain model

Abstract. This article extends the Uncertainty Mode and Effects Analysis (UMEA) to human effects and uses ontologies to connect human driven uncertainty data to the corresponding parts in an aircraft CAD-Assembly. Still, human behaviour is one of the major sources of uncertainty in the product usage phase. Hence, using uncertainty data of human behaviour for product design becomes increasingly important, especially for the control of uncertainty in load carrying systems. In this context, the exchange of semantically enriched uncertainty data between different domains and domain specific applications guarantees the consistency of the data prevents misinterpretation and enables the reuse of existing data for future design decisions.

Introduction

The increasing amount of available data, especially from the production and usage phase, enables but also requires system engineers to cope with new modelling technologies such as Sematic Web Technologies. Connecting the available data in a useful way allows for the reuse and beneficial exploitation of the data for systems that are designed in the cooperation of multiple domains. The necessary data exchange and the interpretation of the meaning behind the data foster the development of standards such as the ISO 10303 series also known as the Standard for the Exchange of Product model data (STEP) or domain specific de facto standards such as the Common Parametric Aircraft Configuration Schema (CPACS). When data complies with these kinds of standards the semantics of the data is given by the commitment to a predefined schema. This entails different problems. Firstly, this only allows the usage of subsets of the standards but no extensions. Secondly, the verification of the encoded semantics is limited to the expressivity of the language in use, e.g. XML, XMI or EXPRESS.

Semantic Web Technologies build up on XML and layer additional semantics in form of RDF(S) and OWL 2 on top of it. OWL 2 applies the well understood semantics of the Description Logic SROIQ which in turn is a decidable subset of First-order predicate logic. The use of OWL 2 allows the machine readable and interpretable definition of new concepts relating to predefined concepts. Thus, uncertainty data capturing e.g. the aircraft acceleration during the touchdown can be semantically connected to the relevant parts of the aircraft assembly and allows a classification according to the behaviour of the pilot.

In the Collaborative Research Centre 805 (CRC 805), the research aims for the control of uncertainty in load-carrying structures. In these systems, uncertainty with respect to loads and strength can have a significant impact on the system behaviour. This article summarises the results of the cooperation between the subprojects A1 “Development of Models, Methods and Instruments for the Acquisition, Description and Evaluation of Uncertainties”, A5 “Information Model for the Representation and Visualisation of Uncertainties” and C6 “Human Factor on the uncertainty of utilization/ Analysis of user behaviour and development of a utilization language". The main contributions of this article are the extension of the UMEA by human factors and the contextualisation of the existing uncertainty information model with regard to the Dolce UltraLite Ontology, the Semantic Sensor Net Ontology, the Quantities and Units Ontology, the domain specific NIST Assembly Ontology and an Aircraft Ontology. This allows the modelling and the exchange of semantic information. Further, conclusions about uncertainty data are enabled by taking into account information of both product and process model.

The article is structured as follows. In section one, the UMEA is introduced and extended both by human factors and by ontological aspects. In section two, the landing process is modelled according to the following subsections. The scope of the inquired data is defined for the landing process of the example aircraft Cessna 172 S and the benefits of an ontology-based representation are explained. Then, the current definition of uncertainty of the CRC 805 is introduced. Further, the landing process model and its location in the process model of the CRC 805 are presented. Next, the human factors as input into the process model and the technical parameters as output from the process model are exposed. Building up on this, uncertainty event chains between the product and the process model are reformulated as requirements for the information model. Section three reviews related works and presents the ontology-based information model in the context of cyber physical systems. Possible future research topics and a summary conclude in the last two sections.

UMEA – Uncertainty Mode and Effects Analysis

The development of the information model is guided by the UMEA methodology.

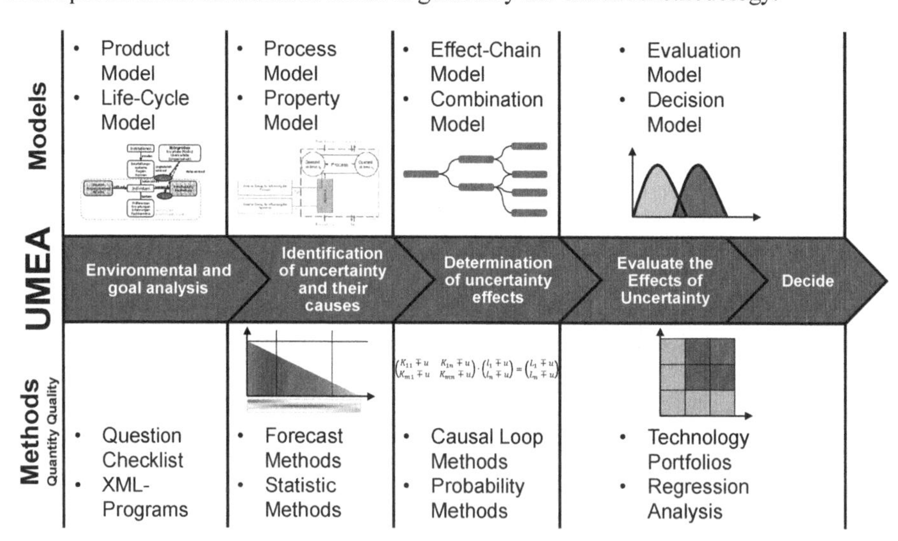

Figure 1: UMEA Methodology cf. Engelhardt (2009)

Figure 1 shows the five steps of the UMEA methodology as well as its methods and models. In the first step of the UMEA, the environmental and goal analysis, a system boundary is determined to separate it from neighboring systems. In addition, the definition of the interactions of the neighboring systems and the specification of evaluation objects is in focus. The effects are associated with institutions like social institutions, organizations or social entities in this context [1]. According to VDI 3780, these institutions are organized in the state, public, technical, scientific and economic sector [2]. The aim is a complete, systematic and comprehensive containment of uncertainties [3]. The identification of uncertainties, the second step of the UMEA, describes and names the single uncertainties and their causes [4]. For subsequent calculations the quantification of uncertainty is necessary during this step. For this purpose, a process analysis of the product life cycle is performed with regard to the process model explained in Chapter 2. The process and product properties are thereby assigned to the individual uncertainties.

In the third step, the detection of the effects of uncertainty, the connection of single uncertainties and their causes throughout the process chain are analyzed. There, the system uncertainty can be determined using the adapted effect chain model. For the calculation of the system uncertainty methods are distinguished between qualitative and quantitative types of uncertainty [4].

The comparison of the boundary values to the system uncertainty leads to the evaluation of uncertainties. There, the evaluation objects are recognized and the system uncertainty is determined. The two possibilities for the evaluation are well-known or established evaluation methods. The aim is the reduction of complexity in order to find a solution to the respective decision problem. The application of the UMEA then concludes with a decision-making process within the design phase based on the prior evaluated uncertainty [4].

Modelling of the landing process

Inquiry of the landing information for the assembly design. The scope of the following analysis focuses on the landing process of a Cessna 172 S aircraft. Similar to Oberle, the landing process starts with the exit of the cruise flight and ends on the runway with the attainment of taxi speed [5]. The information of the landing process, which is gained through a laboratory study using a flight simulator, is attached to the product model of an aircraft. The product model describes the geometric information of the 3D explicit CAD assembly. Additionally, the OWL information model includes a semantic representation of the geometrical and topological CAD information as well as the respective information of the landing process. This allows both the exchange of semantically enriched data between different applications as well as the extension of the data by inference procedures.

Identification of uncertainty. The understanding of uncertainty changed over time from a simple interval related concept to a generic framework capturing numerical methods for engineering applications coping with aleatoric and epistemic uncertainty [6,7]. Uncertainty in load carrying structures can be defined as phenomenon that arises from a system with underdetermined process characteristics [8]. Uncertainty in load carrying systems can be further classified following the uncertainty map of the CRC 805 which builds up on the work of Engelhardt et al. [9]. Currently, parametric and non-parametric uncertainty is distinguished. In the case that all parameters for a verifiable and valid model are known, the parametric uncertainty can be distinguished according to the level of information about their values. Starting from an unknown parameter domain, uncertainty is classified as unknown uncertainty. For known interval boundaries and fuzzy distributions with their corresponding arithmetic uncertainty the term of estimated uncertainty is used. The maximum amount of information about model parameters is attained with known or estimated continuous distribution functions corresponding to the so called stochastic uncertainty.

Here, stochastic uncertainty is taken into account in the form of stochastic sensor data, captured during the landing process. Consecutively, the uncertainty data is linked to the respective parts of the assembly structure. In the next step, the general CRC-process model and the specific landing process model are discussed in order to identify classification criteria for the sensor data.

Process model for the landing manoeuver. The Process model in Figure 2 complies with the process model of Bretz et al. which again extends the process model of Eifler et al. for active and semi-active systems [10,11]. The application corresponds to the Cessna 172 S and the process relates to a landing manoeuvre under visual flight rules. The time dependent operands represent the uncertainty data captured by the sensors. As disturbances wind, sight, topographic conditions and other aircrafts or airborne objects are taken into account. The exchangeable information comprises visual information like the cockpit and instrument view, radio and sensor data. Accordingly, human specific input parameters, such as fatigue or experience, are represented by the system-oriented input from the human user. No resources are exchanged and system quantities are out of scope. For a detailed description of the related experiment it is referred to Oberle [5].

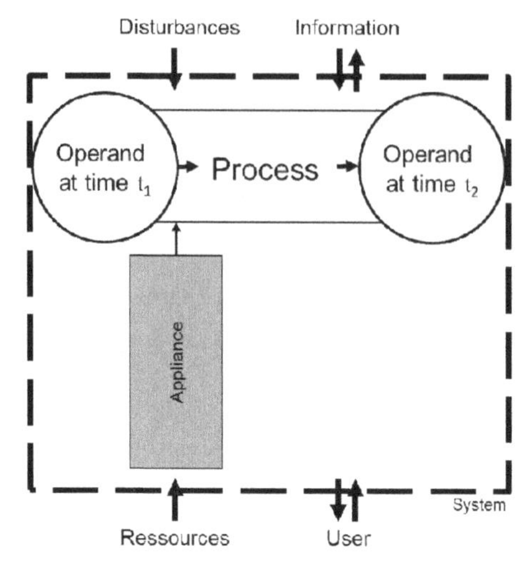

Figure 2: CRC 805 process model cf. [10]

Going more into the details of the landing manoeuvre, the process can be subdivided into the process-chain shown in Figure 3. The landing manoeuver is entered after the cruise flight and ends with the attainment of taxi speed. The main sub-processes are the descending, the final approach, intercept and flare and directional control on the runway. Simultaneous to the change of the trajectory during the landing both mental and physical sub-processes are performed in free order. The accomplishment of the approach briefing and the landing check-up are mental processes. Opposed to these, the setting of the three flap levels and the adaption of the fuel mixture are physical processes. Three additional physical processes can be found in [5]. During the complete landing manoeuvre an amount of 120 technical parameters is stored with a frequency of 9 Hz. The parameters are described in the following paragraph.

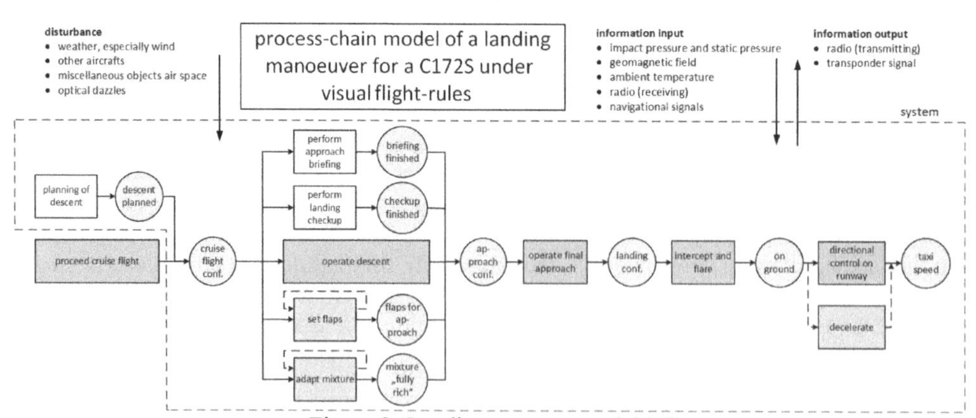

Figure 3: Landing process model [5]

Technical output parameter in the landing process. The operands at time t are operationalized by aircraft and global data. While global data is supposed to be constant for all aircraft components like time or weather conditions, aircraft data can be related to a specific aircraft component like aircraft engine data or aircraft landing gear data. The data is calculated and provided by Microsoft Flight Simulator X SDK and can be accessed via the SimConnect API. For logging Black Box 3 is used [13]. Overall, 120 parameters are recorded during five landing process of 44 pilots. Thus, data of 220 landings is available.

Since the data of all sensors is logged nine times per second, an oversupply of uncertainty data reduces its usability. In order to allow a useful interpretation of the relevant data in a CAD system, it needs to be set into relation with the respective sub-processes. Table 1 shows an exemplary selection of representative uncertainty measures that are, from an engineering perspective, only relevant during a specific phase of the landing process. Other uncertainty measures such as engine data is supposed to be relevant within the whole landing manoeuvre.

Table 1: Process depending stochastic uncertainty measures

Process	Uncertainty measure
Operate descent	Time to touchdown, angle
Set flaps	Time to touchdown of setting flap 1, 2 and 3
Adapt mixture	Time of mixture to touchdown, Mixture in %
Operate final approach	Acceleration and speed in the air
Intercept and flare	Acceleration and speed in the air and on touch down
Directional control on runway	Acceleration and speed, time to taxi speed

Human factors. According to the process model in **Figure 2** the pilot interacts with the aircraft and influences the technical output parameters of the landing process. Typically, pilots can differ in the properties listed in Table 2 which can differ over a continuous domain or in categories.

Table 2: Human factors influencing the landing process model

Human factors	Uncertainty measure
Experience measured in airtime	simulator airtime, real airtime
Fatigue	low, medium, high
Qualification	no motor, motor, commercial licence
Demographic data	age, educational level

Identification of uncertainty event chain. As the human factors influence both the process course and the process output a semantic relation between the process model and the influenced product is necessary. According to the UMEA, these relations can be formalised with respect to the context. The following three perspectives propose connection-rules for the consecutive ontology-based representation.

1. Connection between the sensor data and the product model:
    a. Global data is valid for the whole aircraft, all its subassemblies and parts. Thus the data is supposed to be attached to the top node of the assembly independently from the location of observation.
    b. Aircraft data is only valid for a specific part and is supposed to be attached to the location of observation.
    c. The instantiation of the assembly allows the distinction between the produced assemblies as well as their data.
2. Connection between the sensor data and the process model:
    a. Process independent data, e.g. engine data is important for the whole landing process and is supposed to be attached to all sub-processes.

   b. Process dependent data e.g. the acceleration on touch down is of importance only for a certain sub process and does not need to be attached to all processes.
3. Connection between the data of the human factors and the process model:
   a. Transitive process attributes derive from process independent human factor data, e.g. experience, qualification or demographic data
   b. Non-transitive process attributes derive from human factor data which is supposed to change during the landing process, e.g. fatigue

Ontology based formalization of the information model

The aerospace industry provides a fertile application domain for knowledge reuse and shows interest in ontology-based modelling for a long time [13]. Despite of an even longer tradition of ontology usage in engineering the number of publicly available aircraft ontologies is very limited [14]. A good counterexample is presented by Ast et al. who integrate the qu-rec20 ontology for units and dimensions in an aircraft structure ontology and allows for standard reasoning tasks [15,16]. Glas proposes a collaboration framework for aircraft design which is based on a reference ontology. After matching equivalent concepts changes in domain specific models are supposed to be propagated according to predefined merging rules between domains and by workflows over time [17]. A general ontology-based representation of assembly structures in the Web Ontology Language (OWL) is presented by the National Institut of Standards and Technology [18]. Using ontologies to connect and sort data throughout the product lifecylce is described in [19]. Product Life Cylce Support Library (PLCSlib) uses reference ontologies for the specification of ISO 10303-239 based data exchange between Product Data Management (PDM) Systems. Other related works are provided by Schorr, Verhagen et al. and Wang et al. [20,21,22].

Based on the developed requirements and the presented state of the art a reduced view on the sub-class hierarchy of the OWL based information model is proposed in Fig. 4. According to the UMEA it connects the product model with the process model as well as both of the models to the respective uncertainty information. In order to allow a broad adaption of the information model Dolce UltraLite is used as a reference. This so called top level ontology serves as a repository for general concepts and allows the modelling of interrelation between different domain and application ontologies. All Dolce UltraLite classes are preceded by the namespace abbreviation DUL. Both the Semantic Sensor Net Ontology with the namespace abbreviation ssn and the Quantities and Units Ontology with the namespace abbreviation qu descend from the Semantic Sensor Network Incubator Group of the W3C [23,24]. The Engineering domain specific design patterns of the Assembly Ontology proposed by the Fiorentini et al. with the abbreviation nist constitute the framework for reusing the same artefact repeatedly in the assembly structure without identification and naming problems [18]. This benefit is applied to the domain ontology of Ast et al. [18]. Thus, the artefacts of the Cessna 172 S can be identified by the class nist:Usage. The proposed extension of the CRC805 classes formalises the discussed landing process, connects the sub-processes to the ssn:SensorOutput class and to the crc805:HumanInput class. Additionally, the class ssn:ObservationValue is specialised by the crc805:UncertaintyDataType class with its corresponding sub-classes. Semantic Web Rule Language (SWRL) statements allow the definition of customized inference rules for the connection of ssn:SensorOutput instances with instances of nist:Usage. A particularity is the omittance of a class:Part. Since the Open World Assumption makes it impossible to fulfil the cardinality restriction, saying that a part made of zero assemblies or parts and the Closed World Assumption does not support cardinality constraints, an alleged class:Part would stay unpopulated. Thus all nist:Artifacts which are not classified as nist:Assembly are supposed to be parts.

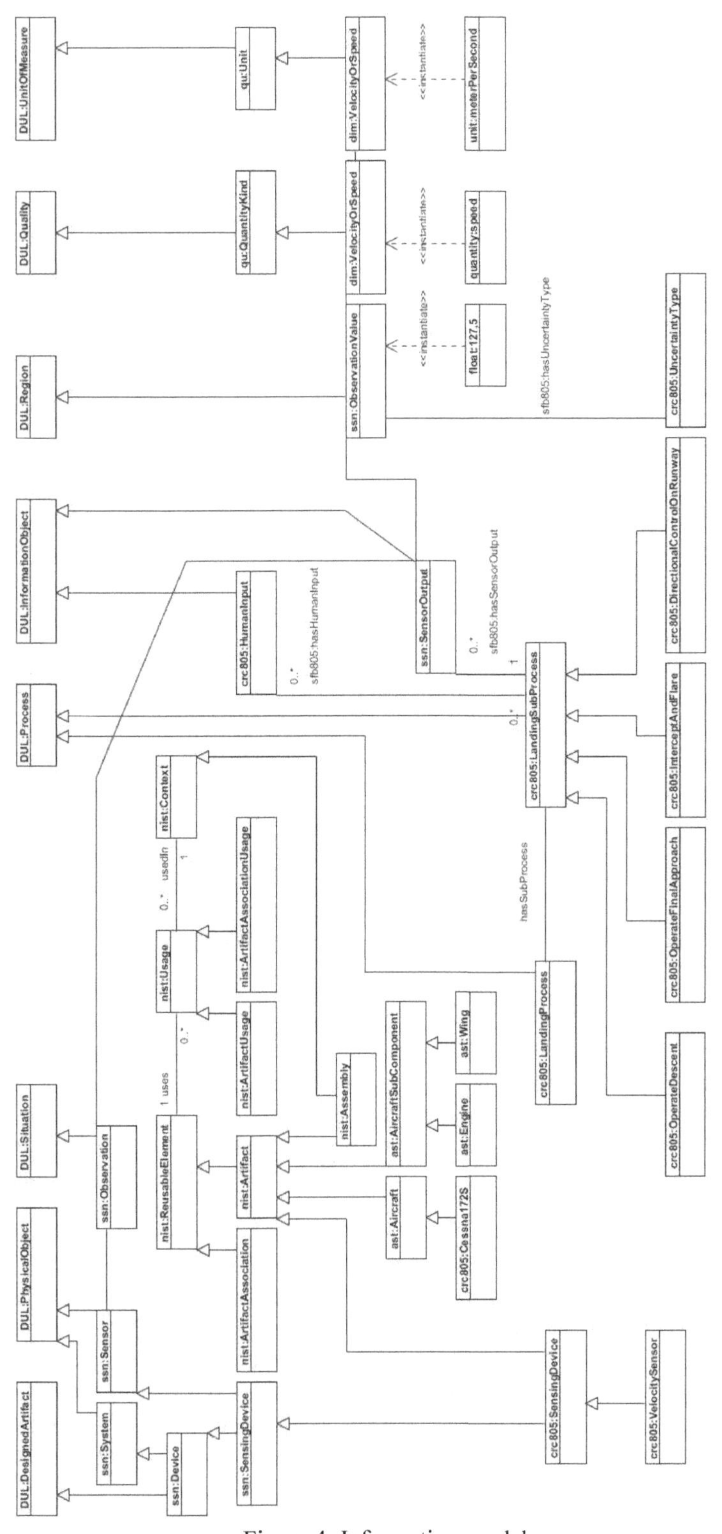

Figure 4: Information model

The information model represents one example for technologies becoming better known in the industrial application of cyber physical systems, also known as Industrie 4.0. The communality between these technologies is a long domain specific development culminating in the recent process to interconnect domain specific systems via internet. This gives rise to new design paradigms that are able to harmonize the determinism of computations and the intrinsically non-deterministic nature of physical systems and processes. In order to realise the communication of domain specific applications and models during the design phase a higher level of abstraction needs to be paired with semantic models [25]. This assertion can be extended to the downstream phases of the product life cycle like production and usage phase as well as to consecutive design phases of new products.

Cyber physical systems can be separated in a hard real-time component for the control of the physical processes and a soft real-time component for the internet protocol based communication [26]. Ontology-based systems are strongly related to the internet orientated part of the cyber physical systems. Firstly, standard reasoning tasks like classification and consistency check in OWL 2 with Direct Semantics are N2EXPTIME-complete with respect to the number of axioms in the T-Box of the knowledge base, thus time consuming [27]. Secondly, the communication between different cyber physical systems takes place on the internet oriented side of the system. Hence Ontologies are needed on the non-real-time side for the interpretation of exchanged data.

For these goals quasi Standards like the Common Data Dictionary of the International Electrotechnical Commission [28], eCl@ss [29] or ETIM [30] are developed. Due to their representation in XML or similar database-like structures the meaning can only be encoded in a traditional way. This means that the human interpretable meaning of a class is stored in a separate description which is not machine-readable. The readability for machines is achieved by mappings between different classes. Compared to the above proposed ontology, these taxonomies use the subclass-axiom; hence use only a fraction of the potential of Semantic Web Technologies. Beside the higher expressivity, OWL knowledge bases benefit from their machine-readable semantics.

Due to the openness of Semantic Web Technologies, distributed and prospective class definitions can be interrelated so that new class definitions can connect different domain interpretations. Contrary to this, classic glossaries or dictionaries are released by a certain interest group. The interrelation between different class definitions or complete repositories is not envisaged or desired. This necessitates a new way of thinking about information models for the exchange of meaningful data between applications in particular and between cyber physical systems in general.

Outlook

Different research areas can be identified for future domain-overlapping cooperation's as well as specialised research efforts. Firstly, the on-going cooperation can be extended to the integration of ontology engineering for the evaluation of uncertainty in the UMEA. Classification rules can enable reasoning over knowledge bases which include more than one system with the corresponding uncertainty data. Another possibility to exploit semantic data representation could be the formalisation of dependencies between human factors and technical data. This would allow situation dependent feedback for the pilot. Secondly, and stronger oriented towards knowledge representation, two major research questions are investigated in the near future. One solution for the clear identification of parts which are opposed to assemblies is needed for the community of engineering ontologists. Simply choosing the Horn-definable part of OWL to evade the question of Open versus Closed World Assumption does not solve this problem since OWL RL does not support cardinality restrictions. A second and much broader research area is the development of semantic models for the exchange of data between specific applications throughout the product life cycle and the exchange of data between cyber physical systems in general need to be promoted strongly. We argue that many enabling technologies like the Semantic Web Technologies already exist and can be used for the development of semantic driven models. Based on this, a visualization concept is developed in the next steps to interpret feedback data from processes meaningful.

Conclusion

Human behaviour is a source of uncertainty which is not enough taken into account in robust system design. In order to present information of the usage phase in the design phase, uncertainty data is connected to the product model. For the example of a Cessna 172 S aircraft uncertainty data is connected from the human pilot through the landing process to the product model of the aircraft. The data such as simulated sensor data is linked to the respective CAD model through an ontology. This allows a semantically consistent data exchange between different applications and the logic-based deduction of new relations between the data. In order to establish the connection between uncertainty data of the sensors on the one hand and the product and process models on the other hand the UMEA methodology is used. The main contributions of this article are the extension of the UMEA by human factors and the integration of the existing uncertainty information model in the context of the top level ontology Dolce UltraLite, the Semantic Sensor Net Ontology, the Quantities and Units Ontology and the domain specific NIST Assembly Ontology and the Aircraft Ontology.

Acknowledgements

The authors like to thank the German Research Foundation DFG for funding this research within the Collaborative Research Center (CRC) 805 "Control of Uncertainties in Load-Carrying Structures in Mechanical Engineering".

References

[1] R. Engelhardt, A model to analyse Uncertainties on Stakeholders' Evaluations in Technical Systems, Proceedings of the TMCE 2010, April 12–16, Ancona, Italy, 2010.

[2] VDI 3780 (2000), Technology Assessment Concepts And Foundations, VDI-Richtlinien, Düsseldorf, 2000.

[3] H. Birkhofer, Analyse und Synthese der Funktionen technischer Produkte, VDI Verlag Düsseldorf, 1980.

[4] R. Engelhardt, Uncertainty-Mode- and Effects-Analysis – an Approach to Analyze and Estimate Uncertainty in the Product Life Cycle, Proceedings of ICED 09, the 17th International Conference on Engineering Design, Vol. 2, Palo Alto, CA, USA, 2009.

[5] M. Oberle, R. Bruder, Process model for the investigation of human uncertainty during the usage of load bearing systems, Proceedings 19th Triennial Congress of the IEA, Melbourne, 2015.

[6] R. Moffat, Describing the Uncertainty in Experimental Results, Experimental Thermal and Fluid Science, New York, pp. 3-17, 1988.

[7] B. Möller, M. Beer, Engineering computation under uncertainty - Capabilities of non-traditional models, Computers and Structures, pp. 1024-1041, 2008.

[8] H. Hanselka, R. Platz, Ansätze und Maßnahmen zur Beherrschung von Unsicherheit in lasttragenden Systemen des Maschinenbaus: Controlling Uncertainties in Load Carrying Systems, VDI-Zeitschrift Konstruktion, pp. 55-62, 11/12 2010.

[9] R. Engelhardt et al., A Model to Categorise Uncertainty in Load-Carrying Systems, 1st MMEP International Conference on Modelling and Management Engineering Processes, Cambridge, 19-20 July 2010.

[10] A. Bretz et al., Darstellung passiver, semi-aktiver und aktiver Maßnahmen im SFB 805-Prozessmodell, 2015. Available on: http://www.sfb805.tu-darmstadt.de/media/sfb805/f_downloads/150310_AKIII_Definitionen_aktiv-passiv.pdf, Accessed on: 1. April 2015.

[11] T. Eifler et al., Approach for a consistent description of uncertainty in process chains of load carrying mechanical systems, Applied Mechanics and Materials, 2012, pp. 133-144.

[12] R. McElrath, BlackBox, 2013, Available on: http://www.robbiemcelrath.com/fs/blackbox/about, Accessed on: 2. April 2015.

[13] M Uschold, R. Jasper, A Framework for Understanding and Classifying Ontology Applications. Proceedings of the IJCAI-99 workshop on Ontologies and Problem-Solving Methods (KRR5), Stockholm, Sweden, pp. 1-12, August 1999.

[14] T. Gruber, G. Olsen, An Ontology for Engineering Mathematics, Fourth International Conference on Principles of Knowledge Representation and Reasoning, Bonn, pp. 258-269, 1994.

[15] M. Ast, M. Glas, T. Roehm, Creating an Ontology for Aircraft Design, Publikationen zum Deutschen Luft- und Raumfahrtkongress 2013, Stuttgart, pp. 1-11. November 2014.

[16] L. Lefort, Ontology for Quantity Kinds and Units: units and quantities definitions, 2010. Available on: http://purl.oclc.org/NET/ssnx/qu/qu-rec20. Accessed on: 26 March 2015.

[17] M. Glas, Ontology-based Model Integration for the Conceptual Design of Aircraft, 2013.

[18] X. Fiorentini et al., An Ontology for Assembly Representation, National Institute of Standards and Technology, 2007.

[19] A. Matsokis, D. Kiritsis, Ontology applications in PLM, International Journal of Product Lifecycle Management (Vol.5, No.1), pp. 84-97, 2011.

[20] H. Schorr, Ontologies, Knowledge Bases and Knowledge Management, University of Southern California, Los Angeles, 2011.

[21] W. Verhagen, R. Curran, An Ontology-based Approach for Aircraft Maintenance Task Support, 20th ISPE International Conference on Concurrent Engineering, pp 494-506, 2013.

[22] Y Wang, et al., Aviation Equipment Fault Information Fusion Based on Ontology, International Conference on Computer, Communications and Information Technology (CCIT 2014), 2014.

[23] W3C, Semantic Sensor Net Ontology, Available on:http://www.w3.org/2005/Incubator/ssn/wiki/Semantic_Sensor_Net_Ontology, Accessed on 8. April 2015.

[24] W3C, Ontology for Quantity Kinds and Units: units and quantities definitions, Available on: http://www.w3.org/2005/Incubator/ssn/ssnx/qu/qu-rec20, Accessed on 8. April 2015.

[25] E.A. Lee, Cyber Physical Systems: Design Challenges, 11th IEEE Symposium on Object Oriented Real-Time Distributed Computing (ISORC), Orlando, pp.363-369, 2008.

[26] F. Mueller, Challenges for cyber-physical systems: Security, timing analysis and soft error protection, High-Confidence Software Platforms for Cyber-Physical Systems (HCSP-CPS) Workshop, Alexandria, Virginia. 2006.

[27] W3C, OWL 2 Web Ontology Language Profiles, http://www.w3.org/TR/owl2-profiles/#Computational_Properties, Accessed on 9. April 2015.

[28] International Electrotechnical Commission, IEC 61360 - Common Data Dictionary, http://std.iec.ch/iec61360, Accessed on 9. April 2015.

[29] eCl@ss e.V., Classification and Product Description, http://www.eclass.de/, Accessed on 9. April 2015.

[30] ETIM Deutschland e.V., Das Klassifizierungsmodell der Elektrobranche, http://www.etim.de/, Accessed on 9. April 2015.

CHAPTER 7:

Uncertainty Quantification

Applied Mechanics and Materials Vol. 807 (2015) pp 195-204
© (2015) Trans Tech Publications, Switzerland
doi:10.4028/www.scientific.net/AMM.807.195

Submitted: 2015-04-29
Accepted: 2015-08-04

Analysis of the effect of uncertain clamping stiffness on the dynamical behaviour of structures using interval field methods

Maurice Imholz[1,a] , Dirk Vandepitte[1,b] and David Moens[2,c]

[1]Department of Mechanical Engineering, Division PMA, Katholieke Universiteit Leuven, Celestijnenlaan 300B, B-3001 Heverlee, Belgium

[2]Department of Mechanical Engineering, Mechanical Engineering Technology TC, KU Leuven, Technology Campus De Nayer, J. De Nayerlaan 5, B-2860 Sint-Katelijne-Waver, Belgium

[a]maurice.imholz@kuleuven.be, [b]dirk.vandepitte@kuleuven.be, [c]david.moens@kuleuven.be

Keywords: interval analysis, interval fields, possibilistic analysis, uncertainty modelling

Abstract. In uncertainty calculation, the inability of interval parameters to take into account mutual dependency is a major shortcoming. When parameters with a geometric perspective are involved, the construction of a model using intervals at discrete locations not only increases the problem dimensionality unnecessarily, but it also assumes no dependency whatsoever, including unrealistic parameter combinations leading to results that probably overestimate the true uncertainty. The concept of modelling uncertainty with a geometric aspect using interval fields eliminates this problem by defining basis functions and expressing the uncertain process as a weighted sum of these functions. The definition of the functions enables the model to take into account geometrically dependent parameters, whereas the coefficients in a non-interactive interval format represent the uncertainty. This paper introduces a new type of interval field specifically tailored for geometrically oriented uncertain parameters, based on a maximum gradient condition to model the dependency. This field definition is then applied to a model of a clamped plate with uncertain clamping stiffness with the purpose of identifying the effects of spatial variability and mean value separately.

Introduction

General concept of uncertainty quantification. Uncertainty Quantification (UQ) refers to the process of identifying and quantifying all possible sources of non-determinism within a design or its intended environment. All engineers will encounter sources of uncertainty at some point during their working life. UQ lies at the base of producing reliable products. Reliability analysis aims at coming to quantitative conclusions on the likelihood of a product to perform its intended action successfully. In practice, this comes down to identifying all possible configurations of a product and its environment that will lead to failure of the product and determining the likelihood of occurrence of such an event. For a designer, both the product and its working environment are subject to all kinds of uncertainties. Some of these are inherent to the non-deterministic character of nature, others are due to a lack of knowledge on the product or the environment. UQ attempts to find the uncertain response characteristics of any product given information on the uncertain nature of it. In numerical modelling, UQ comes down to numerically determining uncertainty on the output quantities of a model, given the uncertainty on the model input or parameters. More specifically, in FE analysis, uncertainty can be present in the model parameters, e.g. the element mass and stiffness values, but also in the model input, e.g. load or excitation characteristics. For numerical UQ to be successful, the modelling of the uncertainty should accurately represent the uncertainty present in real life, and the numerical model needs to propagate it accurately to obtain reliable uncertainty results at the output. Uncertainty can be quantified in many ways, which can be categorized in two groups. Stochastic methods use stochastic parameters such as variance and mean values to model non-deterministic variables. Although well-established in FE analysis (see e.g. [1, 2]), their application in industrial context remains limited because of the increased computational cost of a non-deterministic analysis, but more dominantly because of the high degree of knowledge required to perform it. Usually, stochastic parameters are expensive to obtain and assumptions on the probability distribution are necessary in most cases. Possibilistic methods model uncertainty on model input or model parameters using possibilistic parameters, of which intervals, fuzzy numbers and convex regions are the most common. An interval models uncertainty on a variable as two extreme bounds which are not to be exceeded. A fuzzy number is an extension to this concept, by considering the membership of a parameter to a certain interval as a continuous function, given by a membership function μ that ranges from 0 to 1. A value of 1 indicates the value is certainly part of the interval, whereas a value of 0 indicates that it is definitely not. Convex regions define an uncertainty region in a multidimensional space and attempt to capture uncertainty on parameter combinations by convex shapes such as ellipses. An overview of convex regions is given in [3]. In general, possibilistic modelling of uncertainty requires less extensive knowledge, but consequently, UQ using possibilistic methods provides less information. However, acquiring reliable knowledge on the bounds of the response of a system is of great interest to a designer. Possibilistic methods therefore provide a low-threshold analysis method for UQ compared to stochastic methods.

Non-deterministic fields. When uncertain parameters with a spatial character are involved (e.g. geometric shape uncertainties or material properties), care must be taken. In FE analysis, such a parameter is discretised to element level, leading to localised values in each element. Physically, some degree of dependence always exist within the discretised set. For this, a field representation $x(\mathbf{r}) \in \mathbb{R}$ is suitable, which describes the field variable x as a function of a spatial coordinate **r**. A single field realisation is then comprised of the value of the variable x in each point of the spatial domain considered. Non-deterministic field representations are possible as well. The theory of random fields (RF) [4] defines fields in a probabilistic way, with each field realisation to be a function of both the spatial coordinate and some well-specified random parameters. Next to RF, interval fields (IF) were introduced by Moens et al.[5] to model non-deterministic field variables in a possibilistic context. In its simplest form, interval fields are defined as a sum of weighted basis functions $\phi_i(\mathbf{r})$, in which the weights α_i^I are interval parameters. The general equation of an IF is given in Eq. 1:

$$x^I(\mathbf{r}) = \sum_{i=0}^{n} \alpha_i^I \cdot \phi_i(\mathbf{r}). \tag{1}$$

The functions $\phi_i(\mathbf{r})$ are deterministic functions of the spatial coordinate **r**, α_i^I are simple interval parameters. The objective of interval fields is to define an uncertain set of field realisations that are physically plausible, i.e. obey some geometric dependency criterion. The use of IF can be exploited fully if the definition of α_i^I and $\phi_i(\mathbf{r})$ is done in such a way that the interval parameters are all non-interactive. In this case, the dependency in the field is captured exclusively by the basis functions $\phi_i(\mathbf{r})$, whereas the uncertainty is captured only by the interval parameters α_i^I. This separation enables efficient uncertainty propagation throughout all steps of the analysis to the final output quantities. [6] considers an analysis with a single interval parameter to model the interval field. However, when more interval parameters are necessary to model an uncertain field with some degree of correlation between parameters, a decomposition has to be applied which ensures the interval parameters are truly non-interactive.

This paper will introduce a general IF definition that enables an analyst to model knowledge which he/she possibly has on the uncertainty of the application as straightforwardly as possible, while at the same time ensures this strict division of uncertainty and dependency between the α_i^I and $\phi_i(\mathbf{r})$. Once constructed, the interval parameters α_i^I will span a non-interactive interval space, but the field $x^I(\mathbf{r})$ will nevertheless obey a geometric dependency criterion, defined by a bound on the maximum gradient of the field variable. As an illustration, this field will be applied to a clamped plate model with uncertain clamping conditions, which will be propagated to obtain intervals on the plate's natural frequencies. This will illustrate the possibility to introduce a maximum gradient and show the effect on the set of field realisations.

The paper exclusively discusses intervals to model uncertainty. Throughout this work, an interval parameter x^I with upper bound $\overline{x}$ and lower bound $\underline{x}$ is represented in two ways: using the bounds themselves: $x^I = \langle \underline{x} | \overline{x} \rangle$ or using the centre point $x_c = \frac{\overline{x}+\underline{x}}{2}$ and the interval radius $r_x = \frac{\overline{x}-\underline{x}}{2}$: $x^I = \langle x \rangle_{x_c}^{r_x}$

Uncertainty definition

The starting point of any uncertainty analysis is the knowledge of the analyst. Being an expert at a specific application or product, the analyst may have some knowledge on the uncertainty present. The first section introduces the global uncertainty parameters which are used further on to define an IF decomposition. A second section discusses in more detail the mathematical background of the presented field decomposition method.

Definition of global uncertainty parameters. Consider a field variable $x(\mathbf{r})$. The value in each point is expressed as the sum of the mean field value μ_x and the deviation from the mean value $s_x(\mathbf{r})$ in that point:

$$x(\mathbf{r}) = \mu_x + x(\mathbf{r}) - \mu_x = \mu_x + s_x(\mathbf{r}) \tag{2}$$

where $\mu_x = \frac{1}{\Omega} \int_\Omega x(\mathbf{r}) d\mathbf{r}$ is assumed positive. Ω is a measure for the range of the field domain (e.g. the length, surface, or volume). When the field variable is uncertain, both the mean value and the deviation are subject to uncertainty. Eq. 2 is then left unchanged, with the superscript I added to the variables to indicate they are now interval parameters:

$$x^I(\mathbf{r}) = \mu_x^I + s_x^I(\mathbf{r}) \tag{3}$$

In this description, $\mu_x^I = \langle \underline{\mu_x} | \overline{\mu_x} \rangle$ is a simple interval parameter, but $x^I(\mathbf{r})$ and $s_x^I(\mathbf{r})$ are both interval fields, i.e. a set of possibly interactive intervals in each point of the field domain. It is assumed that the absolute value of the deviation is bounded by a specified value $s_{x,max}$. This indicates the field variable itself never exceeds the interval $\langle \underline{\mu_x} - s_{x,max} | \overline{\mu_x} + s_{x,max} \rangle$ in any point. To introduce the spatial dependency within the field, it is assumed that the difference between two adjacent points of the field is bounded as well, as most physical parameters normally exhibit some smooth evolution. This obliges the field values in adjacent points to vary within reasonable limits. To be able to express this property in a continuous field, *the gradient of the field is constrained.* The uncertainty is assumed to be rotationally symmetric with respect to the spatial dimension, leading to an interval on the norm of gradient, the magnitude of which is constant over the entire domain. The bounds are given by a single value $M = \|\nabla x\|_{max}$ leading to the interval $\langle 0 | M \rangle$.

In summary, the interval field $x^I(\mathbf{r})$ in this section has four major parameters defining the uncertainty:

1. The *lower bound on the mean field value* $\underline{\mu_x}$.
2. The *upper bound on the mean field value* $\overline{\mu_x}$.
3. The maximum absolute value of the *deviation from the mean value* $s_{x,max}$.
4. The maximum norm of the *gradient of the field* $\|\nabla x\|_{max}$.

Implications on local uncertainty in FE meshes. When a field is applied in FE models, the field is discretised to the nodes and elements of the FE model. Normally, the points of the field can be chosen to correspond to the nodes of the FE model. It is important to mention that the definition of a non-zero maximum gradient will lead to an interval of non-zero size in each point of the field. The *dimensionality of the uncertainty is equal to the number of points in the field*, as the interval parameters in each point are not fully dependent of each other (this would mean that the value in one point completely determines the value in another point). However, they are not fully independent either. The field decomposition presented below respects the dimensionality of the uncertainty and only tackles the dependency within the field.

The Local Interval Field Decomposition

Summary of objectives. The starting point for this field decomposition is the explicit interval field description introduced by Moens et al. [5]. Eq. 1 expresses the field $x^I(\mathbf{r})$ as a weighted sum of basis functions, in which the weights are interval parameters. To adapt the field description to the definition of the global uncertainty parameters from the previous section, the IF definition of Eq. 1 is slightly altered to:

$$x^I(\mathbf{r}) = \mu_x^I + \sum_{i=0}^{n} \alpha_i^I \cdot \phi_i(\mathbf{r}). \tag{4}$$

Next to this, we define the set of field realisations X_s as:

$$X_s = \left\{ \tilde{x}(\mathbf{r}) \middle| \tilde{x}(\mathbf{r}) = \tilde{\mu_x} + \sum_{i=0}^{n} \tilde{\alpha}_i \cdot \phi_i(\mathbf{r}), \forall \tilde{\alpha}_i \in \alpha_i^I, \forall \tilde{\mu_x} \in \mu_x^I \right\}. \tag{5}$$

The analyst may specify the four previously defined global uncertainty parameters, which impose conditions on the field that have to be met. Each field realisation that is part of X_s has to obey the following conditions:

1. $0 \leq \underline{\mu_x} \leq \int_\Omega \tilde{x}(\mathbf{r}) d\mathbf{r} = \tilde{\mu}_x \leq \overline{\mu_x}$
2. $\forall \mathbf{r} \in \Omega : |\tilde{x}(\mathbf{r}) - \tilde{\mu}_x| \leq s_{x,max}$
3. $\|\nabla x\| \leq \|\nabla x\|_{max}$

In addition, the interval parameters α_i^I need to be *non-interactive* without violating this conditions for any field realisation $\tilde{x}(\mathbf{r})$ within the uncertain set. The remainder of this section discusses the definition of the field decomposition for a *1D spatial domain*, i.e. a variable which can vary along a line in space. This will suffice for the application on the clamped plate model. However, all properties of the field can be extended to 2D and 3D domains, at the cost of increased mathematical complexity. The spatial coordinate will be indicated by the variable y throughout the remainder of this paper.

Definition of basis functions. In Eq. 4, $\phi_i(\mathbf{r})$ are deterministic basis functions. For the purpose of propagating the interval field to obtain the output uncertainty, it is beneficial that the interval parameters α_i^I span a non-interactive interval space, since existing uncertainty propagation methods are based on non-interactive interval spaces. The basis functions possess the following properties to ensure this condition is met:

1. All ϕ_i are piecewise second order polynomial functions so their first spatial derivative is continuous (see Fig. 1).
2. A single ϕ_i is positioned at the centre point y_i of each element of the FE mesh, centered on this node (see Fig. 2).
3. All ϕ_i are identically shaped
4. All ϕ_i satisfy $\int_\Omega \phi_i(y) dy = 0$, the integral ranging over the entire domain of y.

This leads to basis functions of Eq. (6):

$$\phi_i(y, y_i) = f(|y - y_i|, a, R). \tag{6}$$

In this equation, a is referred to as the *influence strength* of the field, which determines the ability of an individual point to influence the value in nearby points. It directly determines the maximum gradient bound. R is the *influence radius* and determines the distance over which a point can influence the value

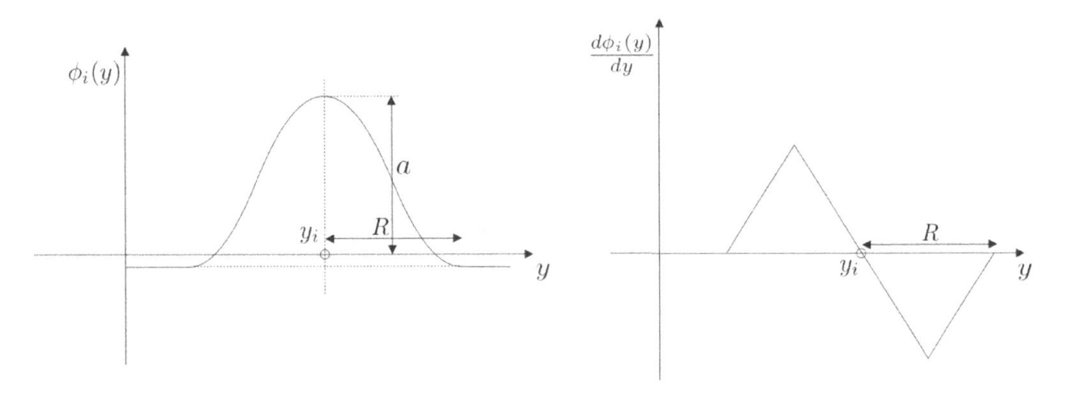

Figure 1: Definition of a basis function and its first derivative.

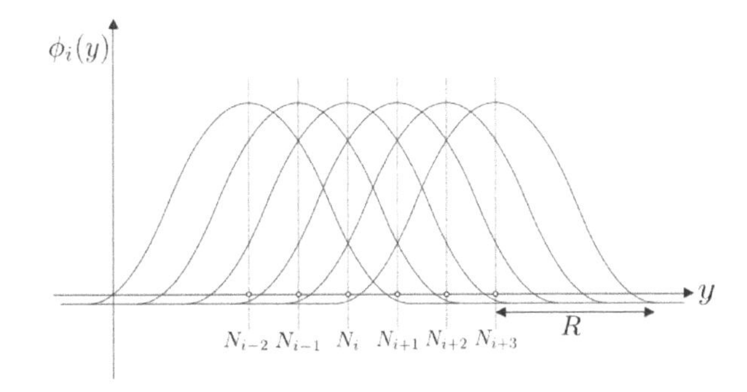

Figure 2: A separate basis function is defined in each field point.

in other points. The basis functions $\phi_i(y)$ are defined in each point along the field, as illustrated in Fig. 2. Fig. 2 shows that the basis functions have a local character: beyond a radius R their ability to change the field value vanishes, hence the name *Local Interval Field Decomposition* (LIFD). Moreover, the choice of piecewise second order polynomials ensures the basis functions decay smoothly to zero at the end of their influence radius with a gradient equal to zero. We emphasize that in points within a distance R of the edges of the domain Ω, the shape of the functions ϕ_i has to be altered slightly to ensure the maximum gradient constraint is constant over the entire domain. The derivation of the shape is omitted from this discussion, but can be found in previous work of the same authors. Using this basis functions, the field is expressed as:

$$x^I(y) = \langle\mu_x\rangle_{C_\mu}^{R_\mu} + \sum_{i=1}^{N_k} \langle\beta_i\rangle_0^1 \cdot \phi_i(y, a, R), \tag{7}$$

with $C_\mu = \frac{\overline{\mu}+\mu}{2}$ and $R_\mu = \frac{\overline{\mu}-\mu}{2}$. The field description of Eq. 7 puts an interval $\langle\beta_i\rangle_0^1$ in each node/element of the FE mesh and therefore it *does not reduce the dimensionality of the uncertainty*. However, *these intervals can now be considered as non-interactive while still obeying the maximum gradient condition*, which was the initial goal of this decomposition method. The $\langle\beta_i\rangle_0^1$ combined with $\langle\mu_x\rangle_{C_\mu}^{R_\mu}$ now span an $(n+1)$-dimensional non-interactive uncertainty region. Fig. 3 illustrates how field realisations are created from random parameter selection within this hypercubic space. The physical feasibility of such a realisation is ensured through the relation between the maximum gradient, the influence radius and the influence strength. This relation is omitted from this paper but can be found in earlier work of the same authors. In Fig. 3, the dotted lines represent basis functions in four points with a random scaling factor within the interval $\langle -a|a\rangle$. Their sum is given by the solid line.

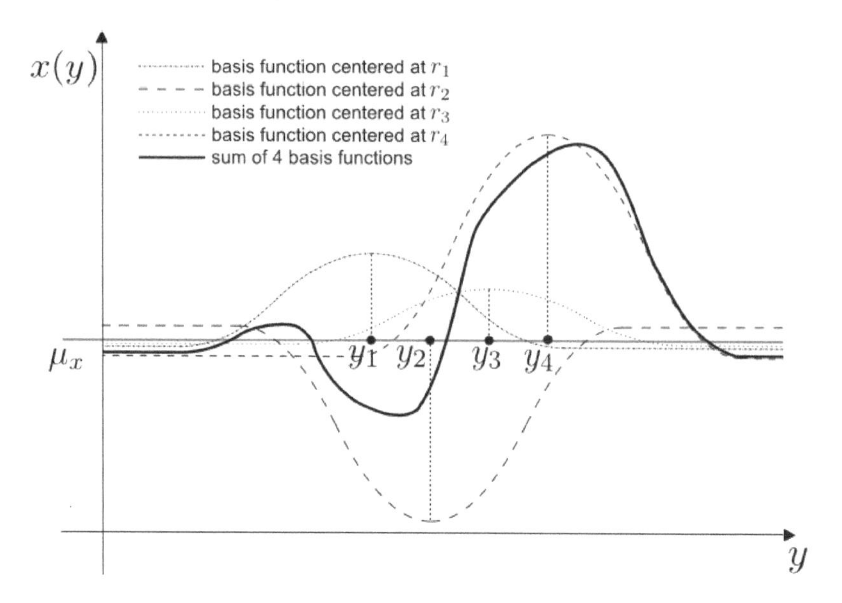

Figure 3: Superposition of 4 basis functions to produce a single field realisation.

Non-deterministic analysis case: clamped plate

The purpose of this numerical example is to illustrate the use of LIFD on a clamped plate FE model with an uncertain clamping stiffness K_θ. For a few values of the maximum gradient, the uncertain clamping stiffness set will be visualised and the effect on the uncertainty on the natural frequencies is assessed.

Description of analysis performed. We examine a homogeneous thin square plate clamped at one edge. The FE model consists of 10x10 quadratic Kirchhoff plate elements. A clamping stiffness value is defined in each element along the clamped edge, leading to a total of 10 clamping stiffness values. Since the LIFD targets spatial uncertainty, we assume the clamping stiffness in each element can be different, and the maximum gradient condition enables the introduction of dependency within the uncertain set. For the purpose of this analysis, the deviation is defined *relative* to the mean value $K_\theta(y) = \mu(1 + s(y))$, leading to Eq. 8. As the influence strength a serves as a scaling factor, it is taken out of the summation.

$$K_\theta^I(y) = \langle \mu_K \rangle_{C_\mu}^{R\mu} \cdot \left(1 + \sum_{i=1}^{N_k} \langle \beta_i \rangle_0^1 \cdot \phi_i(y, a, R)\right) = \langle \mu_K \rangle_{C_\mu}^{R\mu} \cdot \left(1 + a \sum_{i=1}^{N_k} \langle \beta_i \rangle_0^1 \cdot \phi_i(y, R)\right) \quad (8)$$

In this case, $N_k = 10$, the number of elements along the clamped edge. Corresponding to this definition, the maximum gradient condition is normalized to obtain bounds on $\frac{1}{\mu}\frac{\partial K_\theta}{\partial y}$. It can be shown that $\left(\frac{1}{\mu}\frac{\partial K_\theta}{\partial y}\right)_{max} = 20a$, so a choice of the influence strength immediately sets the maximum gradient condition. The influence radius determines the maximum deviation from the mean value. We aim at performing a study on the gradient alone, as a measure for the quality of the clamping stiffness. For this, we assume an interval on the mean value of $\mu_K^I = \langle 10^1 | 10^2 \rangle$. Starting from this, we examine five different values for the maximum gradient. In each case, the influence radius is chosen as a whole number of element lengths, ranging from 5 to only 1. The influence strength is chosen accordingly to keep the maximum deviation constant over all analysis cases, so that the only varying uncertainty parameter is the maximum gradient. Due to this choice of cases, the influence radius decreases linearly, but the influence strength increases non-linearly, however monotonously. The five analysis cases are

Table 1: LIFD parameters for each analysis case.

Global uncertainty parameters	Reference case 0	analysis case 1	analysis case 2	analysis case 3	analysis case 4	analysis case 5
interval on mean value μ_K^I in N/m	$\langle 10^1 \vert 10^2 \rangle$	$\langle 10^1 \vert 10^2 \rangle$	$\langle 10^1 \vert 10^2 \rangle$	$\langle 10^1 \vert 10^2 \rangle$	$\langle 10^1 \vert 10^2 \rangle$	$\langle 10^1 \vert 10^2 \rangle$
relative gradient $\frac{1}{\mu}\frac{\partial K_\theta}{\partial y}$ in m^{-1}	0	3.682	4.044	5.048	6.428	10
relative deviation $\left\| \frac{K_\theta - \mu_K}{\mu_K} \right\|_{max}$	0.9	0.9	0.9	0.9	0.9	0.9
influence strength a in N/m	0	0.1841	0.2022	0.2524	0.3214	0.5
influence radius R (number of elements)	-	5	4	3	2	1

listed in Table 1, including a reference case where $\left(\frac{1}{\mu}\frac{\partial K_\theta}{\partial y}\right)_{max} = 0$, corresponding to a constant clamping stiffness value.

Solution method. Each of the cases defines a set of clamping stiffness configurations $\{K_\theta(y)\}$ for which we will determine intervals on the first 10 natural frequencies of the plate. Since the maximum gradient is monotonically increasing, it is important to notice that every new case expands the uncertain set and that therefore all configurations part of a set corresponding to a certain maximum gradient are also part of all sets corresponding to a larger maximum gradient. It is therefore impossible that the intervals on the natural frequencies decrease in size when the cases 0 to 5 are analysed. Initial simulations show that the natural frequencies are monotonically increasing with respect to increasing mean value. We therefore only consider the extreme values of the mean value. The reference case consists of finding the natural frequencies corresponding to a constant clamping stiffness equal to these extreme values of the mean value. Increasing the maximum gradient creates an uncertain domain of possible clamping configurations in which - for each analysis case - two optimisation problems are solved: at $\mu_K = 10^2$ to find the configuration that provides maximum frequency value, and at $\mu_K = 10^1$ to find configurations that provide minimum frequency values. Each natural frequency is optimised individually, leading to possibly different configurations for each natural frequency. To find the configurations of extreme frequency values, a constant-step gradient based optimisation method is used. The starting configuration in each case is a constant clamping stiffness value equal to the mean value. Experimentation with different starting points indicate that the optimisation problem is convex and the use of gradient-based methods effectively provide the global optimum.

Discussion of results. The results of the optimisation routines are given in Fig. 4, which shows the relative maximum (o) and minimum (x) frequency values for increasing influence strength/maximum gradient. The maxima and minima are given relative to the maximum respectively the minimum frequency value calculated in the reference case, displaying the reference interval as a point at value 1. The initial increase in maximum gradient means a significant increase in interval size for all frequencies. After that, further increase is of much smaller influence. An exception to this seems to be the 10th frequency, for which the interval size continues to grow with increasing maximum gradient. To further analyse this observation, we examine the clamping stiffness configurations that produce the interval bounds on the 10th natural frequency and compare them to the corresponding mode shape. Fig. 5 shows the first 10 mode shapes of the plate, Fig. 6 zooms in the 10th mode shape specifically, determined for a clamping stiffness equal to zero. Looking at Fig. 6, we observe that at the supported edge the angular displacement is maximal in elements 3,4,7 and 8, and very limited in other elements. From Fig. 7, we clearly see large clamping stiffness values at element 3,4,7 and 8 for the maximum frequency value, and small clamping stiffness values in the same elements for the minimum value. The minimum and maximum configurations consists of a low respectively high mean value of the clamping stiffness over elements where it matters for the displacement of the corresponding mode.

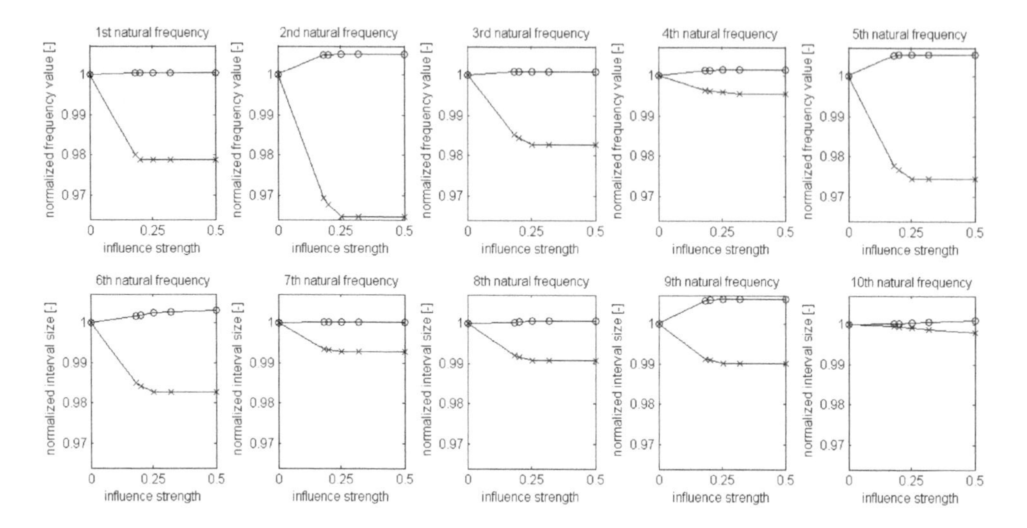

Figure 4: Normalised maximum (o) and minimum (x) frequency values for the 10 lowest natural frequencies for increasing influence radius.

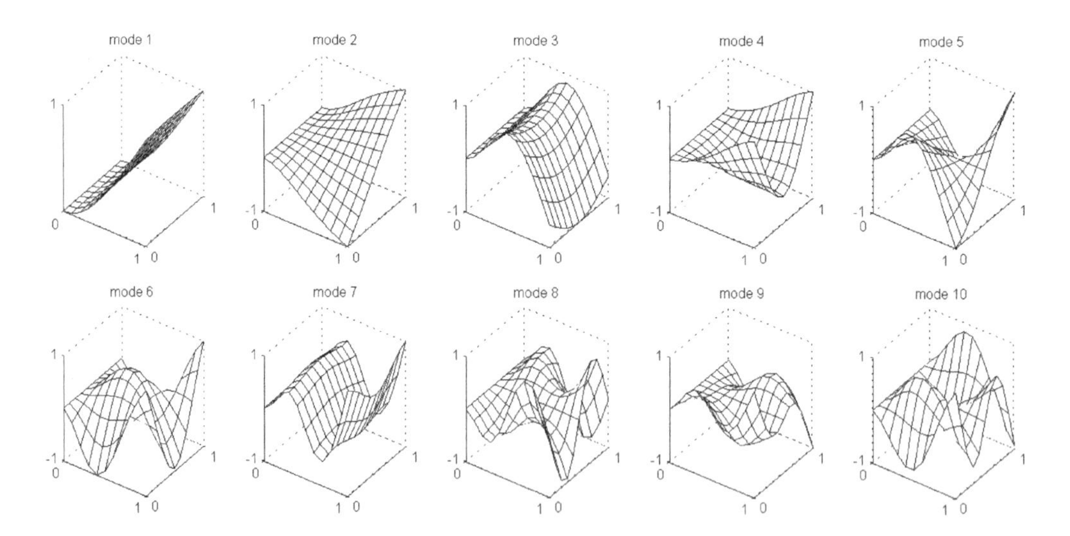

Figure 5: First 10 mode shapes of the clamped plate

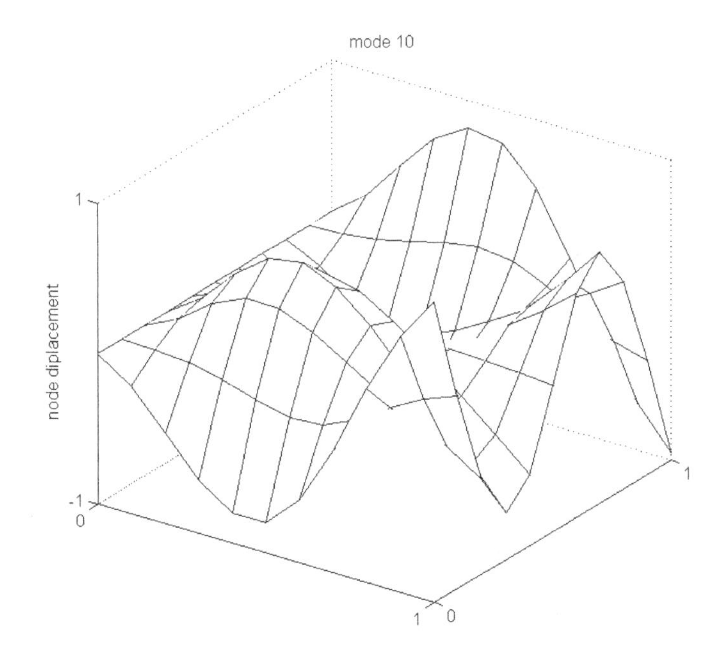

Figure 6: Zoom in on 10th mode shape

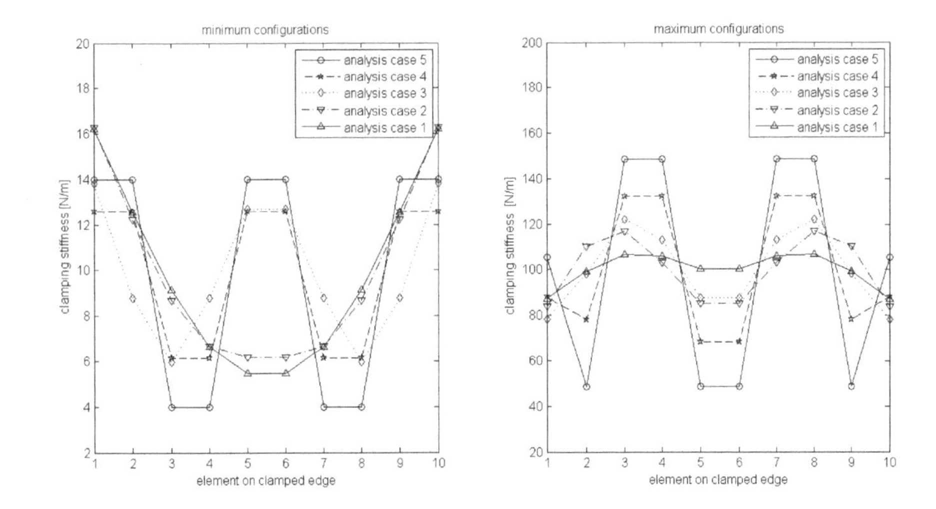

Figure 7: Clamping stiffness configurations that produce the extreme values for the interval on the 10th frequency.

We stress that the maximum deviation is kept constant for all cases, so the maximum and minimum allowable value in any element is constant for all analysis cases. However, for mode 10 specifically, a higher gradient is necessary to attain this extreme values in elements 3,4,7 and 8. In fact, Fig. 7 shows that each subsequent increase in maximum gradient pushes the extreme values further out. For other frequencies, this effect does not show, as the extreme clamping values in the relevant elements are attained at much lower maximum gradient, at which point further increasing the maximum gradient yields no effect. For higher order mode shapes, higher maximum gradients are necessary to reach the extreme clamping values in all relevant points.

Lastly, Fig. 4 proves the usefulness of the selected approach using the LIFD. Analysis case 5 with the influence strength equal to 0.5 corresponds to the case in which all dependency in the input field is neglected, which for certain frequencies can lead to conservative (i.e. too large) output frequency intervals, depending on the true dependency present in the field.

Conclusion

This paper quantifies effects of uncertainty on the clamping stiffness of a clamped plate with possible spatially varying clamping stiffness value. Using the LIFD, which is based on the explicit interval definition, a non-interactive uncertainty region is formed which still obeys a certain maximum gradient condition. By this definition, the effect of the maximal gradient could be separated from influence of the mean value and the deviation of the mean value. For different values of the maximum gradient, the uncertainty was quantified by defining intervals on the natural frequencies of the plate using a basic gradient-based optimisation routine. It was shown that the clamping stiffness configurations that produce these bounds are related to the corresponding mode shape.

Acknowledgement

The Research Fund KU Leuven and the Fund for Scientific Research – Flanders (F.W.O) are gracefully acknowledged for their support.

References

[1] R. Ghanem, P.D. Spanos, Stochastic finite elements: a spectral approach, MIT Press, Cambridge, 2003.

[2] I. Elishakoff, Y.J. Ren, Finite element methods for structures with large stochastical variations, Oxford texts in Applied and Engineering Mathematics, Oxford, 2003.

[3] Y. Ben-Haim, I. Elishakoff, Convex Models of Uncertainty in Applied Mechanics, Elsevier, New York, 1990.

[4] E. Vanmarcke, Random Fields: analysis and synthesis, MIT Press, Cambridge, 1983.

[5] D. Moens, M. De Munck, W. Desmet, D. Vandepitte, Numerical dynamic analysis of uncertain mechanical structures based on interval fields, IUTAM 2009, IUTAM Symposium on the Vibration Analysis of Structures with Uncertainties Vol 27.(2011) 71-83.

[6] W. Verhaeghe, W. Desmet, D. Vandepitte, D. Moens, Interval fields to represent uncertainty on the output side of a static FE analysis, Computational Methods in Applied Mechanical Engineering 260(2013) 50-62.

Applied Mechanics and Materials Vol. 807 (2015) pp 205-217
© (2015) Trans Tech Publications, Switzerland
doi:10.4028/www.scientific.net/AMM.807.205

Submitted: 2015-04-19
Accepted: 2015-08-04

Methodical Approaches to Describe and Evaluate Uncertainty in the Transmission Behavior of a Sensory Rod

Christiane Marianne Melzer[1,a*], Martin Krech[2,b], Lisa Kristl[3,c], Tillmann Freund[4], Anja Kuttich[3], Maximilian Zocholl[5], Peter Groche[2], Michael Kohler[3] and Roland Platz[6]

Technische Universität Darmstadt, Magdalenenstr. 4, 64289 Darmstadt, Germany
[1]Fachgebiet Systemzuverlässigkeit und Maschinenakustik,
[4]Fachgebiet Produktentwicklung und Maschinenelemente

Technische Universität Darmstadt, Otto-Berndt-Str. 2, 64287 Darmstadt, Germany
[2]Institut für Produktionstechnik und Umformmaschinen,
[5]Fachgebiet Datenverarbeitung in der Konstruktion

[3]Technische Universität Darmstadt, Fachbereich Mathematik, Schlossgartenstr. 7, 64289 Darmstadt, Germany

[6]Fraunhofer Institute for Structural Durability and System Reliability LBF, Bartningstr. 47, 64289 Darmstadt, Germany

[a]melzer@szm.tu-darmstadt.de, [b]krech@ptu.tu-darmstadt.de, [c]kristl@mathematik.tu-darmstadt.de

Keywords: Monte-Carlo simulation, fuzzy analysis, interval analysis, sensory rod

Abstract. Load-bearing mechanical structures like trusses face uncertainty in loading along with uncertainty in their strength due to uncertainty in the development, production and usage. The uncertainty in production of function integrated rods is investigated, which allows monitoring of load and condition variations that are present in the product in every phase of its lifetime. Due to fluctuations of the semi-finished parts, uncertainty in governing geometrical, mechanical and electrical properties such as Young's moduli, lengths and piezoelectric charge constants has to be evaluated. The authors compare the different direct methodical approaches Monte-Carlo simulation, fuzzy and interval arithmetic to describe and to evaluate this uncertainty in the development phase of a simplified, linear mathematical model of a sensory rod in a consistent way. The criterion to compare the methodical approaches for uncertainty analysis is the uncertain mechanical-electrical transmission behavior of the sensory rod, which defines the sensitivity of the sensory compound.

Introduction

Within the lifetime of mechanical load-bearing components, there are different sources of uncertainty, accumulating in each phase. The resulting consequences are often detected only in a later usage phase, resulting in unpredictable failures, wide product recalls or high service costs. The German Collaborative Research Centre SFB 805 "Control of Uncertainty in Load-Carrying Structures in Mechanical Engineering" investigates methods and technologies to control uncertainty within the product development, production and utilization of load bearing systems according to the working hypothesis of the SFB 805 that "uncertainty occurs when process properties of a system cannot or only partially be determined" [1]. A requirement for the detection and control of uncertainty is the acquisition of information in the uncertainty afflicted process phases.

A current future-oriented research topic in the field of mechanical engineering is the fusion of mechanical, load-bearing structures with additional functions [2, 3, 4, 5]. By ongoing miniaturization of sensors, electronic components and networking technologies, the possibilities for highly integrated structures with additional mechatronic functions are arising. These so called smart machine elements can monitor their stresses, estimate their remaining lifetime or control maintenance intervals [2]. Uncertainty occurring during the production of these components can already be controlled by process monitoring, utilizing the integrated sensor itself. For a widespread application of products, it is necessary to qualify economic methods of mass production. High outputs can, inter alia, be realized by forming-manufacturing technologies. Some of the current research projects in this field deal with micro-extrusion of piezoelectric fibres [3], deep drawing of Piezo-Metal-Composites [4] or joining by rotary swaging [5]. However, these works do not conduct any investigations about evaluating uncertainty during the manufactoring processes. Within the scope of the SFB 805 the production and utilization of sensory rods is investigated [6]. To estimate uncertainty distributions of the manufactured batch, it is necessary to develop powerful mathematical tools for an efficient computation, which is one of the major aims of the SFB 805. Scientists from mechanical engineering and mathematics are investigating methods in order to evaluate and to reduce uncertainty in the development, production and usage of products. There are several approaches to describe parametric uncertainty with probabilistic and possibilistic methods in literature such as Monte-Carlo simulation [7], interval [8] and fuzzy [9] analysis. In the context of Monte-Carlo simulation, density [10] and quantile [11] estimation are important methods to describe uncertainty.

In this paper stochastic and non-stochastic uncertainty in the production process of sensory rods is investigated with a simplified analytical model of a sensory rod. Fluctuating properties of semi-finished parts result in variations in the mechanical-electrical transmission behavior of the sensory rod. The knowledge of this scatter is important for the engineer in order to define an appropriate amount of calibration effort or to determine adjusted safety factors for the system. The investigation is exemplarily carried out with five varying input parameters from geometrical, material and electrical properties, which are chosen according to their significance and availability of uncertain data. A comparison of several direct estimation methods, parametric and non-parametric density estimation, quantile estimation, direct interval analysis and direct fuzzy analysis, is employed with regard to scatter range and computational cost.

Manufactoring of Sensory Rods by Means of Rotary Swaging

In many phases in the lifetime of a structural load-bearing component, there is a necessity to identify its loads and conditions in order to guarantee its reliability and to gather information about the whole

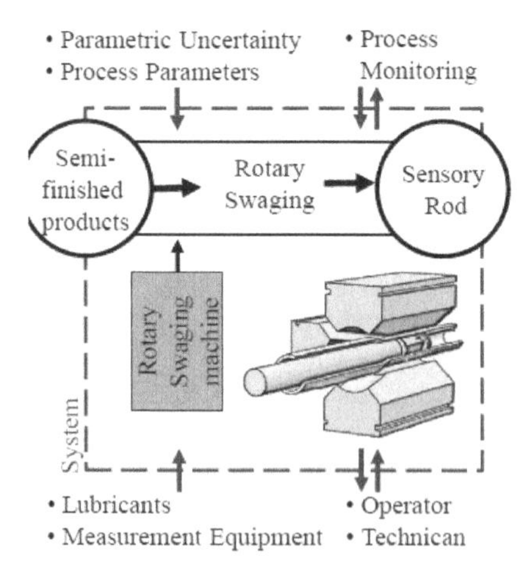

Fig. 1: SFB 805 uncertainty process model for the manufacturing process of sensory rods, [12]

system the component is integrated in. The advanced ability to monitor can be realized by adding sensor elements to the high loaded or sensitive crucial part, which usually is only possible on accessible spots. Unfortunately, the fragile sensors are then exposed to mechanical and environmental influences, which may lead to damages during its lifetime. By integrating sensor elements inside of load-bearing structural components, it is possible to prevent this uncertainty. An adequate manufacturing process for these function-integrated components is the multistage rotary swaging process [6]. Within the scope of the SFB 805, sensory load-bearing components are needed in truss structures with modular passive and active subsystems. The flow of forces within a bar frame can, due to uncertainty, differ noticeable from the prognoses based on analytic and numeric calculations. The production sequence of a sensory rod is illustrated within the uncertainty process model of the SFB 805 in Fig. 1. Rotary swaging is an incremental forming process, which means that the shape change is realized by many small forming steps by four oscillating swaging tools. In a first step, a preform with an inner shoulder and a reduced diameter is produced via infeed rotary swaging and the use of a mandrel. Afterwards, the inner parts, consisting of two end caps and the sensor body, are inserted. In this way, the fragile sensor is placed in an area with already reduced diameter and wall thickness. The high radial forming forces are not acting directly on the sensor, but on the chamfer of the second end cap. The form closure and pretension of the inner parts emerges from the material flow of the tube, which is adapting the chamfer geometry of the second end cap. To control the resulting pretension, it is mandatory to adjust the right lengths, angles and other parameters of preform, tools and end caps. A more detailed description of the joining process is given in [6]. For a large production batch, it can be of interest to estimate the uncertainty distributions in beforehand for a given set-up of tube, end caps and sensor. In this paper, the mechanical-electrical transmission behavior of a piezoceramic sensor is investigated. For this, the piezoelectric sensor is embedded inside a compression-tension loaded tube. The relation between axial forces applied to the passive tube part and the charge displacement of the piezoelectric sensor, the mechanical-electrical transmission behavior, is examined. For a large batch, the question arises whether a costly, individual calibration has to be conducted for each manufactured rod, or if the inaccuracies can be tolerated for a given measurement task. Similarly, the measurement amplifier has to be matched to the sensors output and therefore the measurement range is limited. A simplified mathematical model of the sensory rod, which allows the dimensioning of the joint components and the estimation of uncertainty distribution due to varying geometrical, mechanical and electrical input parameters, is presented in Fig. 2 and Table 1. The left part of the figure shows the design and dimensions of the sensory rod, consisting of a tube, end caps and the sensor body. On the right side, the corresponding spring model with its stiffnesses and its parameters is presented. The stiffness of a section of an elastic part is calculated by the Young's modulus E, an equivalent cross-sectional area A, and a length l. Furthermore, the indices T, S and E represent the tube, the sensor and the end cap. The circumferential stiffnesses of the axially symmetric sections of the tube are summarized in c_{T1} and c_{T2}, even though they are marked only on the upper springs due to symmetry. The cubic sensor body is integrated into the tube under mechanical pretension F_S and is in the flow of forces of externally applied loads. While a tensile load F gradually decreases the pretension of the sensor, a compressive force acting on the tube would lead to the same change of forces inside the sensor with an opposite sign. The equivalent springs of the sensor and end cap sections c_{S1}, c_{E1}, c_{E2} and c_{E3} are compressed, whereas the corresponding tube section is under tensile stress. Compression and tensile stresses are distinguished by drawing the springs accordingly in Fig. 2.

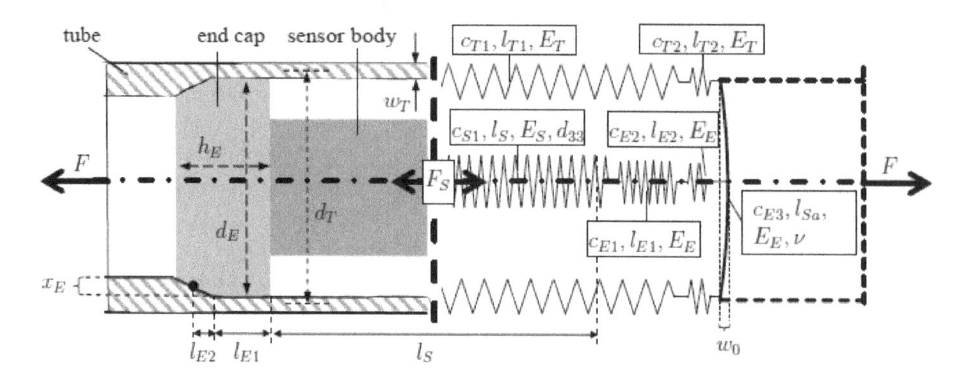

Fig. 2: Design of a sensory rod and mechanical equivalent system

The axial stiffnesses of the joint components determine the mechanical-electrical transmission behavior of axial loads acting on the sensor signal. Similarly to a bolt calculation, different compliances of the clamped parts are calculated and interconnected in parallel and series systems. The subsequent stiffnesses of the mere tube part have no effect on the electrical-mechanical transmission behavior of the sensor and are displayed as dotted lines.

The sensor itself is a uniform block and can be described as c_{S1}, whereas the end cap persists of multiple sections and has to be divided into c_{E1}, c_{E2} and c_{E3}. Also the outer tube is divided into an uniform section c_{T1} and the transition section c_{T2}. For the stiffnesses c_{E2} and c_{T2} in the transition between the end cap and the tube, the surface contact is simplified as a mean line contact along the perimeter. The resulting contact length l_{E2} is taken into account to describe this continuous transition, with the corresponding difference in diameter x_E, see Fig. 2.
The equivalent cross-sectional areas A of the tube sections, the sensor and the end cap sections are given by:

$$A_{T1} = \pi \cdot d_T \cdot w_T, \quad A_{T2} = \pi \cdot d_T \cdot \left(w_T + \frac{1}{2} \cdot x_E \right) \tag{1}$$

$$A_S = l_{Sa}^2 \tag{2}$$

$$A_{E1} = \frac{\pi \cdot d_E^2}{4}, \quad A_{E2} = \frac{\pi \cdot \left(d_E - \frac{1}{2} \cdot x_E \right)^2}{4}. \tag{3}$$

The mean diameter of the reduced tube d_T together with the wall thickness w_T is used to calculate the cross-sectional area of the tube A_{T1}. The lengths of the stiffnesses of the tube sections c_{T1} and c_{T2} result from the lengths of the internal parts. For the uniform section c_{T1}, the length is given by l_S and two times the length of the cylindrical end cap section l_{E1}. Similarly c_{T2} is given by the length of the transition section of the end cap l_{E2}. The resulting axial stiffnesses of the tube, sensor and end cap are given by:

$$c_{T1} = \frac{E_T \cdot A_{T1}}{l_S + 2l_{E1}}, \quad c_{T2} = \frac{E_T \cdot A_{T2}}{l_{E2}} \tag{4}$$

$$c_{S1} = \frac{E_S \cdot A_S}{l_S} \tag{5}$$

$$c_{E1} = \frac{E_E \cdot A_{E1}}{l_{E1}}, \quad c_{E2} = \frac{E_E \cdot A_{E2}}{l_{E2}}. \tag{6}$$

Because of the radial offset in the flux of forces in the end caps, a bending moment occurs in both of the end caps. As indicated by the bended line in the equivalent stiffness model in Fig. 2, an additional axial compliance is added to the stiffness series of the internal parts. The bending of the end cap can be described analytically as a simply supported circular slab [13]. For the investigated model, the deflection at the center of the slab w_0 is taken into account, which related to the change of the sensors pretension force ΔF_S, defines the stiffness c_{E3} as described in the following formulas:

$$c_{E3} = \frac{\Delta F_S}{w_0} = \frac{256\ \pi\ K\ (1+\nu)}{d_E^2(12+4\nu-\beta^2((7+3\nu)+(4+4\nu)\cdot \ln\beta))}, \quad \text{with } K = \frac{E_E \cdot h_E^3}{12\cdot(1-\nu^2)}, \quad \beta = \frac{2\cdot l_{Sa}}{\sqrt{\pi}\cdot d_E}. \tag{7}$$

Thereby, the geometrical factor β defines the ratio of the diameter of the loaded circle compared to the outside diameter of the slab d_E. The square sensor contact area is, for this purpose, transformed to an equivalent circle with the same area. For the Poisson's ratio ν, the value of steel is taken into account and set to be 0.3. The resulting stiffnesses of the springs connected in series are then summarized for the tube sections in c_T and for the internal parts in c_S. Thereby, the end cap stiffnesses as well as the corresponding sections of the tube are accounted twice in the Eq. 8 and 9:

$$\frac{1}{c_T} = \frac{1}{c_{T1}} + \frac{2}{c_{T2}} \tag{8}$$

$$\frac{1}{c_S} = \frac{1}{c_{S1}} + \frac{2}{c_{E1}} + \frac{2}{c_{E2}} + \frac{2}{c_{E3}}. \tag{9}$$

The mechanical-electrical transmission factor Ψ of the compound is then defined as the ratio of charge displacement Q and the applied tube force F. The charge displacement occurs because of the longitudinal piezoelectric effect d_{33} and the change of the sensors pretension force ΔF_S, which results from the parallel system of the stiffnesses c_T and c_S:

$$\Psi = \frac{Q}{F} = \frac{d_{33}\cdot\Delta F_S}{F} = \frac{d_{33}}{F}\cdot\frac{F\cdot c_S}{c_S+c_T} = \frac{d_{33}\cdot c_S}{c_S+c_T}. \tag{10}$$

In this simplified analytical model, holes inside the end caps for wiring the sensor elements, as well as the radial elasticity in the contact zone of the end caps and the tube are neglected.

Direct Monte-Carlo Simulation, Direct Interval Analysis and Direct Fuzzy Analysis

In this section, three direct methods, Monte-Carlo simulation, interval arithmetic and fuzzy arithmetic with δ-cuts are applied to estimate the mechanical-electrical transmission factor Ψ of the sensory rod and compared, regarding the computational cost and the confidence interval. In the section direct Monte-Carlo simulation, parametric and non-parametric density estimation as well as quantile estimation, is presented. Two different density estimation approaches are regarded with the main difference that parametric density estimation assumes that the mechanical-electrical transmission factor Ψ is normally distributed, whereas non-parametric density estimation does not need an assumption about the distribution of Ψ.

Varying Input Parameters. In Table 1, the varying input parameters are shown for this investigation and have been appraised for their significance and availability of uncertain data. For expository purposes, only five varying parameters Young's moduli of the sensor E_S and the tube E_T, the length of the sensor l_S, the diameter of the tube d_T, and the piezoelectric charge constant d_{33} are taken into account. The varying parameters are the basis for the sensitivity analysis and are assumed to be normally distributed with known mean value and standard deviation for direct Monte-Carlo simulation and to have lower and upper limits for direct interval and fuzzy analysis. The difference between upper and lower limit is equal to the sixth standard deviation with a percentage of 99.7 to ensure comparability

Table 1: Varying and non-varying input parameters

Parameter		Deterministic	Normally Distributed	Lower/Upper Limit
Varying	E_S in GPa	46.00	(46.00; 2.30)	[39.10; 52.90]
	E_T in GPa	70.00	(70.00; 0.95)	[67.16; 72.84]
	l_S in mm	18.00	(18.00; 0.01)	[17.97; 18.03]
	d_T in mm	24.40	(24.40; 0.02)	[24.35; 24.45]
	d_{33} in pC/N	640	(640; 32)	[544; 746]
Non-Varying	E_E in GPa	209.00	—	—
	l_{Sa} in mm	14.00	—	—
	l_{E1} in mm	3.00	—	—
	l_{E2} in mm	1.50	—	—
	d_E in mm	23.30	—	—
	x_E in mm	1.36	—	—
	h_E in mm	6.00	—	—
	w_T in mm	1.10	—	—
	ν	0.30	—	—

of the applied methods. The non-varying parameters are the Young's modulus of the end cap E_E, the lengths l_{Sa}, l_{E1} and l_{E2}, the diameter d_E, the heights x_E, h_E and the wall thickness w_T plus the Poisson's ratio ν, shown in Table 1.

Piezoelectric ceramics have a wide variance in their behavior due to uncertainty by sintering manufacturing process [14]. The d_{33} coefficient as well as the Young's modulus E_S of piezoceramics undergo high variations and are assumed to be 15 % of the third standard deviation of the nominal value of 640 pC/N given by the manufacturer [15]. The tolerance of the sensors length l_S in contrast, is very close and is given by $\pm$ 0.03 mm. Also, the Young's modulus E_T of the tube material is highly fluctuating because of uncertainty in the production of the semi-finished extruded tube material. In [16] the standard deviation of EN-AW 6060 aluminium bars is determined to be about 1,3 % the nominal value of 70 GPa. The cross-sectional area of the tube is inter alia determined by the crosssectional area of the tubular semi-finished product before rotary swaging. Despite using the same machine settings, thicker semi-finished tubes will also result in thicker swaged products because of higher forming forces. Therefore, the fluctuation of the diameter d_T is estimated to be around 0.05 mm.

Deterministic Input. For the deterministic approach to calculate the mechanical-electrical transmission factor Ψ, the input parameters Young's moduli E_S and E_T, length l_S, diameter d_T and piezoelectric charge constant d_{33} from Table 1 are illustrated in Fig. 3.

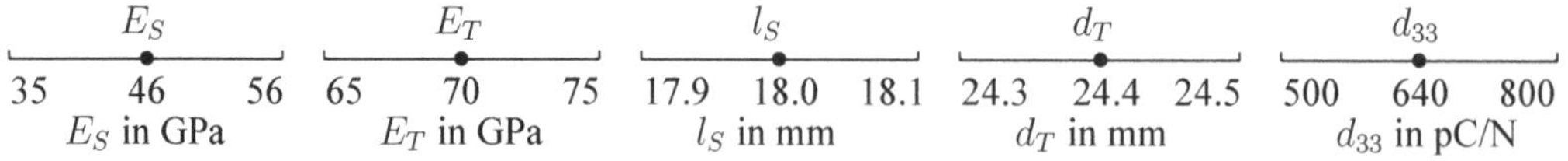

E_S — 35, 46, 56 — E_S in GPa
E_T — 65, 70, 75 — E_T in GPa
l_S — 17.9, 18.0, 18.1 — l_S in mm
d_T — 24.3, 24.4, 24.5 — d_T in mm
d_{33} — 500, 640, 800 — d_{33} in pC/N

Fig. 3: Deterministic parameters Young's moduli E_S and E_T, length l_S, diameter d_T and piezoelectric charge constant d_{33}

Normally Distributed Input. For the Monte-Carlo simulation, the input parameters Young's moduli E_S and E_T, length l_S, diameter d_T and piezoelectric charge constant d_{33} are chosen to be normally distributed as described in Table 1. The corresponding densities are denoted by $p(E_S)$, $p(E_T)$, $p(l_S)$, $p(d_T)$ and $p(d_{33})$. The densities are shown in Fig. 4 with the marked corresponding mean value and sixth standard deviation. The ordinate range is between zero and one, due to the comparability to fuzzy numbers, see *Fuzzy Input*.

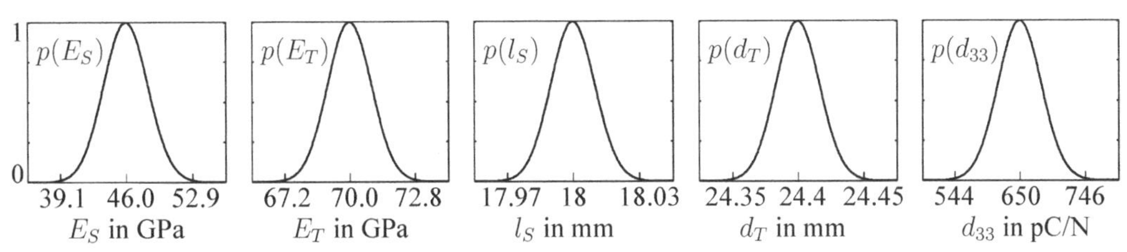

Fig. 4: Densities of the normally distributed parameters Young's moduli E_S and E_T, length l_S, diameter d_T and piezoelectric charge constant d_{33} with mean value and sixth standard deviation

Interval Input. For the direct interval analysis, the lower and upper limits from Table 1 of the input parameters Young's moduli E_S and E_T, length l_S, diameter d_T and piezoelectric charge constant d_{33} are illustrated as intervals $[E_S^-, E_S^+]$, $[E_T^-, E_T^+]$, $[l_S^-, l_S^+]$, $[d_T^-, d_T^+]$ and $[d_{33}^-, d_{33}^+]$ in Fig. 5.

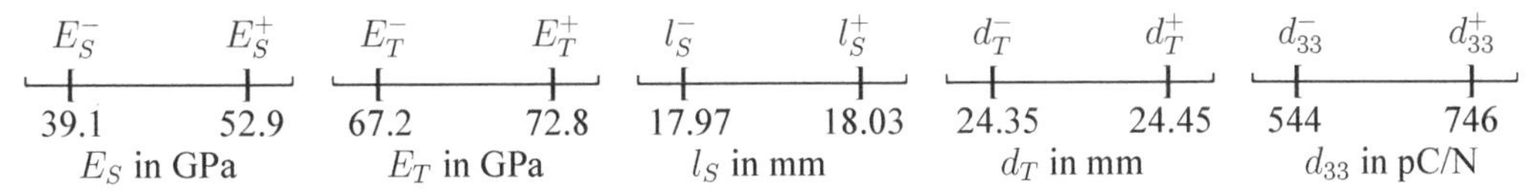

Fig. 5: Interval parameters Young's moduli E_S and E_T, length l_S, diameter d_T and piezoelectric charge constant d_{33} with lower and upper limits

Fuzzy Input. For the direct fuzzy analysis the input parameters Young's moduli E_S and E_T, length l_S, diameter d_T and piezoelectric charge constant d_{33} are assumed as triangular fuzzy numbers. Fuzzy variables are depicted with membership functions and are marked by a maximum ordinate value of 1. The corresponding membership functions are denoted by $\mu_\delta(E_S)$, $\mu_\delta(E_T)$, $\mu_\delta(l_S)$, $\mu_\delta(d_T)$ and $\mu_\delta(d_{33})$. The membership function μ is divided into five δ-cuts, δ = 0, 0.25, 0.5, 0.75, 1. δ-cuts are intervals of the membership function and give the percentage of the possibility that a value is not a member in these intervals. The interval at $\mu_{\delta=0}(E_S)$, $\mu_{\delta=0}(E_T)$, $\mu_{\delta=0}(l_S)$, $\mu_{\delta=0}(d_T)$ and $\mu_{\delta=0}(d_{33})$ represents the input parameters with lower and upper limits and are equal to the sixth standard deviation of the normally distributed input parameters. The deterministic parameters E_S, E_T, l_S, d_T and d_{33} are represented by $\mu_{\delta=1}(E_S)$, $\mu_{\delta=1}(E_T)$, $\mu_{\delta=1}(l_S)$, $\mu_{\delta=1}(d_T)$ and $\mu_{\delta=1}(d_{33})$. Exemplary for $\delta = 0$ and $\delta = 1$ the intervals of the δ-cuts are shown in Fig. 6.

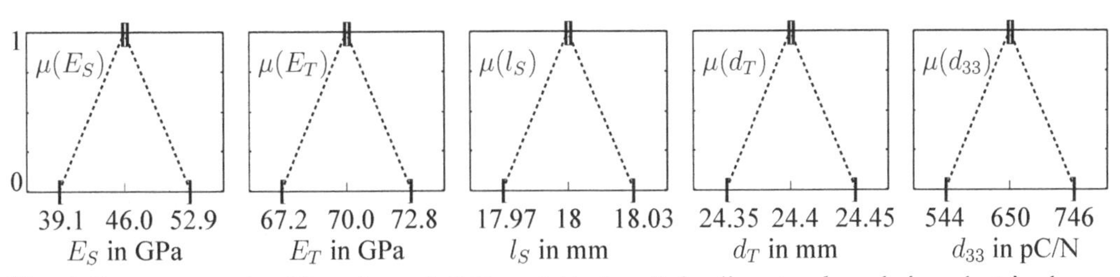

Fig. 6: Fuzzy parameters Young's moduli E_S and E_T, length l_S, diameter d_T and piezoelectric charge constant d_{33} with membership functions

Varying Transmission Factor of a Sensory Rod. Depending on the varying input parameters Young's moduli E_S and E_T, length l_S, diameter d_T and piezoelectric charge constant d_{33} the mechanical-electrical transmission factor Ψ, see Eq. 10, of the sensory rod is estimated by applying the direct Monte-Carlo simulation with parametric and non-parametric density estimation as well as quantile estimation, direct interval analysis and direct fuzzy analysis with δ-cuts. As a reference, the value Ψ is calculated deterministically as $\Psi = 324$ pC/N using Eq. 10. Additionally, the computational cost of each method for the estimation of Ψ is presented.

Direct Monte-Carlo Simulation. In this following subsection, the input quantities Young's moduli E_S and E_T, length l_S, diameter d_T and charge constant d_{33}, as well as the output quantity mechanical-electrical transmission factor Ψ are regarded as random variables. Let $m : \mathbb{R}^5 \to \mathbb{R}$ be the function that computes the random mechanical-electrical transmission factor Ψ corresponding to the input quantities $X = (E_S, E_T, l_s, d_T, d_{33})$, i.e. the function m represents Eq. 10. Using the normal distributions of the Young's moduli E_S and E_T, length l_S, diameter d_T and charge constant d_{33} given in Table 1 respectively the corresponding density functions $p(E_S)$, $p(E_T)$, $p(l_S)$, $p(d_T)$ and $p(d_{33})$ illustrated in Fig. 4 a sample of sample size n of independent and identically distributed (i.i.d.) random variables $X_1, X_2, \ldots, X_n$ distributed as X can be generated and therefore a sample of i.i.d. random variables of $\Psi = m(X)$ of sample size n by setting

$$\Psi_1 = m(X_1), \Psi_2 = m(X_2), \ldots, \Psi_n = m(X_n). \tag{11}$$

Based on this sample, the random mechanical-electrical transmission factor Ψ can be described using *parametric* and *non-parametric density estimation* as well as *quantile estimation.*

Assuming a density g of Ψ exists, it can be estimated via *non-parametric density estimators* such as the kernel density estimator g_n [17, 18]. If Ψ is in addition assumed to be normally distributed, *parametric density estimation* yields a density estimate $\tilde{g}_n$ of a normal distribution using realizations of $\Psi_1, \Psi_2, \ldots, \Psi_n$ to compute the arithmetic mean $\bar{\Psi}$ and the empirical standard deviation σ_Ψ. Since the parametric and non-parametric density estimations in Table 5 are almost identical, it is obvious that the mechanical-electrical transmission factor Ψ is indeed in all cases normally distributed.

Given that Ψ has a density, *quantile estimators*, such as an order statistics quantile estimator [19], can be used to determine an interval $q = [q_{0.0015}, q_{0.9985}]$ that contains Ψ with a probability of $99,7$ %, where $q_{0.0015}$ and $q_{0.9985}$ are the estimated 0.0015- respectively 0.9985-quantiles of the mechanical-electrical transmission factor Ψ. As a result the interval q is smaller than the scatter range of the density estimations g_n and $\tilde{g}_n$.

For the Monte-Carlo simulation, the number of samples n is chosen as 10000 by evaluation of the calculated confidence interval [20] regarding the scatter range (0.07 %) of the mean value of Ψ to ensure accurate results. All input parameters are randomly and independently generated by MATLAB®. For the density estimation the routine *density* in the statistic package *R* was utilized with a rectangular kernel and a bandwidth chosen by crossvalidation.

Table 2: Estimated mechanical-electrical transmission factor Ψ in pC/N for parametric density estimation and quantile estimation

Case	Varying input parameters					Parametric density estimation	Quantile estimation
	$p(E_S)$	$p(E_T)$	$p(l_S)$	$p(d_T)$	$p(d_{33})$	$\tilde{g}_n(\Psi)$ in pC/N	$q(\Psi)$ in pC/N
1	yes	no	no	no	no	(323.9; 3.8)	[311.1; 334.3]
2	no	yes	no	no	no	(324.0; 2.0)	[318.1; 330.1]
3	no	no	yes	no	no	(324.0; 0.02)	[323.7; 324.3]
4	no	no	no	yes	no	(324.0; 0.1)	[323.9; 324.1]
5	no	no	no	no	yes	(323.8; 16.0)	[275.8; 372.2]
6	yes	yes	yes	yes	yes	(323.6; 17.0)	[275.2; 374.0]

In Table 2, the estimated mechanical-electrical transmission factor Ψ via parametric density estimation $\tilde{g}_n(\Psi) = (\overline{\Psi}_g; \sigma_{\Psi_g})$ and quantile estimation $q(\Psi) = [q_{0.0015}(\Psi), q_{0.9985}(\Psi)]$ is shown. As will be shown in Table 5 in the following section *Comparison of Estimated Transmission Factor of a Sensory Rod*, the parametric $\tilde{g}_n(\Psi)$ and non-parametric $g_n(\Psi)$ density estimation yield in all cases almost identical results. The computational cost of parametric density estimation is 4.595 s, of non-parametric density estimation 5.787 s and of quantile estimation 4.958 s. The main part of the computational cost of the Monte-Carlo approaches arise from generating $n = 10000$ samples. In addition, computing the empirical standard deviation and the mean is faster than sorting the samples for the quantile estimation. Therefore, the compuational cost of quantile estimation is higher than the computational cost of parametric density estimation. Non-parametric density estimation has the highest computational cost, thereby especially the selection of a bandwidth by crossvalidation is time consuming.

Direct Interval Analysis. Interval arithmetic can be applied, if the data set has minimum and maximum values. The interval data set for the input parameters Young's moduli E_S=[E_S^-, E_S^+] and E_T=[E_T^-, E_T^+], length l_S=[l_S^-, l_S^+], diameter d_T=[d_T^-, d_T^+] and piezoelectric charge constant d_{33}=[d_{33}^-, d_{33}^+] are shown in section *interval input*, Fig. 5. Interval arithmetic uses four basic mathematical comp-

Table 3: Estimated mechanical-electrical transmission factor Ψ in pC/N for direct interval analysis

	Varying input parameters					Direct interval analysis
Case	[E_S^-, E_S^+]	[E_T^-, E_T^+]	[l_S^-, l_S^+]	[d_T^-, d_T^+]	[d_{33}^-, d_{33}^+]	[Ψ^-, Ψ^+] in pC/N
1	yes	no	no	no	no	[289.79; 358.87]
2	no	yes	no	no	no	[318.11; 330.15]
3	no	no	yes	no	no	[323.43; 324.57]
4	no	no	no	yes	no	[323.70; 324.30]
5	no	no	no	no	yes	[275.40; 372.60]
6	yes	yes	yes	yes	yes	[241.33; 422.00]

utational operations to add, subtract, multiply and divide intervals, [8]. According to [8], the calculation of Ψ as described by Eq. 10 is applied. In Table 3, the estimated mechanical-electrical transmission factor $\Psi = [\Psi^-, \Psi^+]$ via direct interval analysis is shown. The computational cost is 0.025 s for all cases of the estimated transmission factor Ψ.

Direct Fuzzy Analysis. To apply fuzzy arithmetic, a data set of fuzzy numbers is assumed by triangular fuzzy numbers for all input parameters, see section *fuzzy input*, Fig. 6. The calculation of the fuzzy arithmetic by using δ-cuts is based on interval arithmetic. The results of the direct fuzzy arithmetic are presented in Table 4. The possibility for intervals of the mechanical-electrical transmission factor $\Psi = [\Psi^-, \Psi^+]$ at 0, 25, 50, 75 and 100 % (δ = 0, 0.25, 0.5, 0.75, 1) are estimated via direct fuzzy analysis. The computational cost is 0.034 s for all cases of the estimated transmission factor Ψ.

Table 4: Estimated mechanical-electrical transmission factor Ψ in pC/N for direct fuzzy analysis

Case	Direct fuzzy analysis				
	$\mu_{\delta=0}(\Psi)$	$\mu_{\delta=0.25}(\Psi)$	$\mu_{\delta=0.5}(\Psi)$	$\mu_{\delta=0.75}(\Psi)$	$\mu_{\delta=1}(\Psi)$
1	[289.8;358.9]	[298.4;350.0]	[306.9; 341.2]	[315.5; 332.6]	[324.0;324.0]
2	[318.1;330.2]	[319.6;328.6]	[321.0;327.0]	[322.5;325.5]	[324.0;324.0]
3	[323.4;324.6]	[323.6;324.4]	[323.7;324.3]	[323.9;324.1]	[324.0;324.0]
4	[323.7;324.3]	[323.8;324.2]	[323.9;324.2]	[323.9;324.1]	[324.0;324.0]
5	[275.4;372.6]	[287.6;360.5]	[299.7;348.3]	[311.9;336.2]	[324.0;324.0]
6	[241.3;422.0]	[260.8;395.8]	[281.0;370.8]	[302.0;346.9]	[324.0;324.0]

Comparison of Estimated Transmission Factor of a Sensory Rod

In this section the estimated mechanical-electrical transmission factor Ψ of a sensory rod via three direct Monte-Carlo approaches, parametric and non-parametric density estimation as well as quantile estimation, and via direct interval arithmetic as well as direct fuzzy arithmetic with δ-cuts is presented. A sensitivity analysis concentrating on the input parameters Young's moduli E_S and E_T, length l_S of the sensory rod, diameter d_T and piezoelectric charge constant d_{33}, whose significance was stated in section *Varying Input Parameters*, was conducted to detect the influence of the most varying parameters on the scatter range of Ψ. In the development phase of the sensory rod, it is still possible to modify parameters of the joint partners or to claim closer tolerances for the semi-finished parts. Beside using piezoelectric sensors, it is also possible to use strain-gauge based sensor bodies or quartz plates. The assumptions for the varying input parameters are given in Table 1. Once again, in order to compare the applied methods, the difference between the assumed upper and lower limit is equal to the sixth standard deviation of the corresponding assumed normal distribution which includes 99.7% of the possible values. Additionally, the lower and upper quantile are chosen such that the estimated interval $q(\Psi) = [q_{0.0015}(\Psi), q_{0.9985}(\Psi)]$ contains the mechanical electrical transmission factor Ψ with a probability of 99.7%. To compare the methods density estimation and fuzzy arithmetic, the density estimations are normalized such that their maximal values equal one, since the membership functions $\mu_\delta(E_S)$, $\mu_\delta(E_T)$, $\mu_\delta(l_S)$, $\mu_\delta(d_T)$ and $\mu_\delta(d_{33})$ for fuzzy numbers have an ordinate range from zero to one. The results of the estimated mechanical-electrical transmission factor Ψ via direct parametric and non-parametric density estimation, direct quantile estimation, direct interval and direct fuzzy arithmetic compared to the deterministic approach are presented in Table 5. A comparison of the parametric and non-parametric density estimations in Table 5 indicates that the mechanical-electrical transmission factor Ψ is, as assumed for the parametric density estimation, indeed in all cases normally distributed, since the parametric and non-parametric density estimations are almost identical. In all cases the mean value for parametric density estimation $\overline{\Psi}_p$ is at approximately 324 pC/N and thus equal to the deterministic calculation of Ψ. It can be seen, that the scatter range of the quantile estimation is approximately $\pm$ 23 pC/N smaller in case 1, $\pm$ 41 pC/N in case 6 and lesser than $\pm$ 0.5 pC/N in the other cases, in comparison to the results for interval and fuzzy analysis. In comparison to the sixth case of the sensitivity analysis for all applied methods, all input parameters are now varying. Scatter of the five varying input parameters is assumed according to Table 1. The other nine, formerly non-varying, input parameters are assumed to have a scatter range (sixth standard deviation) of 6% regarding the given deterministic values in Table 1. The estimated mechanical-electrical transmission factor Ψ yields similiar results via parametric $\tilde{g}_n(\Psi)$ and non-parametric $g_n(\Psi)$ density estimation as well as for quantile estimation $q(\Psi)$, see case 6 in Table 2. The estimated interval $[\Psi^-, \Psi^+]$ and fuzzy number $\mu_{\delta=0}(\Psi)$ at $\delta = 0$ provide a 7% greater scatter range for the width of the interval compared to case 6 in Table 3.

Table 5: Estimated mechanical-electrical transmission factor for deterministic calculation Ψ (•), parametric $\tilde{g}_n(\Psi)$ $(\cdots)$ and non-parametric $g_n(\Psi)$ (—) density estimation, quantile estimation $q(\Psi)$ (grey), direct interval $[\Psi^-;\ \Psi^+]$ ([ ]) and direct fuzzy analysis $\mu_\delta(\Psi)$ (- -)

Case	Varying input parameters					Transmission factor
	E_S	E_T	l_S	d_T	d_{33}	
1	y	n	n	n	n	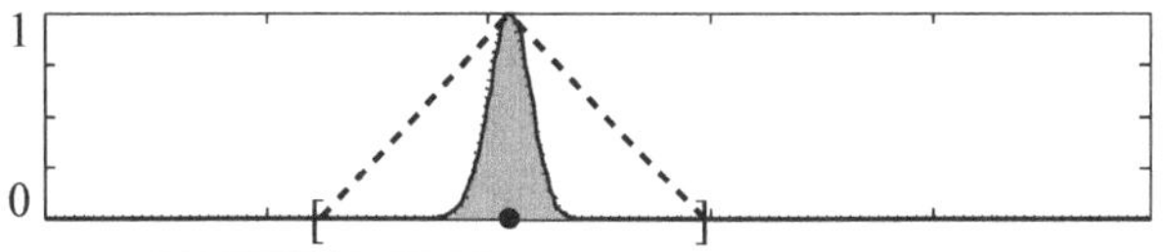
2	n	y	n	n	n	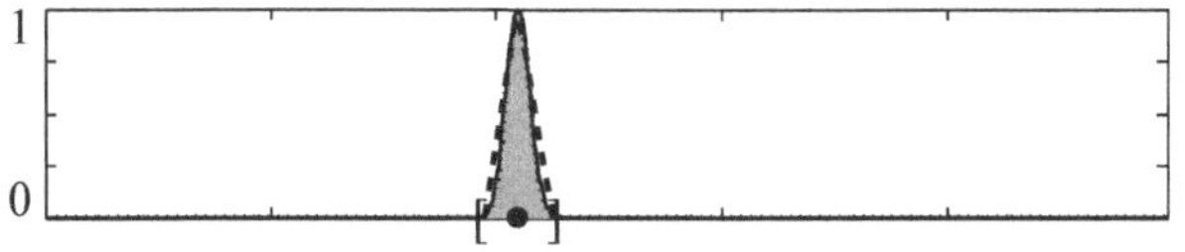
3	n	n	y	n	n	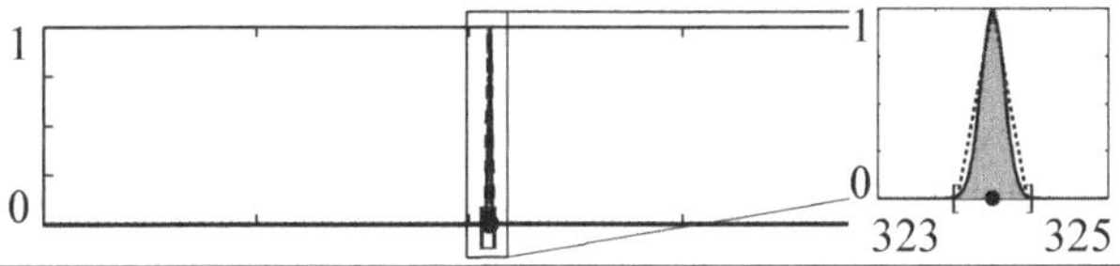
4	n	n	n	y	n	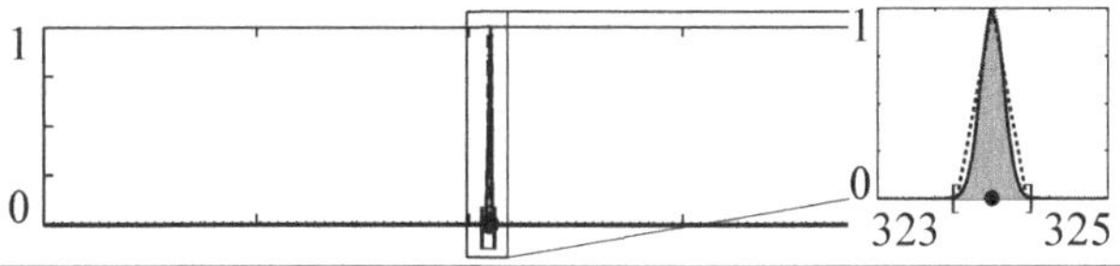
5	n	n	n	n	y	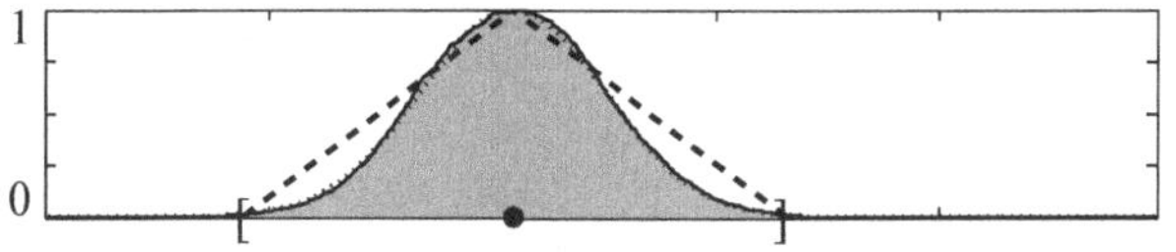
6	y	y	y	y	y	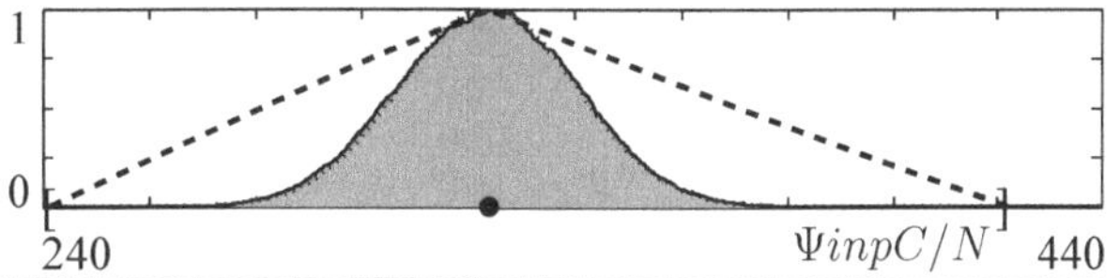

It is shown that, on the one hand, the computational cost of interval and fuzzy analysis is much lower than for the Monte-Carlo simulations with 10000 samples for each input parameter. The sample rate of 10000 samples was proven to be sufficient according to a convergence test against the mean value. On the other hand, however, the implementation of the interval and fuzzy arithmetic is time consuming and the variation of the resulting transmission factor due to input parameters lower and upper limits are overestimated. This is due to the limits of adequately implementing interval arithmetic into mathematical models that use at least one variable of input parameters several times in the calculation.

Conclusions

In this paper, the direct, non-deterministic methods parametric and non-parametric density estimation, quantile estimation, interval and fuzzy arithmetic are compared with each other to describe uncertainty occurring in the mechanical-electrical transmission behavior of a sensory rod, concentrating on input parameters with geometrical, material and electrical properties. For this purpose, a simplified analytical spring model of a sensory rod is considered. Due to fluctuating properties of the semi-finished parts in the production process, the mechanical-electrical transmission behavior is uncertain. A sensitivity analysis examines the individual influence on the selected input parameters Young's moduli of the sensor and the tube, length of the sensor, diameter of the tube and piezoelectric charge constant on the mechanical-electrical transmission factor. The scatter range enables engineers to identify the effective parameters to improve the design of sensory rods, or to design systems with reasonable safety factors. Of the contemplated parameters the piezoelectric charge constant has the largest effect on the mechanical-electrical transmission factor followed by the material parameters Young's moduli of the sensor and the tube. The geometric parameters diameter of the tube and length of the sensor have, in contrast to material and electric parameters, only insignificant influence on the mechanical-electrical transmission factor. The investigation of the combined uncertainty inside a batch of manufactured sensory rods reveals that the standard deviation of the mechanical-electrical transmission factor is approximately 5 % for the considered varying parameters. Due to the high scattering, individual calibration can be neglected only for measuring tasks with a low required accuracy. Comparing the applied methods in terms of computational cost, the results show that direct interval and direct fuzzy analysis are preferable to the three Monte-Carlo approaches having less than 0.8 % of the computational cost of the Monte-Carlo approaches as a consequence of the chosen sample size 10000 for the Monte-Carlo approaches and the number of lower and upper limits for the interval and fuzzy analysis. Notwithstanding computational cost, the implementation of direct interval and direct fuzzy analysis is time consuming, since these methods have to be implemented for each specific model. Moreover, interval and fuzzy analysis estimate a higher variation of the mechanical-electrical transmission factor than the Monte-Carlo approaches using the lower and upper limits of the input parameters. In future work, advanced interval and fuzzy analysis as well as density and quantile estimators could be conducted to reduce implementation cost and to increase accuracy in estimation results in these and in more complex systems containing sensory rods.

Acknowledgements

The authors would like to thank the German Research Foundation (DFG) for funding this research within the Collaborative Research Centre (SFB) 805.

References

[1] H. Hanselka, R. Platz, Ansätze und Maßnahmen zur Beherrschung von Unsicherheit in lasttragenden Systemen des Maschinenbaus - Controlling uncertainty in load-bearing systems, Konstruktion, 11/12, VDI-Verlag Düsseldorf, 2010, 55-62.

[2] B. Denkena, T. Mörke, M. Krüger, J. Schmidt, H. Boujnah, J. Meyer et al., Development and first applications of gentelligent components over their lifecycle, CIRP Journal of Manufacturing Science and Technology 7 (2014) 139–150.

[3] W.-G. Drossel, S. Hensel, M. Nestler, L. Lachmann, A. Schubert, M. Müller, B. Müller, Experimental and numerical study on shaping of aluminum sheets with integrated piezoceramic fibers, Journal of Materials Processing Technology 214 (2014) 217–228.

[4] W.-G. Drossel, S. Hensel, M. Nestler, L. Lachmann, Evaluation of Actuator, Sensor, and Fatigue Performance of Piezo-Metal-Composites, IEEE Sensors J. 14 (2014) 2129–2137.

[5] P. Groche, M. Türk, Smart structures assembly through incremental forming, CIRP Annals - Manufacturing Technology 60 (2011) 21–24.

[6] P. Groche, M. Krech, Integration of functions through rotary swaging – New possibilities for the monitoring of states and loads in tubular machine elements: submitted to New Developments in Forging Technology (2015).

[7] T. Müller-Gronbach, E. Novak, K. Ritter, Monte Carlo-Algorithmen, Springer, Heidelberg, Berlin, 2012.

[8] G. Alefeld, G. Mayer, Interval analysis: theory and applications, Journal of Computational and Applied Mathematics 121 (2000) 421-464.

[9] M. Hanss, Applied Fuzzy Arithmetic, Springer, Berlin Heidelberg New York, 2005.

[10] L. Devroye, L. Györfi, Nonparametric Density Estimation - The L1 View, John Wiley, New York, 1985.

[11] G. Enss, B. Götz, M. Kohler, A. Krzyzak, R. Platz, Nonparametric estimation of a maximum of quantiles, Electronic Journal of Statistics 8 (2014) 3176-3192.

[12] A. Bretz, S. Calmano, T. Gally, B. Götz, R. Platz, J. Würtenberger, Darstellung passiver, semi-aktiver und aktiver Systeme auf Basis eines Prozessmodells (Representation of Passive, Semi-Active and Active Systems Based on a Prozess Model), unreleased paper by the SFB 805 on http://www.sfb805.tu-darmstadt.de/media/sfb805/f_downloads/150310_AKIII_Definitionen_aktiv-passiv.pdf

[13] E. Hake, K. Meskouris, Statik der Flächentragwerke - Einführung mit vielen durchgerechneten Beispielen, Springer-Verlag, Berlin, Heidelberg, 2007.

[14] R. Platz, G. Enss, S. Ondoua, T. Melz, Active stabilization of a slender beam-column under static axial loading and estimated uncertainty in actuator properties, International Conference on Vulnerability and Risk Analysis and Management (2014) 235-244.

[15] Datasheet, Piezomechanik GmbH, Piezostapelaktoren PSt 150, http://www.piezomechanik.com, 07.04.2015.

[16] J. Ferenc, The random variability analysis of the mechanical properties of the selected aluminum alloys, Technical Transaction 3-B (2013) 3-18.

[17] E. Parzen, On the estimation of a probability density function and the mode, The Annals of Mathematical Statistics 33 (1962) 1065–1076.

[18] M. Rosenblatt, Remarks on some nonparametric estimates of a density function, The Annals of Mathematical Statistics 27 (1956) 832–837.

[19] B.C. Arnold, N. Balakrishnan, H.N. Nagaraja, A First Course in Order Statistics, John Wiley & Sons, New York, 1992.

[20] C. Graham, D. Talay, Stochastic Simulation and Monte Carlo Methods, Springer, Heidelberg New York Dordrecht London, 2013.

Applied Mechanics and Materials Vol. 807 (2015) pp 218-225

doi:10.4028/www.scientific.net/AMM.807.218
Submitted: 2015-04-29
Accepted: 2015-08-04

Using Particle Filters to analyse the credibility in model predictions

Peter Lewis Green[1]

[1]Instutute for Risk and Uncertainty, Centre for Engineering Sustainability, School of Engineering, University of Liverpool, Liverpool, UK, L69 3GQ

P.L.Green@liverpool.ac.uk

Keywords: Nonlinear system identification, Bayesian inference, particle filters, model predictions.

Abstract

Models are often used to make predictions far from the region where they were trained and validated. In this paper attempts are made to analyse the credibility that can be placed in such predictions. The proposed approach involves treating a model's parameters as time-variant (even if it is believed that this is not the case), before utilising Bayesian tracking techniques to realise parameter estimates. An example is used to demonstrate that, relative to a Bayesian approach where the parameters are assumed to be time-invariant, treating the parameters as time-variant can reveal important flaws in the model and raise questions about its ability to make credible predictions.

Introduction

Consider a dynamical system for which one has proposed a model structure that depends on a vector of parameters, $\boldsymbol{\theta}$. In the current work the model is based on physical-laws and, as a result, each parameter within $\boldsymbol{\theta}$ carries a physical interpretation (stiffness of a spring element for example). In situations where it is difficult to measure and/or quantify uncertainties about $\boldsymbol{\theta}$, one can choose to infer probabilistic elements of these parameters from a set of training data, D. This can be achieved via the application of Bayes' theorem:

$$p(\boldsymbol{\theta}|D) \propto p(D|\boldsymbol{\theta})p(\boldsymbol{\theta}) \quad (1)$$

which allows one to generate an expression for the posterior parameter distribution, $p(\boldsymbol{\theta}|D)$ (where $\propto$ should be read as 'proportional to'). Such an analysis involves defining the likelihood, $p(D|\boldsymbol{\theta})$, as well as a prior distribution $p(\boldsymbol{\theta})$.

In situations where one's model response varies nonlinearly with $\boldsymbol{\theta}$ (such that analytical expressions for the posterior are impossible to obtain), one can use advanced sampling techniques – usually Markov chain Monte Carlo (MCMC) – alongside model simulations to generate samples of $\boldsymbol{\theta}$ from $p(\boldsymbol{\theta}|D)$. This process is aided somewhat by advanced sampling algorithms such as Transitional MCMC (TMCMC) [1] which are able to sample from distributions with complex geometries and are suitable for parallel processing [2,3] or other algorithms which are suitable for large data sets [4].

The ultimate aim of such system identification varies depending on the application. While sometimes one is simply interested in the values of the parameters themselves, typically the goal is then to use the model to make predictions. Often this involves extrapolating the model into regions where it was not trained. As an example one may use a model of a structure which, having been trained on data from low amplitude ambient vibrations, is used to predict the system's response to an earthquake. Within the Bayesian framework described here, one can evaluate the probability that the system will exhibit the response, x, to a new input, y, using (or at least approximating)

$$p(x|y,D) = \int p(x|y,\boldsymbol{\theta},D)\,p(\boldsymbol{\theta}|D)d\boldsymbol{\theta}. \quad (2)$$

This approach therefore allows one to marginalise one's uncertainty about the model parameters. These predictions are however, by definition, conditional on the data which was used to train the model. There is little to guarantee that the model will perform well in a region far from where it was trained and validated.

In the current paper a numerical example is used to demonstrate this issue. The same problem is then tackled using a particle filter, which can be used to track time-varying parameters of nonlinear systems. It is shown that, by treating the system as if it has time-variant parameters (even if it is believed this is not the case) can reveal much about how the model will perform when it is used to make predictions far from where it was trained.

The analysis described in this paper uses a combination of importance sampling, Markov chain Monte Carlo (MCMC) and Bayesian tracking algorithms (particle filters specifically). As such, the next three sections are devoted to a brief introduction to these methods.

Importance sampling

Consider the situation where one wants to estimate some quantities related to the probability density function, $\pi(\boldsymbol{\theta})$, which is herein referred to as the 'target distribution'. In this case it is assumed that analytical treatment of $\pi(\boldsymbol{\theta})$ is difficult and that, furthermore, the distribution is only known up to a constant of proportionality. At this point it is convenient to define

$$\pi(\boldsymbol{\theta}) = \pi^*(\boldsymbol{\theta})/Z \tag{3}$$

such that $\pi^*(\boldsymbol{\theta})$ is the unnormalised distribution and Z is the normalising constant. With importance sampling, one proceeds by choosing a 'proposal density', $q(\boldsymbol{\theta})$, from which it is relatively easy to generate samples (a Gaussian for example). If the aim is to estimate $\mathrm{E}_\pi[f(\boldsymbol{\theta})]$, one can then make use of the property that

$$\int f(\boldsymbol{\theta})\,\pi(\boldsymbol{\theta})d\boldsymbol{\theta} = \frac{\int f(\boldsymbol{\theta})\,\pi^*(\boldsymbol{\theta})d\boldsymbol{\theta}}{\int \pi^*(\boldsymbol{\theta})\,d\boldsymbol{\theta}} = \frac{\int f(\boldsymbol{\theta})\frac{\pi^*(\boldsymbol{\theta})}{q(\boldsymbol{\theta})}q(\boldsymbol{\theta})d\boldsymbol{\theta}}{\int \frac{\pi^*(\boldsymbol{\theta})}{q(\boldsymbol{\theta})}q(\boldsymbol{\theta})d\boldsymbol{\theta}} = \frac{\mathrm{E}_q[f(\boldsymbol{\theta})w(\boldsymbol{\theta})]}{\mathrm{E}_q[w(\boldsymbol{\theta})]} \tag{4}$$

where $w(\boldsymbol{\theta}) = \pi^*(\boldsymbol{\theta})/q(\boldsymbol{\theta})$ is defined as the 'importance weight'. In this way, using samples $\{\boldsymbol{\theta}^1, \boldsymbol{\theta}^2, \dots\}$ which have been generated from $q(\boldsymbol{\theta})$, one can estimate the quantity of interest using

$$\int f(\boldsymbol{\theta})\,\pi(\boldsymbol{\theta})d\boldsymbol{\theta} \approx \frac{\sum_i f(\boldsymbol{\theta}^i)w(\boldsymbol{\theta}^i)}{\sum_i w(\boldsymbol{\theta}^i)}. \tag{5}$$

(for more details on importance sampling the book [5] is recommended).

Markov chain Monte Carlo

At first sight it may seem that, by defining one's target density as the posterior parameter distribution, importance sampling can be used to estimate quantities associated with Bayesian system identification problems. In practice however, the efficient application of importance sampling relies on the geometry of the proposal density being similar to that of the target density. In Bayesian system identification problems, the posterior is usually very concentrated relative to the prior and, as such, this is difficult to guarantee.

In such situations one can instead choose to sample from the posterior parameter distribution using Markov chain Monte Carlo (MCMC) methods. MCMC involves 'growing' an ergodic Markov chain through one's parameter space while also ensuring that the Markov chain's stationary distribution is equal to (or at least proportional to) the target distribution. If this can be achieved then, once converged to its stationary distribution, the Markov chain will be generating samples from the target. The most well-known MCMC method is the Metropolis-Hastings algorithm [6], which can be used even if the target distribution is only known up to a constant of proportionality.

A full description of modern MCMC algorithms is beyond the scope of the current paper-interested readers may wish to consult the tutorial [7] which is written in the context of structural dynamics.

Particle filters

Particle filters can be used to track the time-varying state of nonlinear dynamical systems. They have been applied to structural dynamics problems since (at least) 2006 [8] and have, for example, been used to aid active control [9], predict the chaotic behaviour of nonlinear systems [10] and track the time-varying parameters present in vehicle-structure interaction problems [11]. Being a well-established technique, only a brief description is given here (see [12] for a tutorial).

In the current work, using n to index a discrete time measurement, the state vector of a system is related to its previous state by

$$\boldsymbol{x}_n = f(\boldsymbol{x}_{n-1}, \boldsymbol{u}_{n-1}, \boldsymbol{v}), \quad \boldsymbol{v} \sim \mathcal{N}(0, \boldsymbol{Q}) \tag{6}$$

where $\boldsymbol{u}$ is a known input. Furthermore, it is assumed that measurements of the system's state are made according to

$$\boldsymbol{z}_n = h(\boldsymbol{x}_n, \boldsymbol{n}), \quad \boldsymbol{n} \sim \mathcal{N}(0, \boldsymbol{R}). \tag{7}$$

(where $\boldsymbol{Q}$ and $\boldsymbol{R}$ are covariance matrices). With a particle filter one is essentially using importance sampling to estimate properties of $p(\boldsymbol{x}_{0:t} | \boldsymbol{z}_{1:t})$. This distribution can, via the application of Bayes' theorem and noting the Markovian nature of equations (3) and (4), be written as

$$p(\boldsymbol{x}_{0:n} | \boldsymbol{z}_{1:n}) \propto p(\boldsymbol{z}_n | \boldsymbol{x}_n) p(\boldsymbol{x}_n | \boldsymbol{x}_{n-1}) p(\boldsymbol{x}_{0:n-1} | \boldsymbol{z}_{1:n-1}). \tag{8}$$

The advantage of this formulation is that, by writing one's proposal density as

$$q(\boldsymbol{x}_{0:n}) \propto q(\boldsymbol{x}_n | \boldsymbol{x}_{0:n-1}) q(\boldsymbol{x}_{0:n-1}), \tag{9}$$

one can calculate current importance weights using those which were realised in the previous iteration of the algorithm:

$$w_n^{(i)} = w_{n-1}^{(i)} \frac{p(\boldsymbol{z}_n | \boldsymbol{x}_n^{(i)}) p(\boldsymbol{x}_n^{(i)} | \boldsymbol{x}_{n-1}^{(i)})}{q(\boldsymbol{x}_n^{(i)} | \boldsymbol{x}_{0:n-1}^{(i)})} \tag{10}$$

(where $w_n^{(i)}$ denotes the weight associated with the ith proposed sample, at the nth iteration of the particle filter). One can then estimate the any quantity of interest ($\mathrm{E}[g(\boldsymbol{x}_n)]$ for example) via

$$\mathrm{E}[g(\boldsymbol{x}_n)] \approx \frac{\sum_i w_n^{(i)} g\left(\boldsymbol{x}_n^{(i)}\right)}{\sum_i w_n^{(i)}}. \tag{11}$$

Example system

In this work measurement data is generated from a simulation of a SDOF Duffing oscillator, whose equation of motion is given by

$$\ddot{x} + c\dot{x} + kx + k_3 x^3 = F(t) \tag{12}$$

where F is a Gaussian white noise force. The training data consisted of 5000 points of displacement time history which was artificially corrupted with Gaussian measurement noise. The parameters were set to $k = 10, c = 0.5, k_3 = 2$ (SI units) and were held constant throughout the simulation. These were chosen so that the effects of the nonlinearity were not particularly obvious in the training data (Fig. 1 shows the effect that setting $k_3 = 0$ has on the system response).

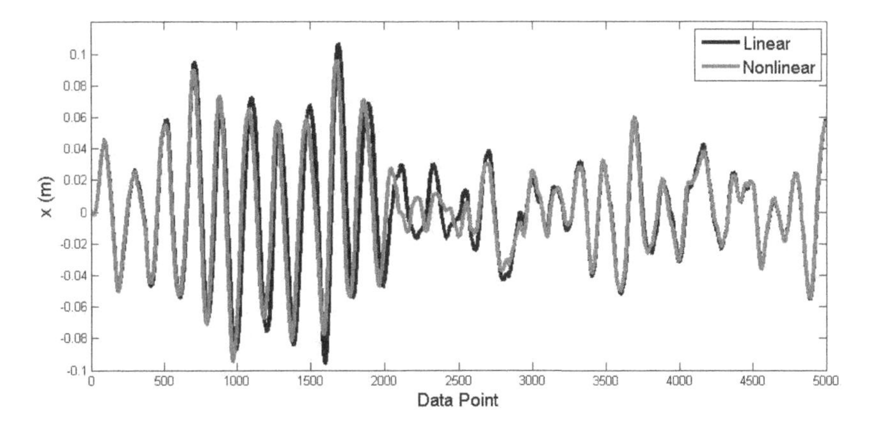

Fig. 1 Response of the simulated system when $\boldsymbol{k_3} = \mathbf{0}$ and $\boldsymbol{k_3} = \mathbf{2}$.

In the following sections a model with no nonlinear stiffness term is fitted to the training data generated from the nonlinear system (the grey line shown in Figure 1). This is a model which, even if fits the training data well, will perform badly if extrapolated to higher amplitudes. The question is: is it possible to tell, using this available data, that the model should not be used to make high-amplitude predictions? In the next section, it is assumed that the system's parameters are time-invariant and Bayesian system identification is performed using MCMC (as described in the introduction).

Assuming time-invariant parameters

To define the likelihood one has to specify a 'prediction model error'. This is designed to answer the question: if it is believed that the chosen model structure (with parameters $\boldsymbol{\theta}$) is a true representation of the real system, what is the probability of witnessing the measurement data? In this case the prediction error model was chosen such the probability of witnessing each measurement was Gaussian, with variance σ^2 and mean given by the model response. It was also assumed that the probability of witnessing separate data points is independent. This choice, having assumed the first 2 moments of $p(D|\boldsymbol{\theta})$, is that which minimises the amount of additional information which must be assumed (according to the principle of maximum entropy [11]). The parameter σ was included alongside k and c in the vector of parameters to be estimated. The prior was chosen to be a multivariate Gaussian

$$p(\boldsymbol{\theta}) = \mathcal{N}(\boldsymbol{m}, \boldsymbol{\Sigma}) \tag{13}$$

where $\boldsymbol{m} = \{10, 0.5, 5^{-6}\}^T$ and $\boldsymbol{\Sigma} = \text{diag}(9, 0.01, 1 \times 10^{-12})$. MCMC results are shown in Fig. 2. while, having used these samples to approximate equation (2), the ability of the model to replicate the training data is shown in Fig. 3. It can be seen that, by increasing the value of the linear stiffness, the model has compensated for the fact that the nonlinear stiffness has not been included. Furthermore, the model appears to have performed well.

For validation purposes, it is common practice to then analyse the ability of the model to replicate some previously 'unseen' data (data which was not used to infer the parameter estimates). Fig. 4. shows that the model also appears capable of replicating a set of validation data (which was also generated using a relatively low amplitude excitation). In here lies the problem with this approach – in the situation where one's training *and* validation data is low amplitude, there is little to suggest that the model will perform poorly at higher amplitudes.

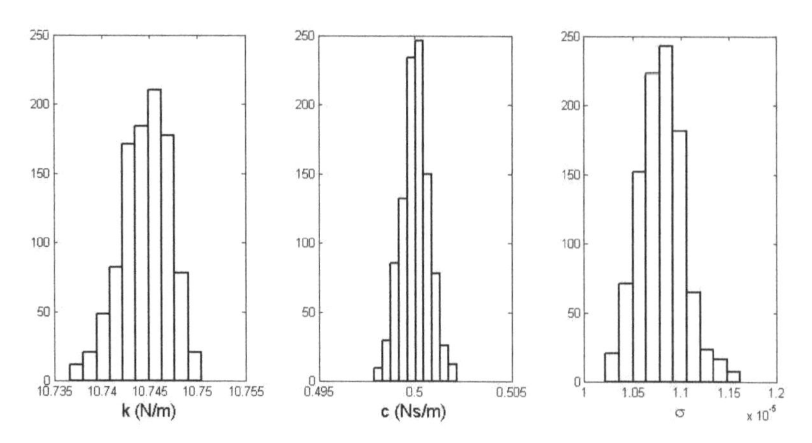

Fig. 2. MCMC samples.

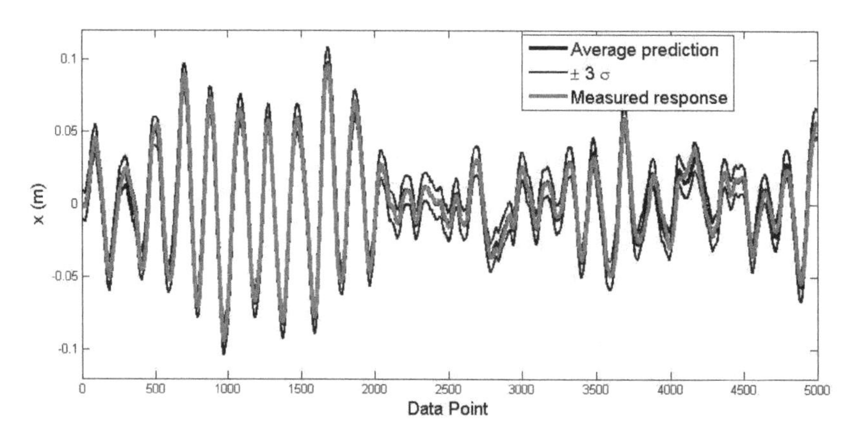

Fig. 3. Propagation of MCMC results – the ability of the model to replicate the training data.

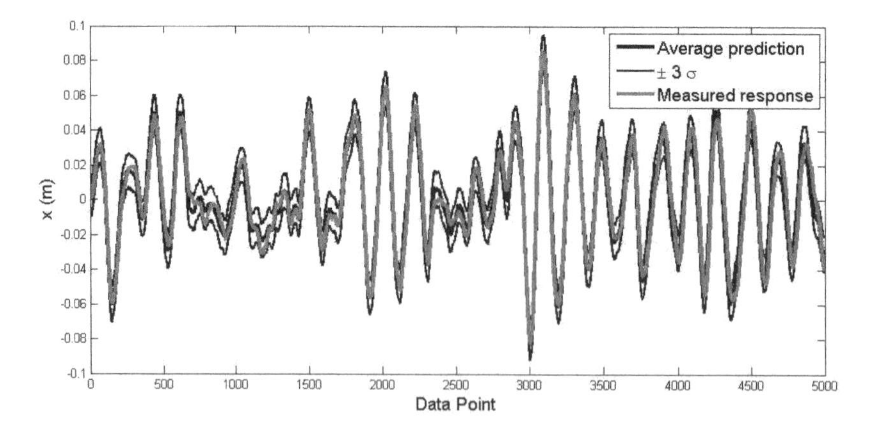

Fig. 4. Propagation of MCMC results – the ability of the model to replicate ‘unseen’ data.

Assuming time-variant parameters

Here the same system is analysed using a particle filter. To account for the possibility that the system's parameters are varying with time, they are included in the state vector. Here then, the system's state at time t is defined as

$$\boldsymbol{x}_t = \{x(t), \dot{x}(t), k(t), c(t)\}. \tag{14}$$

Writing a finite-difference approximation of the equation of motion, the function f maps from $\boldsymbol{x}_{t-\Delta t}$ to $\boldsymbol{x}_t$ according to

$$\begin{Bmatrix} x_1(t) \\ x_2(t) \\ x_3(t) \\ x_4(t) \end{Bmatrix} = \begin{bmatrix} 1 & \Delta t & 0 & 0 \\ -x_3(t-\Delta t)\Delta t & -x_4(t-\Delta t)\Delta t + 1 & 0 & 0 \\ 0 & 0 & 1 & 0 \\ 0 & 0 & 0 & 1 \end{bmatrix} \begin{Bmatrix} x_1(t-\Delta t) \\ x_2(t-\Delta t) \\ x_3(t-\Delta t) \\ x_4(t-\Delta t) \end{Bmatrix} + \begin{Bmatrix} 0 \\ \Delta t\, F(t-\Delta t) \\ 0 \\ 0 \end{Bmatrix} + \boldsymbol{v}. \tag{15}$$

By defining equation (15) it has been hypothesised that each system parameter will be equal the sum of their previous value and a random variable contained in the vector $\boldsymbol{v}$. The covariance matrix, $\boldsymbol{Q}$, was set equal to $\mathrm{diag}(1 \times 10^{-3}, 1 \times 10^{-3}, 1 \times 10^{-2}, 1 \times 10^{-3})$.

With the training data consisting of a time history of displacement measurements, the function h is simply given by $h(\boldsymbol{x}_n) = \{1\ 0\ 0\ 0\}\boldsymbol{x}_n$. In this example R was set equal to variance of the measurement noise. $p(\boldsymbol{x}_n|\boldsymbol{x}_{n-1})$ was used as the proposal density.

The particle filter was used to estimate the mean and variance of the unknown parameters. The resulting parameter estimates (with confidence bounds) are shown in Fig. 5. alongside a plot showing the response of the system with and without the nonlinear stiffness. It is clear that the damping term remains approximately constant during the process, confirming that the energy losses of the system have been modelled appropriately. Fig. 5. also shows several large fluctuations in the linear stiffness term, as the linear model attempts to match the nonlinear training data. It is this information which gives an indication that something is wrong with how the stiffness terms have been modelled – this would be much more difficult to realise using the identification method described in the previous section.

Interpreting the results shown in Figure 5 does, of course, require a certain amount of judgement. As an example, it is possible that one could observe the results in Figure 5 and simply conclude that, for the first 2000 points, the stiffness term is simply converging on its 'true' value. Fortunately, as a result of the particle filter's ability to quickly analyse large sets of data, it would be relatively easy to test this hypothesis on a larger data set. Furthermore, one could also make use of simulated training data – allowing the results of the real system identification problem to be compared with those obtained in a situation where the model is a perfect representation of the process of interest.

Essentially the author would like to stress that, while the method presented here will seldom provide a statement as clear-cut as 'do not trust this model at high amplitudes', it does at least give an indication that something is wrong and that further investigation is needed.

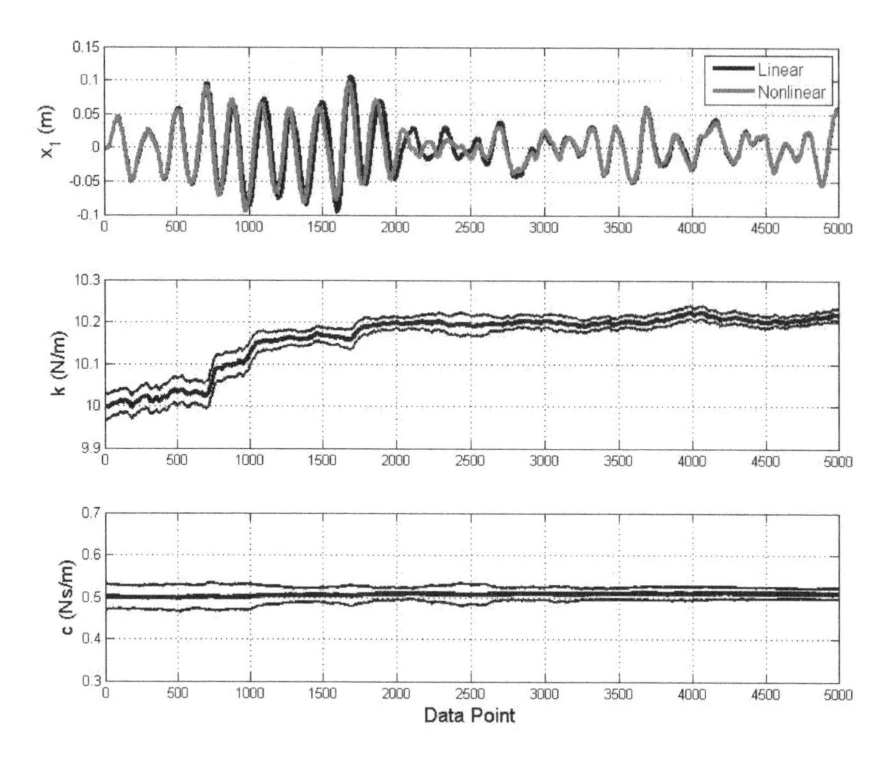

Fig. 5 Particle filter results. For the second and third panels, thick black lines represent expected values while thin black likes are three standard deviations from the mean.

For future work the author is planning to apply a similar methodology to a variety of real scenarios, where the credibility of the 'best' current model is in doubt. It is hoped that the resulting analysis will reveal where it is that these models are struggling to replicate the physics of the system of interest, and how they can be improved.

Conclusions

In this work, measurements from a simulated nonlinear system were used to infer probabilistic parameter estimates of a model. Attempts were made to fit a linear model to the data. It was known that, although able to replicate the training data well, the model would perform poorly at higher amplitudes. Assuming the system's parameters to be time-invariant, parameter estimates were realised using a Bayesian framework combined with MCMC sampling. From this analysis it was difficult to tell that the model will only perform well at low amplitudes. The same problem was then analysed with a particle filter, and the system's parameters were treated as being time-variant. This analysis revealed that relatively large changes in the model parameters were needed to provide a good fit to the training data. This result indicated that the model structure was missing some key components of the system physics and that it may be difficult to trust if used to make predictions far from where it was trained.

References

[1] J. Ching, Y.C. Chen (2007), Transitional Markov chain Monte Carlo method for Bayesian model updating, model class selection, and model averaging. *Journal of engineering mechanics*, *133*(7), 816-832.

[2] P. Angelikopoulos, C. Papadimitriou, P. Koumoutsakos (2012), Bayesian uncertainty quantification and propagation in molecular dynamics simulations: a high performance computing framework. *The Journal of chemical physics*, *137*(14), 144103.

[3] P.E. Hadjidoukas, P. Angelikopoulos, C. Papadimitriou, P. Koumoutsakos (2015). Π4U: A high performance computing framework for Bayesian uncertainty quantification of complex models. *Journal of Computational Physics*, *284*, 1-21.

[4] P.L. Green (2015), Bayesian system identification of a nonlinear dynamical system using a novel variant of simulated annealing. *Mechanical Systems and Signal Processing*, *52*, 133-146.

[5] D.J. MacKay (2003), *Information theory, inference, and learning algorithms* (Vol. 7). Cambridge: Cambridge university press.

[6] N. Metropolis, A.W. Rosenbluth, M.N. Rosenbluth, A.H. Teller, E. Teller. (1953). Equation of state calculations by fast computing machines. *The journal of chemical physics*, *21*(6), 1087-1092.

[7] P.L.Green, K.Worden. Bayesian and MCMC methods for identifying nonlinear systems in the presence of uncertainty. *Philosophical Transactions of the Royal Society A: Mathematical, Physical and Engineering Sciences* (in press).

[8] J. Ching, J.L. Beck, K.A. Porter (2006), Bayesian state and parameter estimation of uncertain dynamical systems. *Probabilistic engineering mechanics*, *21*(1), 81-96.

[9] R. Sajeeb, C.S. Manohar, D. Roy (2007), Use of particle filters in an active control algorithm for noisy nonlinear structural dynamical systems. *Journal of sound and vibration*, *306*(1), 111-135.

[10] M. Khalil, A. Sarkar, S. Adhikari (2009), Nonlinear filters for chaotic oscillatory systems. *Nonlinear Dynamics*, *55*(1-2), 113-137.

[11] H.A. Nasrellah, C.S. Manohar (2010), A particle filtering approach for structural system identification in vehicle–structure interaction problems. *Journal of Sound and Vibration*, *329*(9), 1289-1309.

[12] M.S. Arulampalam, S. Maskell, N. Gordon, T. Clapp (2002), A tutorial on particle filters for online nonlinear/non-Gaussian Bayesian tracking. *Signal Processing, IEEE Transactions on*, *50*(2), 174-188.

[13] E.T. Jaynes (2003), *Probability theory: the logic of science*. Cambridge university press.

CHAPTER 8:

Optimization under Uncertainty

Applied Mechanics and Materials Vol. 807 (2015) pp 229-238
© (2015) Trans Tech Publications, Switzerland
doi:10.4028/www.scientific.net/AMM.807.229

Submitted: 2015-04-15
Accepted: 2015-08-04

Robust Truss Topology Design with Beam Elements via Mixed Integer Nonlinear Semidefinite Programming

Tristan Gally[1,a], Christopher M. Gehb[2,b], Philip Kolvenbach[1,c], Anja Kuttich[1,d*], Marc E. Pfetsch[1,e] and Stefan Ulbrich[1,f]

[1] TU Darmstadt, Optimization, Dolivostr. 15, 64293 Darmstadt, Germany,
[2] TU Darmstadt, System Reliability and Machine Acoustics SzM, Magdalenenstr. 4, 64289 Darmstadt, Germany,
[a] gally@mathematik.tu-darmstadt.de, [b] gehb@szm.tu-darmstadt.de
[c] kolvenbach@mathematik.tu-darmstadt.de, [d] kuttich@mathematik.tu-darmstadt.de,
[e] pfetsch@mathematik.tu-darmstadt.de, [f] ulbrich@mathematik.tu-darmstadt.de

Keywords: beam elements, robust optimization, semidefinite programming, truss topology design

Abstract. In this article, we propose a nonlinear semidefinite program (SDP) for the robust truss topology design (TTD) problem with beam elements. Starting from the semidefinite formulation of the robust TTD problem we derive a stiffness matrix that can model rigid connections between beams. Since the stiffness matrix depends nonlinearly on the cross-sectional areas of the beams, this leads to a nonlinear SDP. We present numerical results using a sequential SDP approach and compare them to results obtained via a general method for robust PDE-constrained optimization applied to the equations of linear elasticity. Furthermore, we present two mixed integer semidefinite programs (MISDP), one for the optimal choice of connecting elements, which is nonlinear, and one for the corresponding problem with discrete cross-sectional areas.

Introduction

Topology optimization of truss structures is an important task in the area of structural design. The goal is to optimize the cross-sectional areas of the bars within a truss structure to find a truss that is both stable and lightweight. A disadvantage of classical TTD, however, is that the trusses are only optimized for given predetermined loads. This leads to the idea of robust TTD, where also small disturbances of the loads are considered to produce a truss that is stable under all expected loads. There is a variety of mathematical models for the optimization of truss structures known in the literature, for example [1, 2, 3]. However, most of them are simple models which are only valid for idealized truss structures. To be able to use these models, two assumptions have to be made [3]: Firstly, the bars are centrically and flexibly connected in the nodes, secondly, the external loads are only applied in the nodes. Under these assumptions all bars are loaded by compression or tension in axial direction only, which excludes mechanical effects such as bending of bars. We avoid the above assumptions by using a finite-element approach so as to take bending into account. We consider a truss as a two-dimensional mechanical structure of single beam elements connected to each other. The beam elements are assumed to be straight bars of uniform cross section.

This article is organized as follows. First, we introduce the robust TTD problem. The main contribution of this article is the extension of the TTD model to trusses with beam elements in the next section. The fourth section discusses a general method for robust PDE-constrained optimization applied to the equations of linear elasticity. We then establish numerical results for these approaches to validate the new TTD model. Afterwards, we give an overview of further extensions of the presented model to also include the optimal choice of connecting elements and discrete sets of possible cross-sectional areas by introducing binary variables. We close with a summary and future research topics.

By extending the applicability of the robust truss models this article contributes considerably to the goals of the Collaborative Research Center 805, which studies uncertainty in load-carrying structures in mechanical engineering.

Robust Truss Topology Design

We first review the TTD problem [3] which deals with the task of finding the stiffest truss for a given amount of total available material. For the optimization model we use the so-called ground structure approach, where a fixed ground structure is given and the goal is to find an optimal material distribution for the bars. The ground structure is defined by a set of nodes $\mathcal{V} = \{v_1, ..., v_n\} \subseteq \mathbb{R}^2$ and a set of potential bars $\mathcal{E} \subseteq \mathcal{V} \times \mathcal{V}$. We distinguish n_f free nodes in $\mathcal{V}_f \subseteq \mathcal{V}$, which are movable, and the remaining fixed nodes, which are supported and cannot move. The optimal design then consists of a subset of nodes and bars of the ground structure with optimal cross-sectional areas for all chosen bars for a given external load.

To evaluate the stability of a truss we use the compliance, which measures the node displacements induced by the external loads. The compliance is defined as $c_f(a) = \frac{1}{2} f^\top u$, where $f \in \mathbb{R}^d$ is the vector of external loads, $a \in \mathbb{R}^m$ are the cross-sectional areas of the bars and $u \in \mathbb{R}^d$ are the node displacements. Here, $d = 2n_f$ represents the degrees of freedom, which in general depend on the quantity of free nodes, degrees of freedom per node and the coupling of the beams by hinges or rigid connections, cf. [4]. The node displacements can be computed via the equilibrium condition $A(a)u = f$ with a symmetric positive semidefinite stiffness matrix $A(a) \in \mathbb{R}^{d \times d}$. Physically, the equilibrium condition states that the reaction forces $A(a)u$ have to compensate the external forces f. The stiffness matrix $A(a) = \sum_{e \in \mathcal{E}} A_e a_e$ is linear in the cross-sectional areas. The single bar stiffness matrices A_e are positive semidefinite rank-1 matrices and can be written as $A_e = b_e b_e^\top$ with

$$b_e = (b_{(v_i,v_j)}(k))_{\{k \le n_f\}} = \begin{cases} \sqrt{E}\frac{v_i - v_j}{\|v_j - v_i\|^{3/2}}, & k = i, \\ \sqrt{E}\frac{v_j - v_i}{\|v_j - v_i\|^{3/2}}, & k = j, \\ 0, & \text{otherwise}, \end{cases}$$

where E is the Young's modulus of the used material.

Minimizing the compliance for a given bound $S_{\max} \in \mathbb{R}$ on the amount of used material leads to the following optimization problem for given lower and upper bounds $\underline{a} \in \mathbb{R}$ and $\overline{a} \in \mathbb{R}$ on the cross-sectional area of each bar:

$$\min_{\tau, a} \quad \tau \quad \text{s.t.} \quad \begin{pmatrix} 2\tau & f^\top \\ f & A(a) \end{pmatrix} \succeq 0, \quad \sum_{e \in \mathcal{E}} a_e \le S_{\max}, \quad \underline{a} \le a_e \le \overline{a} \quad \forall e \in \mathcal{E}.$$

For details we refer the reader to [3]. This semidefinite formulation can be solved efficiently, e.g., with interior-point methods [5].

The resulting trusses are designed to carry a given load and might fail to support even slightly different loads. Therefore, it is necessary to deal with uncertainty in the parameters of the optimization model to obtain robust trusses. Hence, we want to use robust optimization as in [3].

The idea of robust optimization is to optimize the worst-case over a given uncertainty set [6]. In our particular case we aim to find a material distribution for the truss structure that minimizes the worst-case compliance over all loads from a given uncertainty set $\mathcal{U}$. As the uncertainty set we choose an ellipsoid containing the nominal loads given by the vector $\bar{f}$ and a ball of occasional loads of a predefined magnitude. Using a matrix $Q \in \mathbb{R}^{d \times r}$, this ellipsoidal uncertainty set can be written as $\mathcal{U} = \{f \in \mathbb{R}^d \mid f = \bar{f} + Q \cdot \Delta f : \|\Delta f\|_2 \le 1\}$, where r is the dimension of the space of all occasional loads we want to sustain. As in [7], this leads to the SDP

$$\min_{\tau, \tau_1, a} \quad \tau \quad \text{s.t.} \quad \begin{pmatrix} 2\tau - \tau_1 & 0 & -\bar{f}^\top \\ 0 & \tau_1 \mathbb{I}_r & -Q^\top \\ -\bar{f} & -Q & A(a) \end{pmatrix} \succeq 0, \quad \sum_{e \in \mathcal{E}} a_e \le S_{\max}, \quad \underline{a} \le a_e \le \overline{a} \quad \forall e \in \mathcal{E}, \tag{1}$$

where $\mathbb{I}_r \in \mathbb{R}^{r \times r}$ is the identity matrix.

Extension to Beam Elements

In this section, we extend the TTD problem to beam elements. This allows us to model rigid node connections. It also allows to model forces on locations within the beams, as we can add another node with rigid connections and split the beam. Therefore, we could drop the second assumption from the introduction. This would not be possible for pin-connected bars.

We will derive a stiffness matrix which we can then insert into the SDP (1) to include the beam elements. To derive this stiffness matrix, the displacements of the truss due to axial loading and bending are calculated numerically using a finite-element model. Each beam of the truss is discretized by two nodes e_1 and e_2 and a one-dimensional beam element e with element length ℓ_e. Each node has three degrees of freedom – two translational and one rotational, see Fig. 1.

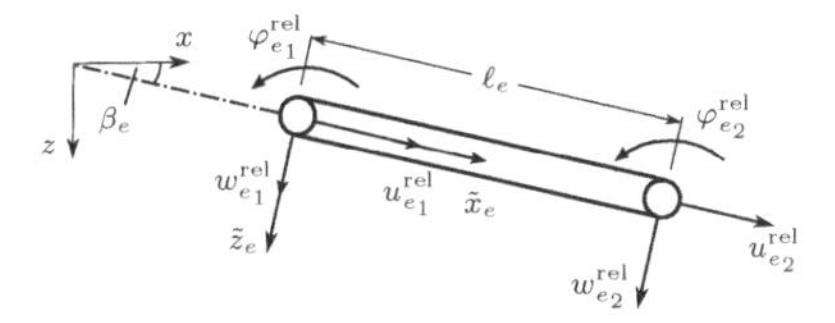

Fig. 1: Finite-element beam model

For each element e, the relationship between axial and lateral element loads and moments f_e^{rel} and axial, lateral and rotational displacements u_e^{rel} is given by $A_e^{\text{rel}}\, u_e^{\text{rel}} = f_e^{\text{rel}}$ with the element stiffness matrix A_e^{rel} in local or relative coordinates $\tilde{x}_e$ and $\tilde{z}_e$. The displacements u_e^{rel} in relative coordinates of one beam element e between nodes e_1 and e_2 can be written as $u_e^{\text{rel}} = (u_{e_1}^{\text{rel}}, w_{e_1}^{\text{rel}}, \varphi_{e_1}^{\text{rel}}, u_{e_2}^{\text{rel}}, w_{e_2}^{\text{rel}}, \varphi_{e_2}^{\text{rel}})^\top$, with the translations $u_{e_{1,2}}^{\text{rel}}$ in axial, translations $w_{e_{1,2}}^{\text{rel}}$ in lateral and rotations $\varphi_{e_{1,2}}^{\text{rel}}$ in angular direction. The element stiffness matrices $A_e^{\text{rel}} \in \mathbb{R}^{6\times 6}$ for each element e are derived from the principle of linear momentum with the Galerkin method [8]. Furthermore, the real displacements are approximated by local shape functions and the nodal degrees of freedom. The local shape functions for each beam e are chosen as

$$\begin{aligned} g_1 &= 1 - \frac{\tilde{x}_e}{\ell_e}, \quad g_2 = 1 - \frac{3\,\tilde{x}_e^2}{\ell_e^2} + \frac{2\,\tilde{x}_e^3}{\ell_e^3}, \quad g_3 = (-\frac{\tilde{x}_e}{\ell_e} + \frac{2\,\tilde{x}_e^2}{\ell_e^2} + \frac{\tilde{x}_e^3}{\ell_e^3}) \cdot \ell_e, \\ g_4 &= \frac{\tilde{x}_e}{\ell_e}, \qquad\quad g_5 = \frac{3\,\tilde{x}_e^2}{\ell_e^2} - \frac{2\,\tilde{x}_e^3}{\ell_e^3}, \qquad\quad g_6 = (\frac{\tilde{x}_e^2}{\ell_e^2} + \frac{\tilde{x}_e^3}{\ell_e^3}) \cdot \ell_e \end{aligned} \tag{2}$$

and the entries of the element stiffness matrices A_e^{rel} are given by

$$A_{e,ji}^{\text{rel}} = \int_0^{\ell_e} E a_e \frac{\partial g_j}{\partial \tilde{x}_e} \frac{\partial g_i}{\partial \tilde{x}_e}\, d\tilde{x}_e,\ j,i \in \{1,4\}, \quad A_{e,ji}^{\text{rel}} = \int_0^{\ell_e} E I_z \frac{\partial^2 g_j}{\partial \tilde{x}_e^2} \frac{\partial^2 g_i}{\partial \tilde{x}_e^2}\, d\tilde{x}_e,\ j,i \in \{2,3,5,6\}. \tag{3}$$

Here, I_z is the moment of inertia of area, which can be computed via $I_z = \pi r_e^4/4 = a_e^2/(4\pi)$ for circular beams. With Eqs. 2–3 the element stiffness matrices can be stated as (compare [9])

$$A_e^{\text{rel}} = \frac{EI_z}{\ell_e^3} \begin{pmatrix} a_e\ell_e^2/I_z & & & & & \\ 0 & 12 & & \text{symmetric} & & \\ 0 & 6\,\ell_e & 4\,\ell_e^2 & & & \\ -a_e\ell_e^2/I_z & 0 & 0 & a_e\ell_e^2/I_z & & \\ 0 & -12 & -6\,\ell_e & 0 & 12 & \\ 0 & 6\,\ell_e & 2\,\ell_e^2 & 0 & -6\,\ell_e & 4\,\ell_e^2 \end{pmatrix}. \tag{4}$$

Before the element stiffness matrices A_e^{rel} for each beam element can be assembled to become the system stiffness matrix A in global coordinates x and z, each element stiffness matrix A_e^{rel} has to be transformed from relative to global coordinates. The transformations for the displacements and for the loads and moments are $u_e = T_e u_e^{\mathrm{rel}}$ and $f_e = T_e f_e^{\mathrm{rel}}$, with the transformation matrix

$$T_e = \begin{pmatrix} \hat{T}_e & 0 \\ 0 & \hat{T}_e \end{pmatrix} \quad \text{with} \quad \hat{T}_e = \begin{pmatrix} \cos\beta_e & \sin\beta_e & 0 \\ -\sin\beta_e & \cos\beta_e & 0 \\ 0 & 0 & 1 \end{pmatrix},$$

where β_e is the angle between the beam element's $\tilde{x}_e$ axis and the global coordinate axis x as shown in Fig. 1. Now it is possible to express the equilibrium condition in global coordinates as

$$T_e\, A_e^{\mathrm{rel}}\, T_e^{-1}\, u_e = A_e^{\mathrm{glb}}\, u_e = f_e.$$

To be able to assemble the system stiffness matrix A from the element stiffness matrices A_e^{glb} in element degrees of freedom, incidence matrices $L_e \in \{0,1\}^{6\times d}$ for each element are needed [8]. These boolean matrices define the relationship between the six element degrees of freedom u_e and the d system degrees of freedom u by connecting each element node with the corresponding global node. Then, we derive the element stiffness matrices $A_e = L_e^\top A_e^{\mathrm{glb}} L_e$ in global degrees of freedom and the system stiffness matrix as $A = \sum_{e\in\mathcal{E}} A_e$ in global coordinates. Finally, the system matrix equation in global coordinates can be written as $Au = f$ with the system displacements u and the load vector f in global coordinates.

Note that in Eq. 4 we again have $I_z = a_e^2/(4\pi)$, so the disadvantage of using beam elements is that our semidefinite program becomes nonlinear.

Robust PDE-constrained optimization applied to the equations of linear elasticity

We validate the newly developed TTD problem with a method for the robust optimization of general nonlinear programs with discretized PDE constraints that are subject to uncertain parameters, which we will apply to the equations of linear elasticity. The method is thoroughly and more generally discussed in the dissertation of Sichau [10]. It has been applied successfully to the robust geometry optimization of different load-carrying structures and it is neither restricted to TTD in specific nor to structural mechanics in general. In the simple case without uncertain inequality constraints, we consider an uncertain, fully discretized nonlinear program

$$\min_{u,a} \quad J(u,a;p) \quad \text{s.t.} \quad e(u,a;p) = 0 \text{ and } c(a) \le 0, \tag{5}$$

where $a \in \mathbb{R}^{n_a}$ is the vector of design variables, $u \in \mathbb{R}^{n_u}$ is the vector of state variables, $p \in \mathbb{R}^{n_p}$ is the vector of uncertain parameters, J is the scalar objective function and $c(a) \le 0$ are vector-valued design constraints. For given a and p the state equation $e(u,a;p) = 0$ is required to yield a unique state u. As in the second section, we assume the uncertain parameter to be restricted to a known ellipsoidal uncertainty set $\mathcal{U}$ of the form

$$\mathcal{U} = \{p \in \mathbb{R}^{n_p} \mid p = \bar{p} + Q\cdot\Delta p,\ \|\Delta p\|_2 \le 1\} = \left\{p \in \mathbb{R}^{n_p} \,\middle|\, p = \bar{p} + \Delta p,\ \|Q^{-1}\Delta p\|_2 \le 1\right\}$$

with an invertible matrix $Q \in \mathbb{R}^{n_p \times n_p}$ and a center $\bar{p} \in \mathbb{R}^{n_p}$ which we call the *nominal* value. This enables us to incorporate the parameter p in the optimization problem by considering the worst-case $\Phi(a)$ of the objective for a given design with respect to the uncertainty set, that is, formally,

$$\Phi(a) = \max_{u,p} \quad J(u,a;p) \quad \text{s.t.} \quad e(u,a;p) = 0 \text{ and } p \in \mathcal{U}.$$

The robust counterpart of problem (5) is to find the design with the best worst-case, i.e.,

$$\min_{a} \quad \Phi(a) \quad \text{s.t.} \quad c(a) \le 0.$$

This min-max problem is difficult to solve for a general nonlinear objective due to its two-level structure. The key idea in [10] is to approximate the objective by a second-order Taylor expansion and to linearize the state equation. We expand these functions around a point $\bar{x} = (\bar{u}, a; \bar{p})$ for which $e(\bar{x}) = 0$ holds. In doing so we arrive at

$$\begin{aligned} \tilde{\Phi}(a) = \max_{\Delta u, \Delta p} \quad & J(\bar{x}) + \partial_{(u,p)} J(\bar{x}) s + \frac{1}{2} s^\top \begin{pmatrix} \partial_{uu} J(\bar{x}) & \partial_{up} J(\bar{x}) \\ \partial_{pu} J(\bar{x}) & \partial_{pp} J(\bar{x}) \end{pmatrix} s \\ \text{s.t.} \quad & e(\bar{x}) + \partial_{(u,p)} e(\bar{x}) s = \partial_u e(\bar{x}) \Delta u + \partial_p e(\bar{x}) \Delta p = 0, \quad \|Q^{-1} \Delta p\|_2 \leq 1, \end{aligned}$$

where $\Delta u \in \mathbb{R}^{n_u}$, $\Delta p \in \mathbb{R}^{n_p}$ and $s = (\Delta u^\top, \Delta p^\top)^\top$. With the linearized state equation we can easily eliminate Δu from the maximization problem and obtain a trust-region problem of the form

$$\tilde{\Phi}(a) = \max_{\Delta p} \quad q(a, \Delta p) \quad \text{s.t.} \quad \|Q^{-1} \Delta p\|_2 \leq 1,$$

with a function q that is quadratic in the optimization variable Δp. Both the approximated worst-case function $\tilde{\Phi}$ and its gradient $\nabla \tilde{\Phi}$ can be computed efficiently, which follows from well-known trust-region theory [11]. For details on a computationally tractable representation of the gradient, we refer to [10]. It is important to note, however, that the first partial derivatives of the state equation have to be available. We conclude that the approximated problem

$$\min_{v} \quad \tilde{\Phi}(a) \quad \text{s.t.} \quad c(a) \leq 0$$

can be solved efficiently with gradient-based algorithms such as standard SQP methods [12].

Next, we apply this method to the robust geometry optimization of an elastic structure under uncertain external loads. We describe the physics of the system with the equations of two-dimensional linear elasticity. After finite-element discretization we obtain for the state equation

$$e(u, a; p) = A(a) u - N p = 0,$$

where $A(a) \in \mathbb{R}^{n_u \times n_u}$ is the stiffness matrix, $N \in \mathbb{R}^{n_u \times n_p}$ is the boundary mass matrix and the uncertain parameter $p \in \mathbb{R}^{n_p}$ defines the surface load. Note that the vector Np corresponds to the load vector f in the previous sections. As in the second section, we again choose the compliance $J(u, a; p) = \frac{1}{2} u^\top N p$ as the objective function. For this specific objective, the second-order Taylor expansion is exact and therefore $\Phi = \tilde{\Phi}$ holds, which means we can optimize for the exact worst-case. We employ the same design constraints as before and obtain $c(a) = (a - \overline{a}, -a + \underline{a}, \sum_{j=1}^{n_a} a_j - S_{\max})$, with lower and upper bounds $\underline{a}$ and $\overline{a}$ both in $\mathbb{R}^{n_a}$ and $S_{\max} \in \mathbb{R}$.

Numerical Results

We now apply the techniques presented in this paper to the optimization of a simple truss structure with six nodes and nine bars, see Fig. 2.

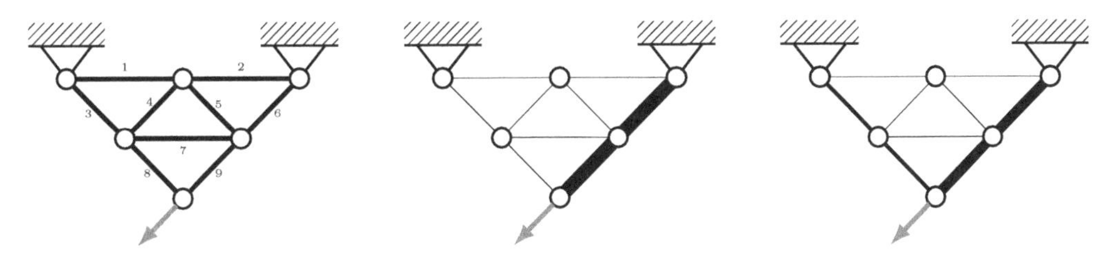

Fig. 2: Groundstructure, nominal and robust solution of the SDP approach (not to scale)

Table 1: Solutions and compliances for the different models

	cross-sectional areas in mm^2									compliance	
model	a_1	a_2	a_3	a_4	a_5	a_6	a_7	a_8	a_9	nominal	worst-case
SDP-N	0.50	0.50	0.50	0.50	0.50	15.75	0.50	0.50	15.75	2.00	4.59
PDE-N	0.50	0.50	0.50	0.50	0.50	15.76	0.50	0.50	15.74	2.00	4.59
SDP-R	0.50	0.50	2.61	0.50	0.50	13.64	0.50	2.61	13.64	2.30	2.95
PDE-R	0.50	0.50	2.51	0.50	0.50	12.90	0.50	2.56	14.54	2.29	2.94

The optimization problem is to find the cross-sectional areas in square millimeters given as a vector $a \in \mathbb{R}^9$ such that the respective objective is minimal, which is the compliance for the nominal problems and the worst-case of the compliance for the robust problems. The length of the horizontal beams is $1.8\,\text{m}$, the length of the diagonal ones $1.2\,\text{m}$. The material of the truss is steel with a Young's modulus $E = 210\,\text{GPa}$ and a Poisson's ratio $\nu = 0.3$. We set $\underline{a} = 0.5\,\text{mm}^2$, $\overline{a} = 18\,\text{mm}^2$ and $S_{\text{max}} = 35\,\text{mm}^2$. The top left and top right nodes of the truss are fixed. The bottommost node is subject to an uncertain force with a nominal magnitude of 75 N with direction $(-1, -1)^\top$. We define the uncertainty set such that disturbances of up to $\pm 5\,\%$ in the direction of the nominal load and up to $\pm 20\,\%$ orthogonal to that direction are permitted in the bottommost node. Hence, we have

$$\bar{p} = \frac{75}{\sqrt{2}} \begin{pmatrix} -1 \\ -1 \end{pmatrix} \quad \text{and} \quad Q = \frac{\|\bar{p}\|_2}{\sqrt{2}} \begin{pmatrix} -0.05 & 0.2 \\ -0.05 & -0.2 \end{pmatrix}$$

and the uncertainty set $\mathcal{U} = \{p \in \mathbb{R}^2 \mid p = \bar{p} + Q \cdot \Delta p \colon \|\Delta p\|_2 \leq 1\}$. For the SDP models, the force vectors p and Δp and the matrix Q have to be augmented with zeros for the three free nodes without load.

All computations were done on a workstation with a quad-core 3.4 GHz Intel i7-4770 CPU. In the following, we refer to the nominal and robust SDP formulation as SDP-N and SDP-R. Analogously, we refer to the nominal and robust problem of the previous section as PDE-N and PDE-R.

For the newly developed TTD problem with beam elements we use a sequential semidefinite programming algorithm based on [13] implemented in Matlab R2014a. We solve the resulting linear SDPs with SeDuMi 1.3 [14]. The runtime for SDP-N was 3.43 s and for SDP-R 1.57 s.

For the PDE-based method we generate a two-dimensional mesh with the Comsol Multiphysics 5.0 software consisting of 2,830 nodes and 3,554 triangle elements. Since we neglect the third dimension, the cross-sectional areas are given as the widths of the beams. We choose linear basis functions for the finite-element discretization. The six nodes are modeled as rigid regular octagons with a side length of 20 mm. We employ a homogeneous Dirichlet boundary condition along the upper edges of the top left and top right node. The load is implemented as a Neumann boundary condition on the bottom left edge of the bottommost node. We use the same robust optimization framework and globalized SQP solver as described in [10], both of which are written in Matlab. The runtime for the robust optimization was 180.78 s and for the non-robust optimization 21.66 s.

The results are given in Table 1. Since the displacements of the zero-dimensional nodes of the SDP models cannot be mapped directly to the displacements of the two-dimensional nodes of the PDE models, the compliance values for the same cross-sectional areas differ slightly between the two models. For this reason, the compliance values given in Table 1 were evaluated using the PDE models for better comparison. The column "nominal" shows the compliance of the truss defined by the cross-sectional areas when the load case is $\bar{p}$. The column "worst-case" shows, for the same truss, the worst compliance for any load $p \in \mathcal{U}$.

We see that for the nominal problems SDP-N and PDE-N the solutions are almost identical. The cross-sectional areas for the robust optimal solutions of SDP-R and PDE-R differ, but the worst-case objective values are almost identical with a relative error of about 0.3 %. For SDP-R, less material is moved to the left bars and the material is evenly distributed between the outer top and the bottom bars, whereas for PDE-R, more material is moved to the left and to the bottom. Apparently, due to the nonconvex nature of the problems, there are multiple local minima that are, even though relatively far apart, of comparable quality and the two solvers under investigation converge to a solution that is slightly favorable for their respective model.

Table 1 shows that both approaches are successful in making the truss more robust to disturbances of the load. At the cost of an increase in the nominal objective of about 15 % the worst-case could be reduced by 36 % from the nominal to the robust solution for both models. The newly developed TTD model with beam elements was more than one hundred times faster then the other approach in the robust case.

Further Extensions

In this section, we present two further extensions of our model using binary variables. First, we want to introduce a possibility for connecting the single beam elements via hinges. We present an optimization model for the optimal selection of the connection type. As a second extension we restrict the cross sectional-areas of the beams to a discrete set. Both of these extensions require the usage of branch-and-bound algorithms to deal with the binary variables. In the first extension, the resulting relaxations are nonlinear SDPs, in the second approach linear SDPs.

Optimal selection of connecting element type. In the context of beam elements it is also possible to model pin-connected beams via hinges instead of rigid connections. For this purpose the stiffness matrix $A(a)$ has to be changed slightly, as it is sufficient to only manipulate the incidence matrices L_e. If our truss should only consist of hinges as connecting elements, we need an additional degree of freedom for the rotation at each connection between a beam and a free node instead of a single rotation per free node. Then we again compute the incidence matrices L_e, which now connect each rotation in the nodes of a beam to the corresponding rotations in the system stiffness matrix.

We now present an optimization model for the optimal selection of the connecting element types. For this purpose, we include stiffness matrices for all four possible combinations of connections for each bar in our optimization problem:

- A_e^{rr} if both nodes of bar e are rigidly connected,
- A_e^{rp} if the first node of bar e is rigidly connected and the second pin-connected,
- A_e^{pr} if the first node of bar e is pin-connected and the second rigidly connected,
- A_e^{pp} if both nodes of bar e are pin-connected.

To derive the different stiffness matrices only the boolean matrices L_e have to be changed. In this case, our system needs the degrees of freedom of both the rigid and pinned connecting elements. We now have to compute four different boolean matrices L_e^{rr}, L_e^{rp}, L_e^{pr} and L_e^{pp}, where the matrices for rigid connections in one of the nodes map that node's rotation to the shared rotation of that node in the system stiffness matrix, while those for hinges map them to the rotation for that node and this beam.

Apart from these additional stiffness matrices, we need n binary variables h_{v_i} that model whether a node v_i is rigidly connected ($h_{v_i} = 0$) or pin-connected ($h_{v_i} = 1$) . Now we need compatibility conditions which enforce that the connections of the beams match those of the nodes. There are different ways to model such constraints (see for example [15]). In our case, we can decide between a Big-M approach or a complementary reformulation.

In the first case, we split the variables for the cross-sectional areas a_e for each beam $e = (e_1, e_2)$ into four different variables a_e^{rr}, a_e^{rp}, a_e^{pr}, a_e^{pp}, enforcing that only one of those may be nonzero, namely the one corresponding to the values of h_{e_1} and h_{e_2}. This can be done with the eight constraints

$$\begin{aligned} a_e^{rr} &\le M(1-h_{e_1}), & a_e^{rp} &\le M(1-h_{e_1}), & a_e^{pr} &\le Mh_{e_1}, & a_e^{pp} &\le Mh_{e_1}, \\ a_e^{rr} &\le M(1-h_{e_2}), & a_e^{rp} &\le Mh_{e_2}, & a_e^{pr} &\le M(1-h_{e_2}), & a_e^{pp} &\le Mh_{e_2}, \end{aligned}$$

where M is a sufficiently large constant, which in our case can be chosen as $S_{\max}$. The first two constraints enforce that a_e^{rr} and a_e^{rp} are zero if the first node is pin-connected ($h_{e_1} = 1$), the third and fourth constraints make sure that a_e^{pr} and a_e^{pp} may not be used if the first node is rigidly connected, and the remaining constraints do the same for the second node. Now we calculate each stiffness matrix using the variable for the cross-sectional area of the corresponding connecting-element type to get the stiffness matrix of the whole truss.

In the second approach, instead of splitting the variables for the cross-sectional areas, we introduce binary variables y_e^{rr}, y_e^{rp}, y_e^{pr}, y_e^{pp} and use

$$\begin{aligned} y_e^{rr} &\le (1-h_{e_1}), & y_e^{rp} &\le (1-h_{e_1}), & y_e^{pr} &\le h_{e_1}, & y_e^{pp} &\le h_{e_1}, \\ y_e^{rr} &\le (1-h_{e_2}), & y_e^{rp} &\le h_{e_2}, & y_e^{pr} &\le (1-h_{e_2}), & y_e^{pp} &\le h_{e_2}. \end{aligned}$$

Then we multiply each stiffness matrix computed with a_e with the corresponding y-variable. The advantage of this approach is that we no longer need $S_{\max}$ as Big-M, which, depending on $S_{\max}$ and the size of the truss, might lead to numerical problems in the branch-and-bound tree. On the other hand, we need more binary variables, although these will not enlarge the branch-and-bound tree, if we branch on the h-variables only, and also get additional nonlinearities due to the products $y_e^{rr} \cdot A_e^{rr}(a)$, $y_e^{rp} \cdot A_e^{rp}(a)$, $y_e^{pr} \cdot A_e^{pr}(a)$ and $y_e^{pp} \cdot A_e^{pp}(a)$.

Discrete cross-sectional areas. The optimal solution of the TTD problem might result in bar sizes which cannot always be produced in practice. Following the approach in [16], we define a discrete set $\mathcal{A}$ of available cross-sectional areas for the beams. Instead of one variable for the cross-sectional area of beam e, we then introduce binary variables x_e^a for all combinations of areas and beams, with the meaning that $x_e^a = 1$ if and only if beam e has cross-sectional area a. This leads to the MISDP (cf. [16])

$$\min_{\tau, \tau_1, x_e^a} \ \tau \text{ s.t.} \begin{pmatrix} 2\tau - \tau_1 & 0 & -\bar{f}^\top \\ 0 & \tau_1 \mathbb{I}_q & -Q^\top \\ -\bar{f} & -Q & \sum\limits_{e \in \mathcal{E}, a \in \mathcal{A}} A_e^a x_e^a \end{pmatrix} \succeq 0, \sum_{e \in \mathcal{E}, a \in \mathcal{A}} a \cdot x_e^a \le S_{\max}, \sum_{a \in \mathcal{A}} x_e^a \le 1 \, \forall e \in \mathcal{E}, \ x_e^a \in \{0,1\}.$$

As we compute the stiffness matrices A_e^a for all possible cross-sectional areas before starting the optimization process, the nonlinearity of the stiffness matrix in a is now hidden in the computation of the stiffness matrices A_e^a, which are constant for the solving process. This means that we can now solve a linear MISDP. But it also means that we are linearizing the stiffness matrices for the relaxations, so the quality of the relaxations is decreased, even though the model is exact for each integer solution. Nevertheless, this allows us to solve the problem globally, while for the nonlinear SDPs, we can only show convergence to a local optimal solution.

To this model we can again add the optimization of the node-connecting elements from the last subsection, which is even more natural here, as we will stay in the same problem class of MISDPs. In this case we should always use the Big-M approach, as we do not want to add nonlinear constraints to this model. In addition, we can even use $M = 1$, as our variables $x_{e,a}^{rr}$, $x_{e,a}^{rp}$, $x_{e,a}^{pr}$ and $x_{e,a}^{pp}$ will be binary anyway, so we do not have the disadvantages of either model.

Numerical Results for the two extensions. To produce the numerical results for the second extension, we used the MISDP code introduced in [16]. The code combines the branch-and-bound framework of SCIP 3.1.1 [17] with interior-point SDP solvers, in our case DSDP 5.8 [18]. For the continuous problem, we again used a sequential SDP algorithm extended by a surrounding branch-and-bound code, which was implemented using Matlab R2014a and SeDuMi 1.3 [14].

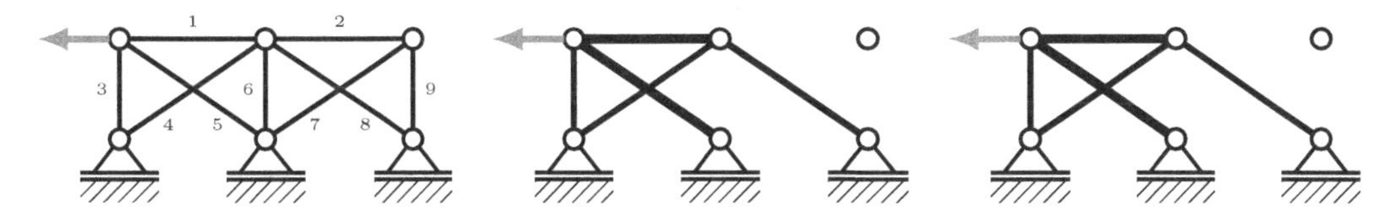

Fig. 3: Groundstructure and solutions for continuous and discrete cross-sectional areas

The types of connecting elements did not have an influence on the example from the last section, because of its ground structure and the position of the force. We therefore use a different example for this section, which can be seen in Fig. 3. In this case we choose $\underline{a} = 0$, $\overline{a} = 2$ and $S_{\text{max}} = 8$. For the discrete problem, we allow six different cross-sectional areas between 0.33 mm^2 and 2 mm^2. Solving took 18.49 s for 515 branch-and-bound nodes. The continuous model took 937.96 s for 48 branch-and-bound nodes. The high computational costs are due to the need to solve a nonlinear SDP in each node. The optimization result is that all free nodes should be connected by rigid elements. The optimal sizes of the beams can be seen in Fig. 3 and Table 2. In contrast to hinges, this improves the objective by 2.3 %. Comparing the results, we see that not all cross-sectional areas of the continuous solution are elements of the discrete set. Therefore the compliance in the discrete model increases by 9.4 %.

Table 2: Solutions and objective values for the optimization of the connecting elements

	cross-sectional areas in mm^2									compliance
model	a_1	a_2	a_3	a_4	a_5	a_6	a_7	a_8	a_9	worst-case
cont.	2.00	0.00	1.23	1.31	2.00	0.00	0.00	1.46	0.00	0.23
discr.	2.00	0.00	1.33	1.33	2.00	0.00	0.00	1.33	0.00	0.26

Conclusion

We presented a nonlinear SDP model for the robust TTD problem with beam elements. We were able to verify our model by comparing it to a general nonlinear robust optimization approach discretized with a standard finite element method. The new model was able to achieve very similar results at less than 1 % of the computational cost of the other approach. The robust TTD model with beam elements reduced the worst-case compliance by as much as 40 %, thus resulting in a truss that is much more robust to uncertain loads. Furthermore, we extended our model to the optimal selection of connection types and restricted the cross-sectional areas to a discrete set.

A possible further extension is to use continuous variables for the stiffness of the connections instead of the binary decision between pinned and rigid connections. This can be achieved by starting from the stiffness matrix of the pin-connected system and adding a further symmetric matrix coupling the rotational degrees of freedom via rotational stiffnesses. These rotational stiffnesses are further parameters accessible for optimization. Another possibility is to add active elements that are able to introduce additional forces to the system or increase the critical buckling load.

Acknowledgement

The authors would like to thank the German Research Foundation (DFG) for funding this research within the Collaborative Research Center 805.

References

[1] M. P. Bendsøe, A. Ben-Tal, J. Zowe, Optimization methods for truss geometry and topology design, Struct. Optimization 7:3 (1994) 141-159.

[2] M. P. Bendsøe, O. Sigmund, Topology Optimization: Theory, Methods and Applications, Springer Science & Business Media, Berlin and Heidelberg, 2003.

[3] A. Ben-Tal, A. Nemirovski, Robust truss topology design via semidefinite programming, SIAM J. Optim. 7:4 (1997) 991-1016.

[4] D. de Klerk, D. J. Rixen, S. N. Voormeeren, General framework for dynamic substructuring: history, review, and classification of techniques, AIAA J. 46:5 (2008) 1169-1181.

[5] H. Wolkowicz, R. Saigal, L. Vandenberghe, Handbook of Semidefinite Programming: Theory, Algorithms and Applications, Kluwer Academic Publishers, Boston, 2003.

[6] A. Ben-Tal, L. El Ghaoui, A. Nemirovksi, Robust Optimization, Princeton University Press, Princeton and Oxford, 2009.

[7] K. Habermehl, Robust Optimization of Active Trusses via Mixed-Integer Semidefinite Programming, PhD thesis, TU Darmstadt, 2014.

[8] B. Klein, FEM - Grundlagen und Anwendungen der Finite-Element-Methode im Maschinen- und Fahrzeugbau, ed. 8, Vieweg+Teubner-Verlag, Wiesbaden, 2012.

[9] J. S. Przemieniecki, Theory of Matrix Structural Analysis, McGraw-Hill, New York, 1968.

[10] A. Sichau, Robust Nonlinear Programming with Discretized PDE Constraints using Second-order Approximations, PhD thesis, TU Darmstadt, 2014.

[11] A. R. Conn, N. I. M. Gould, P. L. Toint, Trust-Region Methods, SIAM, Philadelphia, 2000.

[12] M. Ulbrich, S. Ulbrich, Nichtlineare Optimierung, Birkhäuser Verlag, Basel, 2012.

[13] R. Correa, C. Ramirez, A global algorithm for nonlinear semidefinite programming, SIAM J. Optim. 15:1 (2004) 303-318.

[14] J. F. Sturm, Using SeDuMi 1.02, a MATLAB toolbox for optimization over symmetric cones, Optim. Method. Softw. 11–12 (1999) 625–653.

[15] P. Bonami, A. Lodi, A. Tramontani, S. Wiese, On mathematical programming with indicator constraints, Math. Program. (2015) 1-33.

[16] S. Mars, Mixed-Integer Semidefinite Programming with an Application to Truss Topology Design, PhD thesis, FAU Erlangen-Nürnberg, 2013.

[17] T. Achterberg, SCIP, Solving constraint integer programs, Math. Program. Comput. 1:1 (2009) 1-41.

[18] S. J. Benson, Y. Ye, Algorithm 875: DSDP5 – software for semidefinite programming, ACM Trans. Math. Software 34:3 (2008) 1-20.

CHAPTER 9:

Binary Decisions under Uncertainty

Applied Mechanics and Materials Vol. 807 (2015) pp 241-246
Submitted: 2015-04-15
Accepted: 2015-08-04

doi:10.4028/www.scientific.net/AMM.807.241

Developing a Control Strategy for Booster Stations under Uncertain Load

Philipp Pöttgen[a], Thorsten Ederer[b], Lena C. Altherr[c], and Peter F. Pelz[d]

Chair of Fluid Systems, Technische Universität Darmstadt
Magdalenenstraße 4, 64283 Darmstadt, Deutschland

[a]philipp.poettgen@fst.tu-darmstadt.de, [b]thorsten.ederer@fst.tu-darmstadt.de,
[c]lena.altherr@fst.tu-darmstadt.de, [d]peter.pelz@fst.tu-darmstadt.de

Keywords: Technical Operations Research (TOR), Booster Station, Pump System, Discrete Optimization

Abstract. Booster stations can fulfill a varying pressure demand with high energy-efficiency, because individual pumps can be deactivated at smaller loads. Although this is a seemingly simple approach, it is not easy to decide precisely when to activate or deactivate pumps. Contemporary activation controls derive the switching points from the current volume flow through the system. However, it is not measured directly for various reasons. Instead, the controller estimates the flow based on other system properties. This causes further uncertainty for the switching decision. In this paper, we present a method to find a robust, yet energy-efficient activation strategy.

Introduction

The world's energy consumption keeps rising. Today we are facing a wide range of energy sources and consumers are diversive: Economical sectors as well as private households. About 12 % of the electrical energy in Europe drives pumps [1]. A specific type of pump system is a booster station: A parallel setup of two or more pumps. The individual machines are located close to each other and the whole system is delivered as a package. Booster stations cover a wide performance map and thus are flexible machines with widespread possibilities of usage. Booster stations convey fluid from reservoirs to its destination or support existing networks: If the pressure of a building's water supply is too low, a booster station increases the pressure.

Due to the many applications and thus large energy input, booster stations need to be designed and controlled appropriately. This paper focuses on the control of existing booster stations. Binary decisions are always a discontinuity and hence a burden in the calculation of energy efficient control strategies. We address the following question: How many pumps should cover the uncertain load?

Technical Description

Topological Layout of Booster Stations. Fig. 1 shows the connection scheme of the booster station. Six pumps are connected in parallel. The incoming water flows from the suction pipe, through the single pump units, into the pressure pipe. The pressure at the outlet is always higher than at the inlet. To avoid reverse flow, a check valve is installed behind each pump. The technical topology of the booster station is unquestioned in this paper.

In a pump's field of operation four parameters are important: The volume flow Q, the pressure increase ΔH, the rotational speed n and the power consumption P. Any two of these parameters describe the point of duty. Following industrial standards we measure the pressure increase in meter by scaling the pressure increase Δp in bar with the density ϱ and the specific gravitational constant g:

$$\Delta H = \frac{\Delta p}{\varrho g}. \tag{1}$$

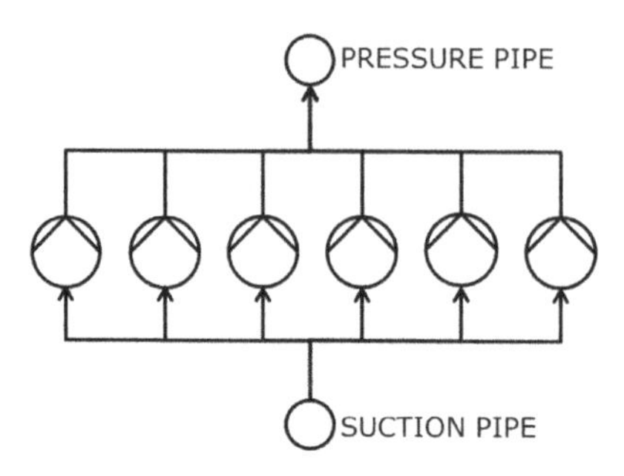

Fig. 1: Topology of a booster station composed of six pumps.

Table 1 shows the characteristics of the considered pump as sample points of the reference curve at maximal rotational speed $n = n_{max}$.

Table 1: Pump Characteristics

Volume flow in $m^3\,h^{-1}$	0	2.05	4.33	6.5	8.7	11	14
Pressure head in m	95.535	94.206	90.937	87.360	82.660	74.486	59.569
Power demand in kW	1.41	1.85	2.45	2.95	3.40	3.75	3.9

For other rotational speeds, the well known scaling laws hold:

$$Q(n) = \left(\frac{n}{n_{max}}\right) Q(n_{max}), \tag{2}$$

$$H(n) = \left(\frac{n}{n_{max}}\right)^2 H(n_{max}), \tag{3}$$

$$P(n) = \left(\frac{n}{n_{max}}\right)^3 P(n_{max}). \tag{4}$$

In case of a booster station the number of active pumps is an additional parameter of operation.

Operational mode. The control variable for the operation of the booster station is the pressure in the pressure pipe. With the premise of a constant supply pressure, this means that the pressure increase of the booster station is controlled to be constant. The controller adjusts the rotational speed of the pumps and switches pumps on and off. The six pumps are of the same type, so the rotational speed in all running pumps is the same. While the rotational speed is a continuous variable which could be controlled, e.g. by a PID controller, the operational status is a discrete variable that causes a discontinuous transition. To react on this discontinuity the designer might use one of the two following control layouts:

1. The first option is a simple layout with few input parameters: Whenever the rotational speed of the active pumps is set to maximum, the controller turns on an additional pump and adjusts the rotational speed. The set-point of the continuous part of the control becomes an input parameter for the discontinuous part. This simple rule guarantees the functionality of the system, but does not consider the energy consumption. The point of duty remains unknown as the total volume flow is unknown.

2. The second option considers the energy consumption, but needs more information on the actual working point of the booster station: The volume flow becomes an input parameter of the control. Volume flow and pressure increase define the working point of the booster station. Two parameters remain to set the working point in any pump. These are (i) the actual number of working pumps and (ii) the rotational speed of the pumps. We optimize the setting of these two parameters, so that the energy consumption of the system becomes minimal.

While the measurement of the pressure in pumps is state of the art, one cannot measure the volume flow in the booster station easily. The conventional measurement devices are way too expensive or would cause an additional flow resistance. However, the controller can make a proper guess for the current volume flow from the already known parameters: With the knowledge about the machine characteristics, the currently measured pressure and the control signal for the rotational speed, the corresponding volume flow is calculated and the exact point of duty of the station is found. With the additional assumption that the volume flow in each pump is the same, we know the point of duty for all pumps.

The estimation of the volume flow assumes stationary flow and exactly measured pump characteristics. Thus it has an error, which leads to a binary decision under uncertainty. We want to apply the second control strategy in technical system to reduce the energy consumption. Our optimization allows to find a control strategy for this case and considers the uncertainty in the calculation of the volume flow.

Optimization task. Following the the TOR-Methodology [2], we have to define (1) the function of the systems, (2) the aim, and (3) the playing field for the optimization:

1. The function of a booster station is to increase pressure to establish a volume flow.

2. The aim of this optimization is to fulfill the function as energy efficient as possible.

3. The playing field is the control strategy. In several duty points the controller can use two degrees of freedom to find a set-point to cover the load. On the one hand, the rotational speed of the pumps, and on the other hand, the number of working pumps.

Based on this, we create an optimization model, which includes both degrees of freedom and covers the uncertain load of the booster station. The mathematical model consists of two stages: First, find a discrete decision for the number of running pumps and secondly, find a continuous decision for the rotational speed.

Mathematical Model

Basic Model. The model for the optimization of the booster station follows the description of [3] with modifications and uses the linearization techniques of [4] to generate a Mixed-Integer Linear Program (MILP). The constraints of the optimization program contain the physical and technical description of the problem. The booster station is modeled as a graph $G(V, E)$. The edges either represent the technical components or simple connections within the graph. Each edge has a variable for the volume flow Q and the head difference ΔH between the vertices. For simple connections the pressure difference equals 0.

The only technical components in this model are centrifugal pumps R: The characteristics within the dependent variables for volume flow Q, pressure head H, rotational speed n and power consumption P describe the flow in every active pump (cf. table 1). We model the dependence of the control variables Q, H, n, P by linearization of the max-curve and the scaling laws as a constraint in the optimization program.

To deactivate a pump, the volume flow in the pump is necessarily set to zero and the pressure of the two connection ports must be uncoupled. This leads to a Big-M formulation with a binary decision for the activation of a pump. The set V represents the connection ports of all technical components plus sources and sinks of the system. In every vertex the volume flow conservation

$$\sum_{(i,v)\in E} Q_{i,v} - \sum_{(v,j)\in E} Q_{v,j} = 0 \qquad \forall v \in V \tag{5}$$

holds as a constraint. Two exceptions from this rule are the vertices for the source s and the sink t: The pressure p in the vertex is given as a boundary parameter of the system's load and volume flow conservation does not hold. Instead of this, the volume flow demand becomes part of the flow conservation in source and sink:

$$\sum_{(s,j)\in E} Q_{s,j} = \sum_{(i,t)\in E} Q_{i,t} = Q_{\text{Load}}. \tag{6}$$

The objective of the optimization is to use as less energy as possible to cover the load. Thus, we minimize the total energy consumption of all pumps:

$$\min \sum_{(i,j)\in R} P_{i,j}. \tag{7}$$

Integration of the uncertain load. We assume that the estimation of the volume flow Q based on the pressure head p and the rotary speed n has no systematic error. The statistical error can be modeled by the probability density function

$$f(Q) = \frac{1}{\sigma_Q\sqrt{2\pi}} \exp\left(\frac{(Q-\mu_Q)^2}{2\sigma_Q^2}\right) \tag{8}$$

of the normal distribution with the mean volume flow μ_Q and the standard deviation σ_Q. A large standard deviation indicates a fluctuating flow or a high uncertainty in the measured pump characteristic.

A continuous distribution is difficult to incorporate in our optimization under uncertainty setting: On the one hand, we have discrete and continuous decision variables depending on the random variable. On the other hand, we want to be able to use the model of this paper in a multi-stage setting. Therefore, the continuous distribution has to be approximated by a discrete distribution.

A good discrete approximation of a continuous probability distribution preserves the lower moments. The idea is to reproduce decisive factors of the distribution's shape like mean, variance, skew and kurtosis. According to [5], we can preserve $2N-1$ moments using N value-probability pairs. The calculation of the discrete approximation is straight-forward. The discrete distribution replaces the system's deterministic load in the model. This results in a stochastic model. In the following computations, we consider $N = 5$ load scenarios:

$$p(Q = \mu_Q) = 53.33\,\%, \tag{9}$$

$$p(Q = 104.07\,\%\,\mu_Q) = p(Q = 95.93\,\%\,\mu_Q) = 22.21\,\%, \tag{10}$$

$$p(Q = 108.57\,\%\,\mu_Q) = p(Q = 91.43\,\%\,\mu_Q) = 1.13\,\%. \tag{11}$$

The total power consumption in the objective function is replaced by the average total power consumption over all scenarios. The determinstic equivalent program is generated and optimized with the commercial solver gurobi.

Results and Discussion

Results for model without uncertainty The application of linearization techniques enables us to use the commercial optimization solver *Gurobi*. It allows one to include numerous discrete decision variables into the model and still finds a solution in a reasonable time span. In order to find control guidelines for the whole field of operation, we have to run the optimization model many times. Every combination of pressure and volume flow demand needs one optimization run. We discretize the pressure and the volume flow with a step size of $\delta H = 1$ m and $\delta Q = 1$ m^3/min and calculate the optimal number of running pumps. Thus we obtain the fields of equal numbers of running pumps within the field of operation as shown in Fig. 2.

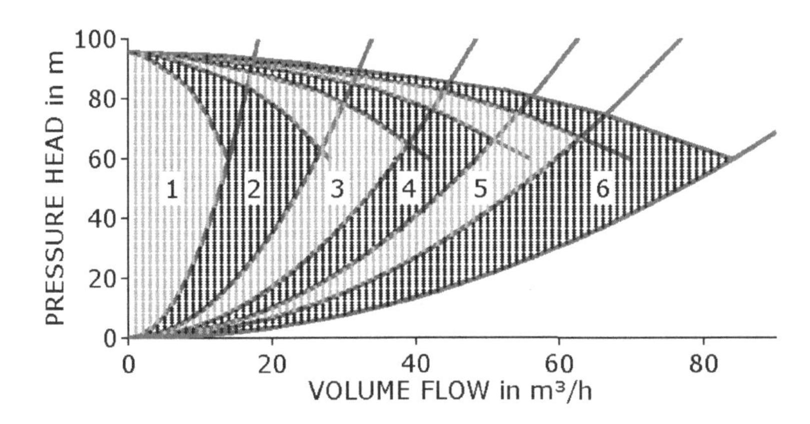

Fig. 2: Optimal control stategy for certain load.

We can identify two reasons for the switch of a pump: (1) For high constant pressure and increasing volume flow the controller increases the rotational speed of the active pumps. Once the maximum is reached, an additional pump is switched on. The decreasing gray lines show the maximum volume flow for fixed pressure head for one, two, three...pumps. (2) For low constant pressure and increasing volume flow, we find a efficiency argument: An additional pump is switched on, because it reduces the total power consumption, even though more pumps are working. We identify the switching line as a parabolic function with

$$H = aQ^2 . \tag{12}$$

The slope of the parabola a is calculated with a fit as shown in Table 2. These parabolas should be integrated into the control strategy of a booster station.

Decisions under uncertainty We apply the same technique of operating field discretization to the model under uncertainty. The same two reasons for the switch of a pump apply to the case with uncertain load: (1) For high pressure the optimization problem reduces once again to a feasibility problem. Due to the assumption of uncertain load, the optimization algorithm activates the next pump at a slightly smaller volume flow than before. (2) For low pressure the switching lines are parabolas again, but with different slopes compared to the results of the model without uncertainty. An uncertain load also leads to efficiency-caused pump activations at lower volume flows.

Table 2: Slopes a of the parabolas in h^2m^{-5}

	certain load	uncertain load
1 - 2	0.3017	0.3565
2 - 3	0.08724	0.08934
3 - 4	0.04264	0.04281
4 - 5	0.02547	0.02557
5 - 6	0.01687	0.01699

Summary and Outlook

In this paper we showed how to lay out a robust control strategy for booster stations: For high pressure rises the optimization problem reduces to a feasibility problem. The described first option for the operational mode is the best solution for the optimization problem. For low pressure rises the measurement of the volume flow in the station is a necessary additional input parameter for the control. The control strategy for a closed loop network should be based on the calculated parabolas to drive the booster station with optimal energy consumption even for low pressure rises.

Acknowledgment

This work is partially supported by the German Federal Ministry for Economic Affairs and Energy funded project "Entwicklung hocheffizienter Pumpensysteme" and by the German Research Foundation (DFG) funded SFB 805.

References

[1] P. F. Pelz: 250 Jahre Energienutzung: Algorithmen übernehmen Synthese, Planung und Betrieb von Energiesystemen, Festvortrag anlässlich der Ehrenpromotion von Hans-Ulrich Banzhaf, Universität Siegen, 23.01.2014.

[2] P. F. Pelz, U. Lorenz, G. Ludwig: Besser geht's nicht. TOR plant das energetisch optimale Fluidsystem, Chemie and more, issue 1, 2014.

[3] P. F. Pelz, U. Lorenz, T. Ederer, S. Lang, G. Ludwig: Designing Pump Systems by Discrete Mathematical Topology Optimization: The Artificial Fluid Systems Designer (AFSD), International Rotating Equipment Conference, 2012.

[4] J. P. Vielma, S. Ahmed, G. Nemhauser: Mixed-integer models for nonseparable piecewise-linear optimization: unifying framework and extensions, Operations research, volume 58, number 2, 2010.

[5] A. C. Miller III, T. R. Rice: Discrete approximations of probability distributions, Management science, volume 29, number 3, 1983.

Applied Mechanics and Materials Vol. 807 (2015) pp 247-256
© (2015) Trans Tech Publications, Switzerland
doi:10.4028/www.scientific.net/AMM.807.247
Submitted: 2015-04-15
Accepted: 2015-08-04

Multicriterial Optimization of Technical Systems Considering Multiple Load and Availability Scenarios

Lena C. Altherr[1,a*] , Thorsten Ederer[1,b], Philipp Pöttgen[1,c], Ulf Lorenz[2,d], and Peter F. Pelz[1,e]

[1]Technische Universität Darmstadt, Chair of Fluid Systems, Magdalenenstr. 4, 64289 Darmstadt, Germany

[2] Universität Siegen, Lehrstuhl für Betriebswirtschaftslehre insb. Technologiemanagement, Hölderlinstr. 3, 57076 Siegen, Germany

[a]lena.altherr@fst.tu-darmstadt.de, [b]thorsten.ederer@fst.tu-darmstadt.de, [c]philipp.poettgen@fst.tu-darmstadt.de, [d]ulf.lorenz@uni-siegen.de, [e]peter.pelz@fst.tu-darmstadt.de

Keywords: sustainability, availability, energy efficiency, mixed-integer linear programming, system synthesis.

Abstract. Cheap does not imply cost-effective – this is rule number one of zeitgeisty system design. The initial investment accounts only for a small portion of the lifecycle costs of a technical system. In fluid systems, about ninety percent of the total costs are caused by other factors like power consumption and maintenance. With modern optimization methods, it is already possible to plan an optimal technical system considering multiple objectives. In this paper, we focus on an often neglected contribution to the lifecycle costs: downtime costs due to spontaneous failures. Consequently, availability becomes an issue.

Introduction

In times of planned obsolescence private and corporate consumers call for more sustainable products and systems. Energy efficiency is key to ecological sustainability. To achieve *economical* sustainability investment costs and availability have to be considered as well. Thus, system designers are confronted with a multi-criterial optimization problem.

The decision whether to buy a specific component or not affects the economic value of the system: A component can be cheap regarding investment costs. However, if its energy consumption is high, it might be worthy to invest in a more efficient one. Given a diverse load spectrum, investing in several components which share the load might also be well spent money if the reduction in energy costs in a given deprication period outweighs the additional investment costs. If components have to be renewed or repaired often due to high failure rates, the downtime of the system increases. In this case, it might be beneficial to invest in more expensive, but also more robust components.

Uncertainty in the load and in the availability of the components due to random failures impede the assessment of expected energy and downtime costs during system planning. In this paper, we present a method for the design of technical systems named Technical Operations Research (TOR). With TOR we are able to find the global optimal system layout which fulfills multiple load scenarios and consists of the optimal combination of optional components. Given different availability scenarios with probabilities which depend on the purchase decisions, this layout incurs the minimum total cost of investment, energy, and failure.

Technical Application and Load Scenarios

We illustrate our method by designing a pump system. The planner has to find the optimal combination of pumps which fulfills the required function while minimizing costs. In this example, optional pumps with characteristic curves as illustrated in Fig. 1 are given. We want to find the parallel connection of up to four pumps which incurs the minimum sum of investment, energy, and failure costs.

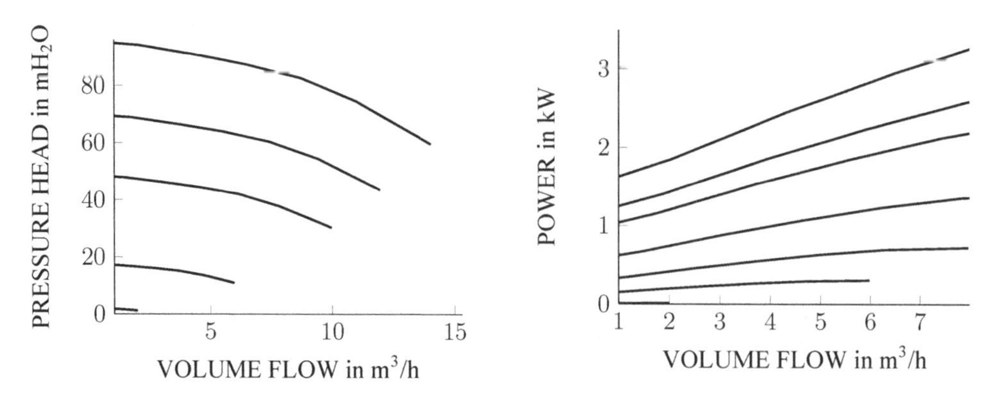

Figure 1: Characteristic curves of a pump in the construction kit of optional components.

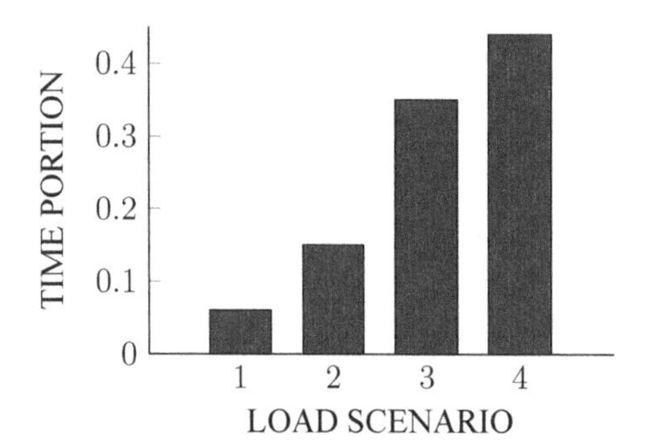

Load Scenario	Q in m^3/h	H in mH_2O
1	6.25	33.13
2	12.50	36.25
3	18.75	39.38
4	25.00	42.50

Figure 2: Load scenarios with probabilities corresponding to their respective assumed time portion define the required function of the system. In each scenario, a different volume flow Q and pressure head H is required.

The required function of the pump system is a distribution of different loads, subject to uncertainty. We assume a set of discrete load cases $\lambda \in L$, which we call load scenarios, cf. Fig. 2. To each of them a probability p_λ is assigned. It represents the assumed portion of the system's life time in which the load corresponding to λ is supposed to occur. The power consumption of the system depends on a given load scenario. The overall power consumption is the weighted sum over the different scenarios. To calculate the energy costs of the system, the power consumption $P_{\lambda,\text{pump}}$ of each pump in load scenario λ is taken into account. Given a deprication period τ and electricity costs ζ the energy costs are

$$U = \tau \cdot \zeta \cdot \sum_{\lambda \in L} \sum_{\text{pumps}} p_\lambda \cdot P_{\lambda,\text{pump}}. \tag{1}$$

However, the power consumption and thus the energy costs of the system are not only dependent on the load scenario, but also on the number of available pumps. On the one hand, if one of the purchased pumps fails, the other pumps may have to run in an operating point which is less energy efficient in order to still fulfill the load. On the other hand, given low penalty costs during system downtime, one could also deliberately switch off the pumps if they had to run in a highly energy consuming operating point. Therefore, we consider not only different load scenarios, but also the different system states which arise if some of the components fail.

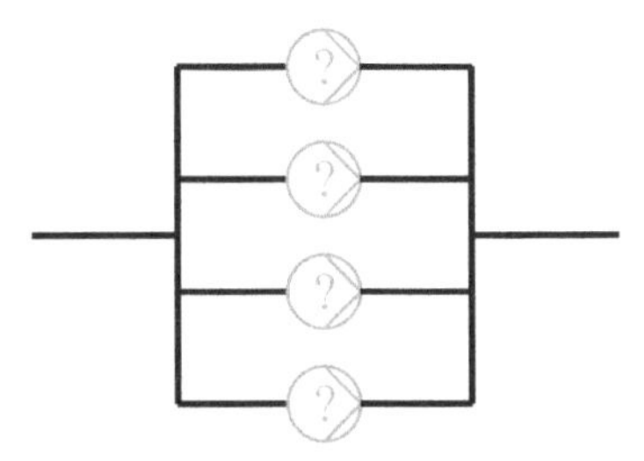

Figure 3: Planning the optimal pump system. How many and which optional pumps should be bought to build the system which incurs the minimum sum of investment, energy and failure costs?

System Availability

If we put n new, statistically identical and independent items into operation at time $t = 0$, at the time t a portion $\tilde{v}(t)$ of these items have not yet failed.

$$\hat{R}(t) = \frac{\tilde{v}(t)}{n} \tag{2}$$

is the empirical reliability which converges to the reliability function $R(t)$ for $n \rightarrow \infty$ [1]. This equation describes the probability that the item will perform its required function under given operating conditions, i.e., that it will not fail in the interval $(0, t]$. For an arbitrary time interval $(t, t + \delta t]$, the empirical failure rate $\hat{\lambda}(t)$ is given by [1]

$$\hat{\lambda}(t) = \frac{\tilde{v}(t) - \tilde{v}(t + \delta t)}{\tilde{v}(t)\delta t}. \tag{3}$$

The combination of Eq. 2 and 3 yields

$$\hat{\lambda}(t) = \frac{\hat{R}(t) - \hat{R}(t + \delta t)}{\hat{R}(t)\delta t}. \tag{4}$$

Given $R(t)$ differentiable, Eq. 4 converges for $n \rightarrow \infty$ to the (instantaneous) failure rate

$$\lambda(t) = \frac{-\mathrm{d}R(t)/\mathrm{d}t}{R(t)}. \tag{5}$$

For a large population of new, statistically identical and independent items, often a so-called bathtub curve with three phases is observed (cf. Fig. 4). In phase 1, early failures due to randomly distributed weaknesses in the items occur and the failure rate $\lambda(t)$ decreases rapidly with time. Stochastic failures are observed in phase 2 and $\lambda(t)$ is approximately constant. In phase 3, failures due to wearout or aging occur and $\lambda(t)$ increases with time.

In case of stochastic failure and constant time independent failure rate $\lambda(t) = \lambda$, the reliability is given by

$$R(t) = e^{-\lambda t}. \tag{6}$$

The mean of the expected failure free time, i.e., the Mean Time To Failure ($MTTF$), is

$$MTTF = \int_0^\infty R(t)dt. \tag{7}$$

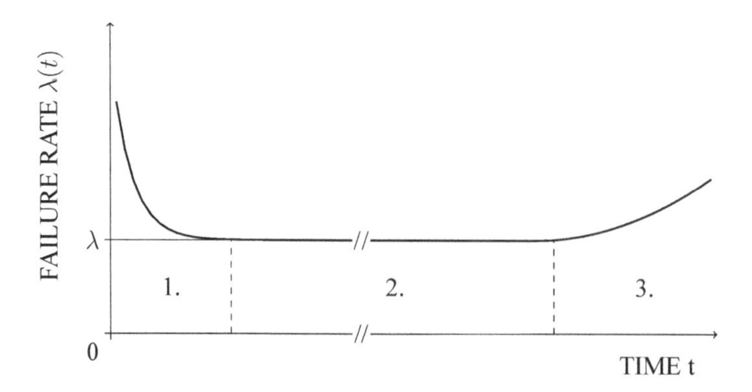

Figure 4: The time-dependency of the failure rate λ can often be represented by a bathtub curve with three phases. For most of the item's life the failure rate is constant and its reliability can be described by an exponential distribution.

For a constant failure rate this yields

$$MTTF = \frac{1}{\lambda}. \tag{8}$$

A repairable item goes through a sequence of failure and repair events. Instead of $MTTF$, the term Mean Time Between Failures ($MTBF$) is used. When only considering two states of an item (good/failed), and assuming that it is as-good-as-new after each repair action, successive failure-free times are independent random variables. Given a constant failure rate λ, the mean of the exponentially distributed successive failure-free times is given by

$$MTBF = \frac{1}{\lambda}. \tag{9}$$

The Point Availability $PA(t)$ describes the probability that an item performs its required function at a given time t, i.e., that it is not in a failure state or undergoing a repair action when requested for use. High reliability promotes high availability, but does not imply so. Even a highly reliable item can still be poorly available if the required repair is very time-consuming. If we assume the maintenance time to be exponentially distributed according to $e^{-\mu t}$ with constant repair rate μ, the Mean Time To Repair ($MTTR$) is given by:

$$MTTR = \frac{1}{\mu}. \tag{10}$$

The point availability $PA(t)$ of an item converges to the stationary value of the average availability

$$A = \frac{MTBF}{MTBF + MTTR}. \tag{11}$$

A gives the expected value for the fraction of time during which the item is operational. Assuming the availability of the components of a system to be independent, the fraction of time during which two components P_1 and P_2 are both available can be calculated by

$$A_{sys} = A_1 \cdot A_2 = \frac{MTBF_1}{MTBF_1 + MTTR_1} \cdot \frac{MTBF_2}{MTBF_2 + MTTR_2}. \tag{12}$$

Based on this assumption, different availability scenarios arise which are dependent on the purchase decision. If we buy three components P_1, P_2 and P_3 with corresponding availabilities A_1, A_2, A_3 there are 2^3 possible system states with different probabilities which we call availability scenarios:

Table 1: Availability scenarios for the purchase decision to buy components P_1, P_2 and P_3.

avail. scen.	P_1	P_2	P_3	time fraction	description
1	0	0	0	$(1-A_1)\cdot(1-A_2)\cdot(1-A_3)$	P_1, P_2, P_3 are down
2	1	0	0	$A_1\cdot(1-A_2)\cdot(1-A_3)$	P_1 is working, P_2, P_3 are down
3	0	1	0	$A_2\cdot(1-A_1)\cdot(1-A_3)$	P_2 is working, P_1, P_3 are down
4	0	0	1	$A_3\cdot(1-A_1)\cdot(1-A_2)$	P_3 is working, P_1, P_2 are down
5	1	1	0	$A_1\cdot A_2\cdot(1-A_3)$	P_1, P_2 are working, P_3 is down
6	1	0	1	$A_1\cdot A_3\cdot(1-A_2)$	P_1, P_3 are working, P_2 is down
7	0	1	1	$A_2\cdot A_3\cdot(1-A_1)$	P_2, P_3 are working, P_1 is down
8	1	1	1	$A_1\cdot A_2\cdot A_3$	P_1, P_2, P_3 are working

High availability at the system level is often achieved by redundancy. By providing more components than actually needed in order to fulfill the required function, the reliability is increased. In a parallel connection of n identical pumps, of which $k \leq n$ are necessary to provide the demanded volume flow, $n-k$ are in reserve. This structure is called a k-out-of-n redundancy [1]. If we assume the pumps to be as-good-as-new after a repair action we can describe the system by states $S_1, S_2, \ldots,$ S_{n-k}, S_{n-k+1} where S_i represents the system state in which i of the identical pumps have dropped out, cf. Fig. 5. For a constant failure and repair rate and statistically identical and independent pumps, this system can be described by a Markov process. In case of $n-k+1$ failed pumps, the system is down. Since this state can only be left if one of the pumps is repaired, the probability to stay in this state is $1-\mu\,\delta t$. The transition probabilities for the other states are given by $v_k\,\delta t$, where $v_k = \lambda^{n-k+1}\frac{n!}{(k-1)!}$ for $k = 1,\ldots,n-k$ [1].

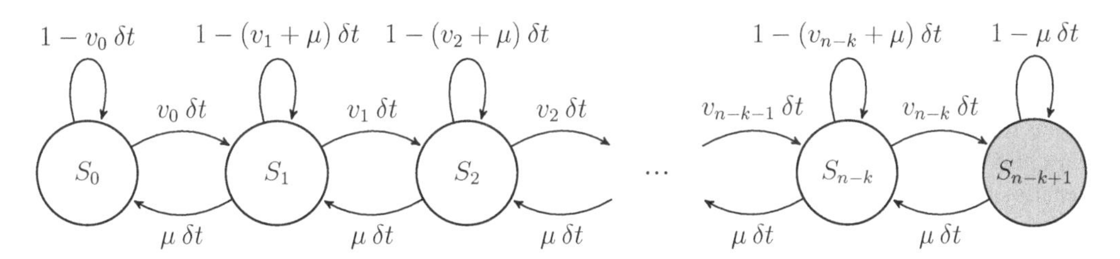

Figure 5: A system of n parallel pumps out of which k are necessary to fulfill the load is an example for a k-out-of-n redundant system. Under certain assumptions, it can be described by a Markov process.

Systems of pumps which have different failure rates λ, cannot be described by the above process. A growing number of states makes the representation of the system's availability more difficult. In this paper, we deal with pumps with different $MTBF$ and thus we do not only consider $n-k$ states, but all possibilities which can arise if different combinations of the pumps fail.

Mixed Integer Linear Program

We model the multicriterial optimization problem by a Mixed Integer Linear Program consisting of two stages: First, find a low-priced investment decision in an adequate set of optional pumps. Secondly, find energy-efficient operating settings for the selected pumps and ensure low failure costs. However, instead of optimizing investment, failure and energy costs separately, we compare all possible systems and minimize the weighted sum of investment costs I, of system downtimes D and of power consumption U simultaneously, given a deprication period τ, electricity costs ζ and downtime penalty costs per hour ξ:

$$\min \quad I + \tau \cdot \xi \cdot D + \tau \cdot \zeta \cdot U. \tag{13}$$

The pumps available in the construction kit have different MTBF and prices:

Table 2: Pumps available in the construction kit.

pump	MTBF in years	MTTR in weeks	price in €
P1	0.5	2	3000
P2	4	2	4000
P3	2	2	3500
P4	2	2	3800

According to [2], all possible systems are modelled by the complete graph $G = (V, E)$ with edges E corresponding to the optional pumps and pipes, and vertices V representing the connection points between these components. In this study we want to compare all possible parallel connections which can be built with N given pumps, see Fig. 3. A binary variable $b_{i,j}$ for each optional pump (i, j) in the set of pumps $\mathcal{P} \subseteq V$ indicates whether it is purchased ($b_{i,j} = 1$) or not ($b_{i,j} = 0$).

Out of a set of N optional pumps, 2^N different systems can be built. These purchase possibilities B are represented by the power set $\mathfrak{P}(\mathcal{P})$. Each set-up $\beta \in B$ results in a set of availability scenarios A_β. If we decide to buy n pumps, 2^n availability scenarios exist. Note that the number of availability scenarios depends on a first stage decision. This dependency can be resolved by a redundant enumeration of all possible purchase decisions in the scenario set. In general, this results in a set of combined purchase and availiability scenarios $A = \{(\beta, \alpha) : \beta \in B, \alpha \in A_\beta\}$. The number of combined scenarios equals the cardinality of the binary relation S on the power set of pumps, such that xSy iff $x \subseteq y$ for all $x, y \in \mathfrak{P}(P)$. Haye [3] calculated this to be $|A| = 3^N$. For instance, for the case of four optional pumps which is considered in this study, 81 different scenarios exist. Given a set of load scenarios L this results in a set of scenarios $Sc = A \times L$, cf. Fig 6.

For each scenario $sc = (\beta, \alpha, \lambda) \in Sc$ and each pump $(i, j) \in \mathcal{P}$, a binary variable $a_{sc,i,j}$ is introduced. This makes it possible to deactivate purchased pumps during operation in each scenario separately.

A physical constraint imposed on the fluid system is the conservation of the volume flow. In each scenario sc the sum of the flow $Q_{sc,i,v}$ going into vertex v must be equal to the sum of the flow $Q_{sc,v,j}$ going out of vertex v:

$$\forall\, sc \in Sc, \forall\, v \in V: \quad \sum_{(i,v) \in E} Q_{sc,i,v} = \sum_{(v,j) \in E} Q_{sc,v,j}. \tag{14}$$

An additional constraint with an adequate upper limit Q_{max} ensures that only components (i, j) which operate in scenario sc contribute to the volume flow conservation:

$$\forall\, (i,j) \in E: Q_{sc,i,j} \leq Q_{\text{max}} \cdot a_{sc,i,j}. \tag{15}$$

Another physical constraint is the pressure propagation

$$\forall\, sc \in Sc,\ \forall\, (i, j) \in E: H_{sc,j} \leq H_{sc,i} + \Delta H + M \cdot a_{sc,i,j}, \tag{16}$$

$$H_{sc,j} \geq H_{sc,i} + \Delta H - M \cdot a_{sc,i,j}. \tag{17}$$

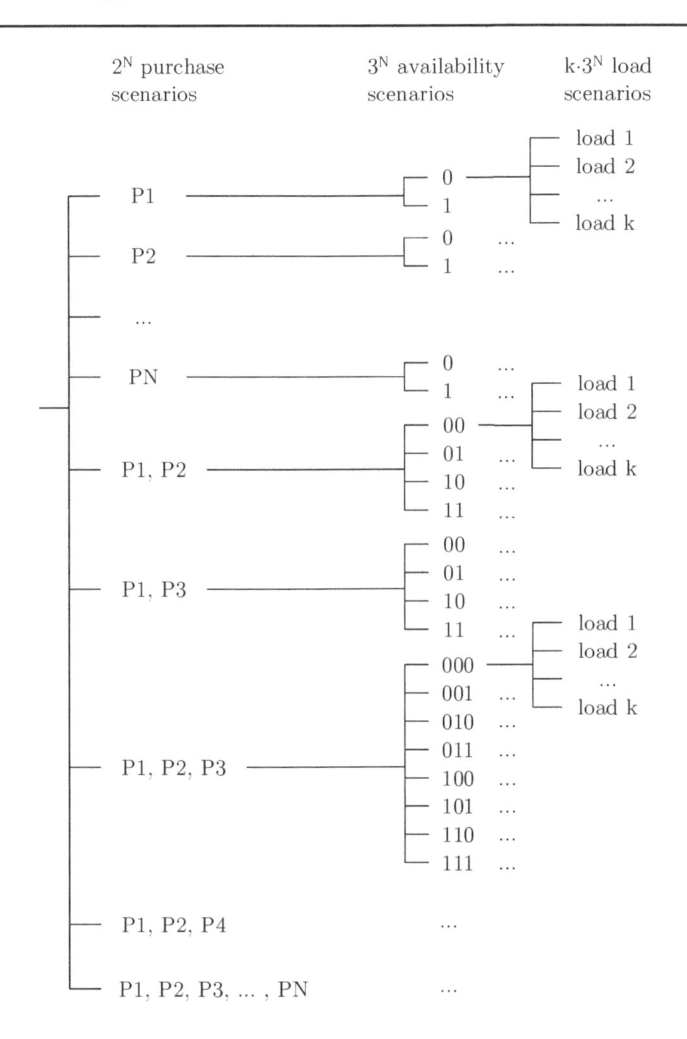

Figure 6: k different load scenarios and N optional pumps yield $k \cdot 3^N$ different scenarios.

which has to be fulfilled along each edge (i, j) if the component is active in scenario sc. The parameter M is an adequately chosen number allowing to activate the constraint if the component is operating ($a_{sc,i,j} = 1$) and to deactivate the constraint if the component is switched off ($a_{sc,i,j} = 0$). The variable ΔH represents the change in pressure when passing through component (i, j). For pumps, the increase of pressure ΔH depends on the rotational speed n of the pump and on the volume flow Q which is conveyed. This dependency is given for each pump by its head curve, cf. Fig. 1. For pipes, a negative ΔH represents the pressure loss due to friction, increasing with the volume flow to the power of two. The head curves of the pumps and the pipes have been piecewise linearly approximated and represented by a convex combination formulation [4]. The second characteristic curve in Fig. 1 represents the relationship between power consumption P of the pump and the volume flow Q conveyed at a given rotational speed n. It is also piecewise linearly approximated with the help of a convex combination formulation and is zero for $a_{sc,i,j} = 0$.

For each scenario $sc = (\beta, \alpha, \lambda)$, a binary variable f_{sc} indicates whether a given combined purchase and availability scenario (β, α) fulfills the load specified by load scenario λ. If the load is not fulfilled in scenario sc, all pumps $(i, j) \in \mathcal{P}$ shall be deactivated:

$$\forall\ sc \in Sc, \forall\ (i, j) \in \mathcal{P} : a_{sc,i,j} \leq f_{sc}. \tag{18}$$

Once the purchase decisions are made, i.e., one purchase possibility $\tilde{\beta}$ is chosen, all scenarios corresponding to other purchase possibilities are to be excluded. This is achieved by the following constraints: Firstly, if a pump is part of the set-up of a given purchase possibility $\tilde{\beta}$, the load in any scenario $(\tilde{\beta}, \alpha, \lambda)$ can only be fulfilled if this pump is bought:

$$\forall\ (i,j) \in \mathcal{P}, \forall \tilde{\beta} \in B \mid (i,j) \in \tilde{\beta} : f_{\tilde{\beta},\alpha,\lambda} \leq\ b_{i,j}. \tag{19}$$

Secondly, the load can only be fulfilled by one purchase possibility:

$$\forall\ (\beta_1, \alpha_1, \lambda_1), (\beta_2, \alpha_2, \lambda_2) \in Sc \mid \beta_1 \neq \beta_2 : f_{\beta_1,\alpha_1,\lambda_1} + f_{\beta_2,\alpha_2,\lambda_2} \leq\ 1. \tag{20}$$

Each scenario $sc = (\beta, \alpha, \lambda)$ occurs during a time fraction $p_{sc} = p_\lambda \cdot p_\alpha$. We calculate the system's power consumption U by summing up the power consumption $P_{sc,i,j}$ of all pumps $(i,j) \in \mathcal{P}$ in every scenario:

$$U = \sum_{(i,j)\in\mathcal{P}} \sum_{sc} p_{sc} \cdot P_{sc,i,j}. \tag{21}$$

With Eqs. 18, 19 and 20 we ensure, that only pumps which have been bought, are available, and active contribute to the energy costs. The failure costs are calculated by

$$\tau \cdot \xi \cdot D = \tau \cdot \xi \cdot \sum_{\lambda} p_\lambda \cdot \left(1 - \sum_{\beta,\alpha} p_\alpha \cdot f_{\beta,\alpha,\lambda}\right). \tag{22}$$

Optimization Result

Table 4 shows the optimal purchase and operating decisions for different magnitudes of downtime penalties. Once the purchase decisions are made, this results in specific availability scenarios. For each availability scenario the operating strategy in load scenario λ_i is given.

For downtime penalties of 0.13 €/hour, the purchase decision which incurs the minimum sum of investment, energy and failure costs is to buy none of the pumps. In this case, the initial investment and the operation of the system cause more costs than simply accepting downtime during the whole deprication period.

If we double the downtime penalties to 0.26 €/hour, the optimal decision is to buy the cheapest pump P1 and to operate it during the scenario with the highest time portion and the minimum energy costs, λ_1. Pump P1 is not able to fulfill the load in scenarios λ_3 and λ_4, but it can provide the volume flow required in λ_2. However, it is not operated in λ_2. Reducing the downtime costs is not profitable due to resulting energy costs. This yields an average uptime of ca. 41 %, as given in the following Table:

Table 3: System uptime for different magnitudes of downtime penalties.

downtime penalty in €/hour	purchase decision	avg. uptime in load scenario λ_i in %				avg. uptime during deprication period in %
		λ_1	λ_2	λ_3	λ_4	
0.13	-	0.00	0.00	0.00	0.00	0.00
0.26	P1	92.8753	0.00	0.00	0.00	40.8651
0.30	P1	92.8753	92.8753	0.00	0.00	73.3715
0.65	P2	99.0502	99.0502	0.00	0.00	78.2497
0.91	P1, P2	99.9323	99.9323	91.9932	91.9932	98.2651
26.00	P1, P2, P4	99.9987	99.9987	99.8493	99.7829	99.9634

Table 4: Optimal purchase and operating decisions depend on the magnitude of downtime penalties.

downtime penalty in €/hour	purchase decision	availability scenario				pumps operating in load scenario			
		P1	P2	P3	P4	λ_1	λ_2	λ_3	λ_4
0.13	-	0	0	0	0	-	-	-	-
0.26	P1	0	-	-	-	-	-	-	-
		1	-	-	-	P1	-	-	-
0.30	P1	0	-	-	-	-	-	-	-
		1	-	-	-	P1	P1	-	-
0.65	P2	-	0	-	-	-	-	-	-
		-	1	-	-	P2	P2	-	-
0.91	P1, P2	0	0	-	-	-	-	-	-
		1	0	-	-	P1	-	-	-
		0	1	-	-	P2	-	-	-
		1	1	-	-	P1, P2	P1, P2	P1, P2	P1, P2
26.00	P1, P2, P4	0	0	-	0	-	-	-	-
		1	0	-	0	P1	P1	-	-
		0	1	-	0	P2	P2	-	-
		0	0	-	1	P4	P4	P4	-
		1	1	-	0	P1	P1	P1, P2	P1, P2
		1	0	-	1	P1	P1	P4	P1, P4
		0	1	-	1	P2	P2	P4	P2, P4
		1	1	-	1	P2	P2	P1, P2	P1, P2

When further raising downtime penalties, pump P2 is bought. This pump has the same characteristic curve as P1, but a higher price and eight times higher $MTBF$, cf. Table 2. Pump P2 is also only operated in scenario λ_1, but due to its lower failure rate, the average uptime of the system reaches about 73 %, cf. Table 3.

For downtime penalties of 0.91 €/hour, pumps P1 and P2 are bought. Now, the required volume flow can also be provided in scenarios λ_3 and λ_4 by operating P1 and P2 in parallel. Whenever the system can fulfill the load in an availability scenario, it does so. The average uptime of the system rises to more than 98 %.

For the highest downtime penalties of 26.00 €/hour, pumps P1, P2 and P4 are bought. Since two of the three pumps can fulfill the load in each scenario λ_i, this results in a 2-out-of-3 redundancy and the average uptime increases slightly to ca. 99.96 %. In this case, higher investment costs are profitable, regarding the otherwise high failure costs.

Summary

Downtime costs can account for a substantial share of the lifecycle costs of a technical system. We showed how to consider its availability in a multicriterial optimization. The magnitude of the failure costs determines the optimal system topology. We observed three distinct cases:

If downtime costs are small, the system's availability is not the optimization's prime directive. The resulting topology consists of few components, even though the load cannot be fulfilled in certain scenarios. This saves investment and energy costs. For average downtime costs, the optimal system and operating strategy is similar to one which results when optimizing without considering failures at all. All load scenarios are fulfilled with minimum energy costs, higher investment costs are balanced by energy savings. If system downtime incurs high costs, it is worthwhile to invest in more components than crucially necessary. The system's redundancy is increased.

We solved our MILP with the standard optimization software CPLEX based on the branch and cut algorithm. For further research it would be interesting to formulate the model as a two stage stochastic program and to evaluate if dual decomposition methods [5] provide a speed-up.

Acknowledgment

This work is partially supported by the German Federal Ministry for Economic Affairs and Energy funded project "Entwicklung hocheffizienter Pumpensysteme" and by the German Research Foundation (DFG) funded Collaborative Research Centre SFB 805.

References

[1] A. Birolini, Reliability engineering, Vol. 5, Berlin: Springer, 2007.

[2] P. F. Pelz, U. Lorenz, T. Ederer, S. Lang, G. Ludwig, Designing Pump Systems by Discrete Mathematical Topology Optimization: The Artificial Fluid Systems Designer (AFSD), International Rotating Equipment Conference, 2012.

[3] R. La Haye, Binary relations on the power set of an n-element set, Journal of Integer Sequences, 12(2), 3.

[4] J. P. Vielma, S. Ahmed, G. Nemhauser, Mixed-integer models for nonseparable piecewise-linear optimization: unifying framework and extensions, Operations research, volume 58, number 2, 2010.

[5] C. C. Carøe, R. Schultz, Dual decomposition in stochastic integer programming, Operations Research Letters 24.1 (1999): 37-45.

Keyword Index

Author Index